Advances in Rare Diseases Biomarkers

Advances in Rare Diseases Biomarkers

Andrea Bernini

MDPI

Basel • Beijing • Wuhan • Barcelona • Belgrade • Novi Sad • Cluj • Manchester

Editor
Andrea Bernini
Department of Biotechnology,
Chemistry and Pharmacy
University of Siena
Siena
Italy

Editorial Office
MDPI AG
Grosspeteranlage 5
4052 Basel, Switzerland

This is a reprint of articles from the Special Issue published online in the open access journal *International Journal of Molecular Sciences* (ISSN 1422-0067) (available at: www.mdpi.com/journal/ijms/special_issues/Z239P02Z4M).

For citation purposes, cite each article independently as indicated on the article page online and as indicated below:

Lastname, A.A.; Lastname, B.B. Article Title. *Journal Name* **Year**, *Volume Number*, Page Range.

ISBN 978-3-7258-2412-0 (Hbk)
ISBN 978-3-7258-2411-3 (PDF)
doi.org/10.3390/books978-3-7258-2411-3

Contents

About the Editor

Andrea Bernini

Andrea Bernini is an Associate Professor of Biochemistry at the Department of Biotechnology, Chemistry, and Pharmacy at the University of Siena. He obtained his Ph.D. in Biotechnology in 2001 after completing his training at the Oxford Centre for Molecular Science, Oxford University, UK. Andrea Bernini has authored nearly 100 papers in international peer-reviewed scientific journals. His scientific expertise lies in nuclear magnetic resonance of biomolecules, structural biology, and bioinformatics. His research interests center on the multi-omics investigation of rare metabolic disorders and rare malignancies, especially in metabolomics and structure determination using magnetic resonances and mass spectrometry. Their current research projects focus on alkaptonuria, an ultra-rare inherited metabolic disease, and soft tissue sarcomas (rare mesenchymal tissue neoplasms).

Preface

A rare disease is a health condition with a lower prevalence than common diseases. The World Health Organization defines a rare disease as one that strikes fewer than 65 per 100,000 people. However, their combined effect is significant: around 7,000 rare diseases affect approximately 350 million people worldwide.

Biomarkers play a crucial role in diagnosing and monitoring rare diseases, which are often challenging to detect and understand due to their low prevalence and diverse clinical manifestations. Biomarkers serve as measurable indicators of biological processes or conditions in rare diseases, offering valuable insights into disease mechanisms and progression. These markers may include genetic mutations, protein levels, or other molecular signatures unique to a rare condition. The discovery and validation of such biomarkers contribute to early detection and the development of targeted therapies, allowing for more effective and personalized treatment approaches.

As technology advances, the integration of omics technologies, such as genomics, proteomics, and metabolomics, has further expanded the repertoire of potential biomarkers, fostering a deeper understanding of rare diseases and paving the way for innovative diagnostic and therapeutic strategies. Biomarkers enhance our ability to navigate diagnostic challenges and promise to improve patient care and foster breakthroughs in treatment modalities.

Andrea Bernini
Editor

International Journal of
Molecular Sciences

MDPI

Review

Biomarker Landscape in RASopathies

Noemi Ferrito [1,2,3,†], Juan Báez-Flores [1,2,3,†], Mario Rodríguez-Martín [1,2,3], Julián Sastre-Rodríguez [1], Alessio Coppola [1,2,3], María Isidoro-García [3,4,5,6], Pablo Prieto-Matos [2,3,7,8] and Jesus Lacal [1,2,3,*]

1 Laboratory of Functional Genetics of Rare Diseases, Department of Microbiology and Genetics, University of Salamanca (USAL), 37007 Salamanca, Spain; noemi.ferrito@usal.es (N.F.); alumni.jbaez@usal.es (J.B.-F.); juliansastre@usal.es (J.S.-R.); acoppola@usal.es (A.C.)
2 GIR of Biomedicine of Rare Diseases, University of Salamanca (USAL), 37007 Salamanca, Spain; pprieto@saludcastillayleon.es
3 Institute of Biomedical Research of Salamanca (IBSAL), 37007 Salamanca, Spain; misidoro@saludcastillayleon.es
4 Clinical Biochemistry Department, University Hospital of Salamanca, 37007 Salamanca, Spain
5 Clinical Rare Diseases Reference Unit DiERCyL, 37007 Castilla y León, Spain
6 Department of Medicine, University of Salamanca (USAL), 37007 Salamanca, Spain
7 Department of Pediatrics, University Hospital of Salamanca, 37007 Salamanca, Spain
8 Department of Biomedical and Diagnostics Science, University of Salamanca (USAL), 37007 Salamanca, Spain
* Correspondence: jlacal@usal.es
† These authors contributed equally to this work.

Abstract: RASopathies are a group of related genetic disorders caused by mutations in genes within the RAS/MAPK signaling pathway. This pathway is crucial for cell division, growth, and differentiation, and its disruption can lead to a variety of developmental and health issues. RASopathies present diverse clinical features and pose significant diagnostic and therapeutic challenges. Studying the landscape of biomarkers in RASopathies has the potential to improve both clinical practices and the understanding of these disorders. This review provides an overview of recent discoveries in RASopathy molecular profiling, which extend beyond traditional gene mutation analysis. mRNAs, non-coding RNAs, protein expression patterns, and post-translational modifications characteristic of RASopathy patients within pivotal signaling pathways such as the RAS/MAPK, PI3K/AKT/mTOR, and Rho/ROCK/LIMK2/cofilin pathways are summarized. Additionally, the field of metabolomics holds potential for uncovering metabolic signatures associated with specific RASopathies, which are crucial for developing precision medicine. Beyond molecular markers, we also examine the role of histological characteristics and non-invasive physiological assessments in identifying potential biomarkers, as they provide evidence of the disease's effects on various systems. Here, we synthesize key findings and illuminate promising avenues for future research in RASopathy biomarker discovery, underscoring rigorous validation and clinical translation.

Keywords: RASopathies; biomarkers; molecular signatures; precision medicine; personalized diagnostics

Citation: Ferrito, N.; Báez-Flores, J.; Rodríguez-Martín, M.; Sastre-Rodríguez, J.; Coppola, A.; Isidoro-García, M.; Prieto-Matos, P.; Lacal, J. Biomarker Landscape in RASopathies. *Int. J. Mol. Sci.* **2024**, *25*, 8563. https://doi.org/10.3390/ijms25168563

Academic Editor: Maurizio Margaglione

Received: 6 July 2024
Revised: 28 July 2024
Accepted: 2 August 2024
Published: 6 August 2024

1. Introduction

RASopathies represent a group of rare genetic diseases affecting fewer than 1 in 1000 people worldwide, characterized by a germline mutation in one of the genes encoding components of the RAS/MAPK pathway [1,2]. These mutations may be de novo or inherited, and the associated phenotypic features and congenital anomalies are diverse [3]. These conditions share overlapping features, including developmental delays, craniofacial dimorphism, cardiac malformations, cutaneous manifestations, musculoskeletal abnormalities, neurologic issues, and an elevated risk of cancer, making patients' lifespan shorter [3]. These diverse and often severe symptoms significantly affect patients' quality of life, impacting multiple organ systems. For this reason, it is important to advance research in this area to improve diagnosis, treatment, and support for affected individuals. The spectrum of RASopathies encompasses diverse disorders such as neurofibromatosis type 1 (NF1), Noonan syndrome (NS), Noonan syndrome with multiple lentigines (NSML), Noonan syndrome-like with loose anagen hair (NS-LAH), neurofibromatosis–Noonan syndrome (NFNS), Noonan syndrome-like (NSL), Legius syndrome (LS), cardiofaciocutaneous syndrome (CFC), Costello syndrome (CS), and capillary malformation–arteriovenous malformation syndrome (CM-AVM) [4]. NF1 results from various types of loss-of-function mutations in the tumor suppressor gene *NF1*. NS presents the highest locus heterogeneity: approximately half of NS patients carry mutations in *PTPN11*, while the other half present mutations in *SOS1, SOS2, SPRY1, CBL, LZTR1, RIT1, RRAS2, NRAS, HRAS, RRAS, KRAS, MRAS, PPP1CB, CRAF, RAF1, MAP3K8*, and *ERK2* [3,5]. While NS arises from gain-of-function mutations in RAS pathway proteins, NSML is due to loss-of-function mutations [6]. NS-LAH arises from mutations in *SHOC2* or *PPP1CB* [7,8], whereas NFNS arises from mutations in *NF1* and may involve the co-occurrence of variants in *NF1* and another gene associated with NS [9]. LS is caused by mutations in *SPRED1* [10]; CFC results from mutations in *KRAS, BRAF, MEK1*, and *MEK2* [11]; CS is caused by mutations in *HRAS* [12]; and CM-AVM is most frequently caused by mutations in *RASA1*, although recently mutations in *EPHB4* have been identified [13]. RASopathy patients exhibit alterations in the expression and activation of proteins in several signaling pathways, including RAS/MAPK, PI3K/AKT/mTOR, Rho/ROCK/LIMK2/cofilin, cAMP/PKA, JAK/STAT, Hippo, Wnt/β-catenin, and TGF-β [14–17]. Given the overlapping clinical presentations among RASopathies alongside their unique characteristics, leveraging biomarkers becomes imperative to enhance diagnostic accuracy, prognostic assessment, and treatment strategies. Moreover, there is a need for continued exploration and validation of these biomarkers to better understand their roles and potential clinical applications.

A biomarker is a measurable and quantifiable indicator of normal or pathological processes or a response to exposures or interventions. Biomarkers can be divided into different categories based on the evaluated indicator: molecular (DNA, RNA, proteins, and metabolites), histologic (tissue samples), or physiologic (blood pressure, heart rate, electrocardiography, echocardiography, and energy X-ray absorptiometry). An ideal biomarker should be either binary (absent or present) or objectively quantifiable, and it should be sensitive and specific. For this reason, developing a biomarker is a lengthy process that requires several steps, including identification, validation, and characterization [18]. Depending on their specific use, biomarkers can be categorized into diagnostic, monitoring, pharmacodynamic/response, and prognostic categories, and for their different uses, they assume great biological relevance [19]. A diagnostic biomarker identifies and confirms the presence and subtype of disease, playing a pivotal role in redefining disease classification in the precision medicine era. Early disease diagnosis enables earlier patient treatment, making the design of early diagnostic biomarkers crucial. Additionally, a prognostic biomarker predicts the likelihood of clinical events, disease recurrence, or progression and is essential for assessing the risk of poor outcomes. Monitoring biomarkers are indispensable for tracking disease progression, as they reflect disease status and exposure to or effects of medical agents. They also aid in treatment tailoring by providing evidence of off-target effects and ensuring drug safety through stable biomarker levels [19]. Furthermore, biomarkers enable an understanding of the molecular basis of physiological or pathological mechanisms, and they can also be essential in drug development and personalized medicine, as they make it possible to tailor treatments to the molecular profiles of patients [20].

Historically, NF1 was the first described RASopathy, diagnosed through clinical analysis and characteristic phenotypical features such as café-au-lait spots, intertriginous freckling, neurofibromas, and skeletal dysplasia, among other physiological and histological biomarkers [2]. The development of genetic testing techniques in the 2000s enabled precise identification of various mutations, leading to more accurate diagnoses despite overlapping characteristics. Today, the advent of next-generation sequencing (NGS) technologies allows a deeper analysis of variants causing RASopathies [20]. Currently, biomarkers used in RA-

Sopathies include genetic and transcriptional analysis of the RASopathy-associated genes, activation levels of RAS/MAPK proteins, measurement of serum levels of immune markers involved in the RAS/MAPK pathway, cardiac biomarkers, neurocognitive assessments, and physical features [3,21–24]. Nonetheless, only a few biomarkers are used in the detection and monitoring of RASopathies, and furthermore, there is a lack of specific therapies.

Even today, a significant number of RASopathy patients test positive in clinical analysis but not in genetic testing, due to the presence of mutations in non-coding regions or mosaicism [25]. Furthermore, despite the extensive literature on RASopathies and their molecular signatures, very few of these findings are employed as biomarkers in clinical practice, not only for diagnosis but also for tailoring precision medicine. The validation of a new biomarker for clinical use is a large, multi-step process ensuring its reliability, accuracy, and utility, involving analytical validation to establish its characteristics and biological relevance, followed by clinical validation and utility through retrospective and prospective clinical trials to confirm its ability to predict clinical outcomes reliably [18]. For RASopathies, the primary challenge for a biomarker is achieving high sensitivity and specificity in distinguishing between closely related disorders reproducibly, especially given the absence of RASopathy-specific treatments and the need for drugs targeting comorbidities or specific pathological features, necessitating a wide array of biomarkers [2]. The aim of this review is to consolidate existing knowledge on biomarkers utilized in the diagnosis, monitoring, and treatment of RASopathies, while also proposing avenues for future research to identify novel biomarkers. To achieve this goal, we have summarized recent studies and findings on RASopathies, categorizing biomarkers into molecular, metabolic, histological, and physiological sections.

2. Molecular Biomarkers of RASopathies

Molecular biomarkers are defined as one or a group of individual molecules measured by differential expression or concentration between a disease state and a normal control [26]. Molecular RASopathy biomarkers used thus far include genetic variants evaluated by genetic testing, mRNAs and ncRNAs assessed through both qualitative and quantitative analysis, and protein activity and expression levels [26].

2.1. Genes as Biomarkers

The most employed method of diagnosis of RASopathies is clinical examination; nonetheless, genetic testing is the most reliable method of detection and a well-established source of diagnostic molecular biomarkers, even if it is not routinely conducted for all patients [26]. This includes genomic DNA analysis of the RASopathy genes (Figure 1A) to find the pathogenic genetic variants by Sanger sequencing as well as whole-genome sequencing (WGS) and whole-exome sequencing (WES) for more comprehensive genetic testing [22,27]. Identified variants are normally classified following the criteria of the Human Genome Variation Society as benign, likely benign, uncertain clinical significance, likely pathogenic, or pathogenic, so that they can be used as genetic biomarkers of the disease [28,29]. Nonetheless, genetic screenings may sometimes prove ineffective in detecting the disease, as seen in the case of NS, where approximately 10–20% tested negative [25]. Furthermore, mosaicism, a condition in which individuals have two or more populations of cells with different genotypes due to spontaneous new mutations occurring during early embryonic or fetal development, can also contribute to the challenges in diagnosing RASopathies, as detection in blood is rarely successful [30]. In such cases, it is necessary to directly test the affected tissue by somatic sequencing to identify the mutation [31].

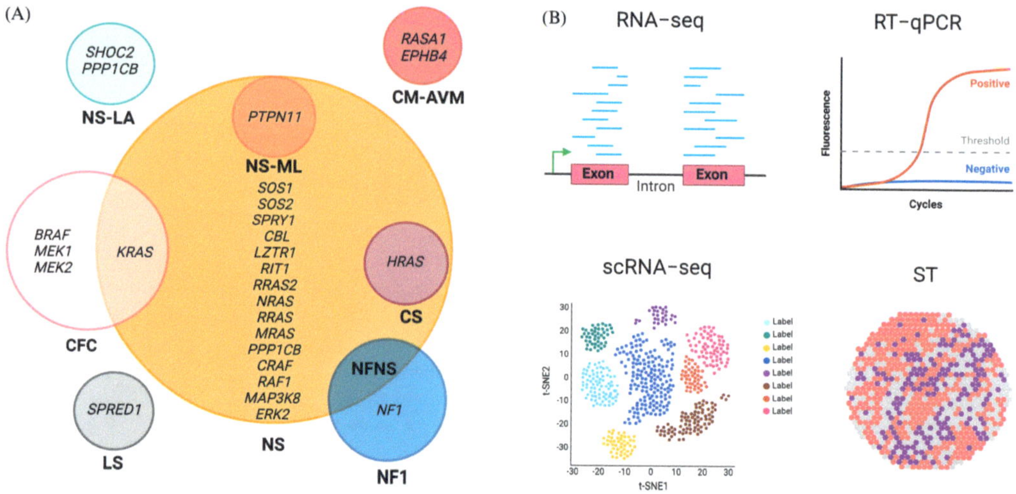

Figure 1. Molecular biomarkers of RASopathies. (**A**) Each circle represents a specific RASopathy and includes the associated genes. (**B**) mRNA detection methods: Overview of the main mRNA detection methods used to study the RASopathies. Created with BioRender.com (accessed on 3 August 2024).

2.2. mRNA Biomarkers

The second most common type of biomarker used in RASopathies is at the RNA level, involving qualitative and quantitative analysis of mRNA and non-coding RNA (ncRNA), such as microRNAs (miRNAs) and long non-coding RNAs (lncRNAs) [26,32]. In various studies, mRNA has been utilized as a diagnostic and prognostic biomarker and as a tool to uncover the molecular mechanisms underlying RASopathies, employing diverse transcriptomics techniques (Figure 1B, Table S1) [32–35].

In NF1, certain pathogenic variants can impact pre-mRNA splicing, and they are typically missed by genetic testing and bioinformatic prediction. For this reason, mRNA analysis has been crucial for identifying these variants and diagnosing the disease [33,34]. mRNA has also been used as a prognostic biomarker in CS. RT-qPCR studies demonstrated that the severity of the phenotype and the frequency of cancer in this syndrome may result from the splicing efficiency of exon 2 inclusion due to activating HRAS mutations. This also suggests that therapeutic interventions for CS could potentially target exon 2 skipping [35]. Furthermore, mRNA studies enable the observation of differential gene expression in tissues and cell types, potentially revealing gene expression profiles of both affected and unaffected tissues in RASopathies. Although these differences have not yet been used as biomarkers of RASopathies, they could be crucial for designing new biomarkers in the future (Table S2) [26,36]. In NS, RNA-seq profiles of WT and *LZTR1*-deficient induced pluripotent stem cell-derived cardiomyocytes (iPSC-CMs) have already been analyzed in a preclinical model of CRISPR repair for NS-associated cardiomyopathy, demonstrating that such therapy could potentially offer a means to normalize the hypertrophic phenotype commonly observed in NS patients [37].

Gene expression analysis in cardiomyocytes from an NSML patient revealed more than 400 downregulated genes and 200 upregulated genes, with the latter being genes associated with muscle development [38]. Furthermore, RNA-seq analysis has been utilized in CM-AVM to assess whether genetic variants causing the disease could result in a reduction in *RASA1* or RAS/MAPK gene expression [39]. Regarding NF1, it was discovered that the extracellular matrix of cutaneous neurofibromas (cNFs) in NF1 patients is largely comprised of collagen VI, a collagen type associated with a pro-tumorigenic role, in contrast to the typically pro-fibrogenic collagen I [40]. Furthermore, a study analyzing mRNA profiles of the plexiform neurofibroma (pNF) tumor environment demonstrated that these tumors are enriched with fibroblasts and non-myelinated Schwann cells [32]. Moreover, through the analysis of differential gene expression using scRNA-seq, it was revealed that orbitofacial NFs exhibit heightened activation of pathways related to cell proliferation, interferons, and immunity, suggesting a potential explanation for their increased local aggressiveness compared to other types of NFs [41]. All this evidence highlights the potential of incorporating mRNA into clinical practice and diagnosis.

2.3. ncRNA Biomarkers

The majority of the human genome is transcribed into non-coding RNAs (ncRNAs), which can be divided into classes according to their function, shape, and length [42]. Among these, two important categories found to have an impact on RASopathies are microRNAs (miRNAs) and long non-coding RNAs (lncRNAs) [43,44]. Indeed, up to 49 ncRNAs in total have been found to be involved in RASopathies (Table S3). Nonetheless, these are not yet widely used in clinical practice for diagnosing or monitoring RASopathies.

miRNAs serve as fine-tuning negative regulators of gene expression and, when mutated, can lead to RASopathies [43]. A study involving five patients with previously diagnosed RASopathies identified novel variants in mature miRNAs, pinpointing these as the underlying causes of the diseases. Therefore, both genomics and transcriptomics can be employed to analyze miRNAs and discover new variants associated with RASopathies, which are typically not identified through gene sequencing [43]. Furthermore, differential expression of miRNAs can be observed in RASopathies and has been crucial for understanding the molecular signature of malignant peripheral nerve sheath tumors (MPNSTs) compared to pNFs, as well as high-grade gliomas versus low-grade gliomas in NF1 [45–47]. Similarly, it has been valuable in clarifying the progression of juvenile myelomonocytic leukemia (JMML) in patients with NS caused by mutations in *PTPN11* [48]. These promising findings highlight the important role of miRNAs in RASopathies and their potential as biomarkers, underscoring the need for further experimental research.

Less is known about lncRNAs; they interact with proteins and nucleic acids to regulate gene expression, yet research into their role in RASopathies remains limited [49]. The lncRNA ANRIL is the only currently reported NF1 lncRNA, and the polymorphism rs2151280 has been associated with pNFs in NF1 patients. Interestingly, this polymorphism has also been identified in optic gliomas in NF1 patients, and some tumors from NF1 patients have exhibited loss of heterozygosity (LOH) at the ANRIL locus [50]. In another study, it was used as a diagnostic biomarker: a boy who was clinically diagnosed with NS but had negative genetic testing was found to have a deletion in lncRNA-*Dnm3os*, as determined by oligonucleotide-based array comparative genomic hybridization [44]. This lncRNA, which encodes two *N-Ras*-regulating miRNAs, was demonstrated to be required for maintaining the proliferative potential of articular chondrocytes by triggering Nerve Growth Factor (NGF) signaling [44]. This evidence not only resolved a particular case of NS but also highlighted a new class of lncRNAs, which, apart from producing miRNAs, form a regulatory network that maintains a proper pool of proliferating chondrocytes to support bone growth, dysregulated in NS.

2.4. Protein Biomarkers

As RASopathies result from specific genetic mutations that can alter protein expression or activation levels directly or indirectly, the use of proteins as biomarkers is crucial in these diseases. To date, 20 proteins have been utilized as biomarkers in RASopathies, with post-translational modifications (PTMs) including sixteen phosphorylations, two dephoshorylations, five ubiquitinations, two SUMOylations, three methylations, two acetylations, one palmitoylation, one ADP-ribosylation, one O-GlcNAcylation, and one S-nitrosylation being analyzed (Figure 2, Table S4).

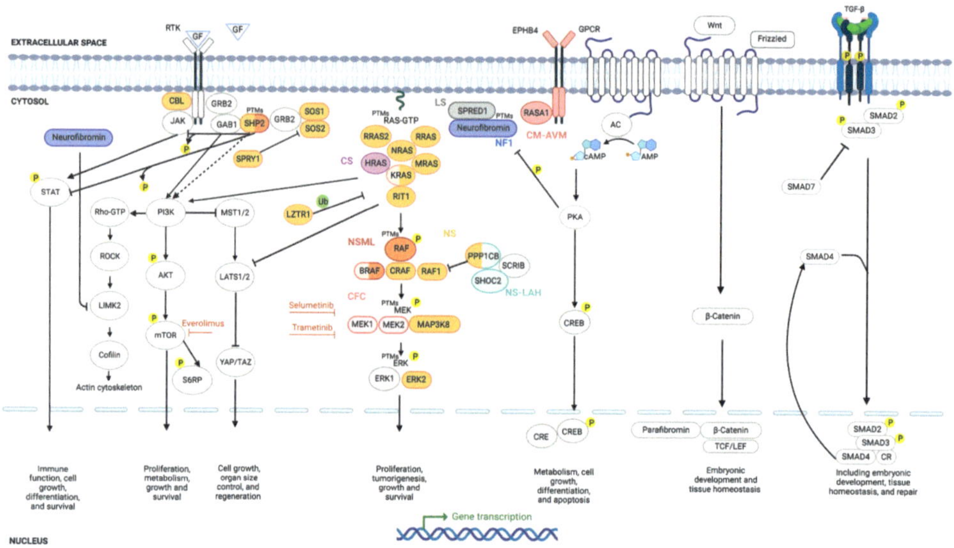

Figure 2. Protein biomarkers of RASopathies. Each color represents a disease: NF1 (blue), LS (gray), CM-AVM (pink), NS (brown), CS (violet), NSML (orange), and CFC (light pink). Inhibitors are indicated in red. Yellow circles denote phosphorylation, and green circles denote ubiquitination. PTM indicates the presence of multiple types of posttranslational modifications on that protein. Created with BioRender.com (accessed on 3 August 2024).

2.4.1. RAS/MAPK Biomarkers

The RAS/MAPK pathway biomarkers, including HRAS, KRAS, NRAS, ERK, MEK, RAF, and AKT, are common biomarkers [51] (Figure 2). This pathway is also intricately regulated by PTMs such as phosphorylation, ubiquitination, SUMOylation, and methylation (Figure 2, Table S4) [52]. These modifications are crucial for precise control of RAS signaling, and their dysregulation can contribute to RASopathies and tumorigenesis. RAS protein levels are regulated by ubiquitination and SUMOylation, which play a crucial role in regulating this pathway [53,54]. Additional PTMs regulating RAS degradation include phosphorylation of HRAS by GSK3β on residues T144 and T148 [55]. Other PTMs include methylation of K5 and K147 of KRAS [56] and palmitoylation of C181 and C184 of HRAS and NRAS, respectively, affecting the subcellular location and activity of RAS [57]. PTMs, facilitated by various proteins, play a fundamental role in regulating the degradation of RAS proteins, ensuring tight control cover their signaling functions. The interplay of these PTMs is vital for proper RAS signaling, and disruptions in these modifications can lead to sustained RAS activation, contributing to RASopathies and tumorigenesis. Understanding these PTMs provides valuable insights into the molecular mechanisms underlying RAS-driven diseases and highlights potential therapeutic targets for intervention.

Beyond RAS proteins, neurofibromin levels have been studied as biomarkers in NF1 [58–60]. Additionally, mutants such as KRAS^{G13D} exhibit impaired NF1 binding and are specific biomarkers for predicting sensitivity to EGFR-targeted therapies [61]. Interestingly, neurofibromin also undergoes phosphorylation at site S2808 by PKC-ε, which is a crucial regulatory process that impacts the protein's function in nuclear import and chromosome congression during mitosis [62]. Neurofibromin has long been implicated using various cell types and animal models as a positive or negative regulator of cAMP levels, although identifying the key molecules involved in this coupling has proven challenging, and the role of Ras has been reported to be variable [62]. Additionally, neurofibromin responds to growth factor stimuli, which modulate its interaction with regulatory proteins such as 14-3-3, thereby affecting its GAP activity towards Ras [63]. In NF1, the loss of neurofibromin results in defective regulation of RAS ubiquitination, leading to sustained RAS activation and contributing to tumorigenesis [58]. Regarding PTMs, neurofibromin undergoes methylation [64] and SUMOylation, with 15 SUMO consensus motifs and two SIM sites predicted by JASSA [65]. Palmitoylation also occurs in the SPRED1–neurofibromin–KRAS complex [66]. Other proteins with PTMs include BRAF, c-RAF (RAF1), MEK1, MEK2, and ERK [53–56,67–77] (Table S2).

2.4.2. PI3K/AKT/mTOR Biomarkers

Due to the crosstalk between the PI3K/AKT/mTOR and RAS/MAPK pathways, phosphorylation levels of AKT, S6RP, and mTOR have been previously used as biomarkers in research on RASopathies [16,78] (Figure 2). To the best of our knowledge, these are not currently used as diagnostic biomarkers for RASopathies. Nonetheless, based on the following evidence, we propose their use in clinical practice for the study of affected tissues. In a case study of a patient with *PTPN11*-caused NS, immunohistochemistry of a lesion from a glioneuronal neoplasm showed positive staining for pmTOR [16]. Similar results were obtained in an NSML patient with hypertrophic cardiomyopathy caused by a mutation in *PTPN11*, where the levels of pAKT and pS6RP (a downstream target of mTOR) in the patient's skin fibroblasts demonstrated enhanced PI3K/AKT/mTOR pathway activity. Furthermore, treatment with everolimus, an mTOR inhibitor, resulted in an improvement in heart failure risk and a reduction in brain natriuretic peptide levels [78]. Nonetheless, the functional link between SHP2 (encoded by *PTPN11*) and mTOR is not fully understood, although it has been found that the regulatory subunit p85 of PI3K interacts directly with SHP2 [79]. High levels of pAKT, pmTOR, and pS6RP have also been found in MPNSTs, indicating their role in the aggressive clinical behavior of these tumors in NF1 patients [80]. Preclinical and clinical trials have demonstrated the effectiveness of mTOR inhibitors in treating cNFs and low-grade gliomas, underscoring the potential of pmTOR as a treatment response biomarker for these conditions in NF1 patients [81–83]. Furthermore, it would be valuable to investigate whether pmTOR could serve as a predictive biomarker for progression from benign to malignant fibromas in NF1.

2.4.3. Rho/ROCK/LIMK2/Cofilin Biomarkers

The Rho/ROCK/LIMK2/cofilin pathway is a key pathway for various cellular processes such as cytoskeletal dynamics, cell migration, cell morphology, and cell adhesion [84] (Figure 2). This pathway has not been extensively studied in the context of RASopathies [85,86], but in NF1, ROCK plays a crucial role in regulating actin cytoskeleton dynamics and cell contractility [85]. Dysregulation of this pathway can lead to cardiac abnormalities and developmental defects, as seen in conditions such as NS and LS, due to gain-of-function SHP-2 mutations that induce hyperactivity of ROCK [87].

2.4.4. cAMP/PKA Biomarkers

Up to three biomarkers have been identified in the cAMP/PKA pathway in RASopathies, namely, cAMP [88,89], PKA [60], and the ratio between neurofibromin isoforms I and II [90]. These molecules influence the balance between the cAMP/PKA and RAS/MAPK pathways, which can be used as a therapeutic strategy for cNFs [89] (Figure 2). The cysteine-serine-rich domain of neurofibromin may regulate adenylate cyclase (AC) activity; thus, the deletion of neurofibromin results in dysfunctional cAMP signaling [88,91]. On the other hand, PKA phosphorylates neurofibromin, impairing its GAP activity and thus dysregulating the RAS/MAPK pathway [60]. The ratio between neurofibromin isoforms I and II contributes to this regulation, as isoform II is a weaker negative regulator of RAS than isoform I [90].

2.4.5. JAK/STAT Biomarkers

The JAK/STAT pathway is composed of non-receptor tyrosine protein kinases (JAKs) and signal transducers and activators of transcription (STATs) [92]. This pathway plays integral roles in various cellular processes, including mitosis, differentiation, apoptosis, hematopoiesis, the development of the immune system, and the functioning of exocrine glands [93]. In RASopathies, this pathway's components have not yet been used as biomarkers. Given the interplay between the JAK/STAT and RAS/MAPK pathways [94] (Figure 2), it was suggested that this pathway should be explored as a biomarker for RASopathies. Indeed, SHP2 directly dephosphorylates STAT5 and JAK1, thereby attenuating their activity through inhibition of dimerization [95,96]. Additionally, STAT3 is identified as a critical factor in the initiation of neurofibromas [97].

2.4.6. Hippo Pathway Biomarkers

The Hippo tumor suppressor pathway is an evolutionarily conserved signaling cascade that regulates numerous biological processes, such as cell growth, organ size control, and regeneration [98,99] (Figure 2). Although, to the best of our knowledge, the components of the Hippo pathway have not yet been used as diagnostic or monitoring biomarkers in RASopathies, there is evidence highlighting its impact in NF1. WES studies in NF1 patients assessing the role of acquired somatic mutations in the growth of cutaneous neurofibromas (cNFs) have revealed significant dysregulation of Hippo signaling, suggesting its involvement in cancer progression [100]. Further investigations using WES datasets of cNFs from NF1 patients identified 30 somatic mutations in the Hippo pathway genes, along with elevated expression levels of YAP and TAZ [15]. Moreover, evidence suggests that the transformation of MPNSTs from SCs is influenced by deregulation of the Hippo pathway [68,101]. These studies, along with the interplay between the Hippo pathway and the RAS/MAPK and PI3K/AKT/mTOR pathways mediated by core Hippo kinase proteins (MST1/2 and LATS1/2) [14], suggest that investigating Hippo pathway components as biomarkers could yield novel diagnostic and monitoring tools for RASopathies, improving disease management and treatment outcomes.

2.4.7. Wnt/β-Catenin Biomarkers

The Wnt/β-catenin pathway, the canonical pathway involving the Wnt cascade [102], plays crucial roles in embryonic development and tissue homeostasis [103]. The key molecule in this pathway is β-catenin, which acts as a nuclear effector of the pathway and is also an important component of the cytoskeleton [104]. Up to four biomarkers have been related to the Wnt/β-catenin pathway in RASopathies, namely, β-catenin [17,105], the Wnt ligand Frizzled (FZD) [17], neurofibromin [17], and parafibromin [106] (Figure 2). These biomarkers collectively provide insights into the interplay between the Wnt/β-catenin and RAS/MAPK pathways, crucial for understanding the molecular mechanisms underlying RASopathies.

2.4.8. TGF-β Pathway Biomarkers

Transforming growth factor-β (TGF-β) represents an evolutionary conserved family of secreted proteins with cell-type-specific and developmental-stage-specific actions, playing roles in embryogenesis, differentiation of most cell lineages, and adult tissue homeostasis [107,108]. Three biomarkers have been identified for the TGF-β pathway in RASopathies (Figure 2). TGF-β itself is the primary biomarker, as the frequency of CD4+ cells expressing it is increased in NF1 patients, suggesting an immunosuppressed status [109]. SHP2 is a biomarker for NS, since mutations excessively activate the TGF-β pathway, producing an impairment in early neuroectodermal development in NS-iPSCs [110]. Finally, regarding CS, hyperactivation of SMAD3 signaling during osteogenic differentiation of CS-patient-derived mesenchymal stem cells leads to aberrant expression of extracellular matrix remodeling proteins [73]. Understanding the TGF-β signaling pathway might clarify the molecular pathogenesis of tumor development in NF1 and how neurodevelopment and osteogenic differentiation are affected in NS and CS, respectively. The integration of molecular biomarkers with clinical data, including blood tests, cardiac and immune biomarkers, and imaging techniques such as MRI and PET, enhances diagnosis, monitoring, and treatment tailoring for patients. This approach is crucial in clinical practice, providing a more holistic understanding of RASopathies.

3. Metabolite Biomarkers

Studying the metabolome can reveal the organism's metabolic response to a pathological stimulus and provide information on the molecular pathways involved in the development and progression of a certain disease, helping to bridge the genotype–phenotype gap [111,112]. Very few studies have been conducted to correlate metabolomics and RASopathies, and only a limited number of relevant metabolites have been identified. Metabolomic studies on RASopathies have identified up to five different potential biomarkers: adipokines, glucose levels, high-density lipoprotein (HDL) cholesterol, triglycerides (TGs), and urinary catecholamine. It has been demonstrated that individuals with NF1 exhibit increased metabolic levels of adipokines—cytokines regulated by adipocytes—compared to controls [113]. NF1 patients also exhibit reduced fasting blood glucose and enhanced glucose clearance compared to matched controls [113,114]. This enhanced metabolism of glucose and other energy substrates may participate in tumor growth and transformation [115]. Furthermore, other research suggests that both children and young adults affected by NS or Noonan-related diseases have an unfavorable metabolic profile with low HDL cholesterol, a tendency toward elevated TGs, and an impairment in glucose metabolism despite presenting a lean phenotype [116]. Elevated urinary catecholamine metabolites were observed in a patient suffering from progressive neonatal hypertrophic cardiomyopathy (HCM) and dysmorphic features in whom a well-known NSML-associated *PTPN11* mutation (c.1403 C>T; p.T468M) and a novel, potentially pathogenic missense *SOS1* variant (c.1018 C>T; p.P340S) were found. In particular, mild elevations of vanillylmandelic acid (VMA) and homovanillic acid (HVA) raised concerns of neuroblastoma that, along with the progressive HCM, suggested a RASopathy [117]. This altered metabolic profile has also been described in patients with CS, with increased levels of VMA, HVA, epinephrine, norepinephrine, dopamine, and other metabolites [118]. The detection of such metabolites is feasible in terms of accessibility, as the measurements are taken from urine or blood, for example, and the alterations are determined with precision by using methods such as analytical liquid chromatography–mass spectrometry (LC-MS) [119]. Alterations in glucose metabolism and lipid profiles significantly impact disease progression in RASopathies. Increased glucose clearance can support tumor development and exacerbate disease symptoms, while abnormal lipid profiles, such as low HDL cholesterol and high triglycerides, contribute to cardiovascular issues and metabolic complications. These metabolic disruptions fuel tumor growth, elevate cardiovascular risks, and disrupt overall metabolic balance, complicating patient management and outcomes.

4. Biomarkers in Histology and Molecular Tissue Characterization

In pathology, various techniques identify potential biomarkers that are essential for diagnosis, prognosis, and treatment response. These biomarkers play a pivotal role in clinical trials for disease prediction and treatment monitoring [120]. Histological and molecular biomarkers of RASopathies have been utilized (Figure 3).

These include cellular and tissue structure changes, such as alterations in cell shape and size or extracellular matrix modifications including fibrosis [40,121]. Tissue abnormalities, such as hypertrophic cardiomyopathy, skeletal muscle myopathy, and cutaneous papillomas, are also significant [121–123]. Molecular markers involve the presence or absence of specific proteins detectable through immunohistochemistry [124], including Ki-67, p53, AXL, ERK phosphorylation, growth factors and hormone receptors, VEGF, Interleukin-6, CXCR4/CXCL12, calbindin D, S100, p16, and CDKN2A, among others, providing insights into tissue structure, function, and pathological states (Figure 3).

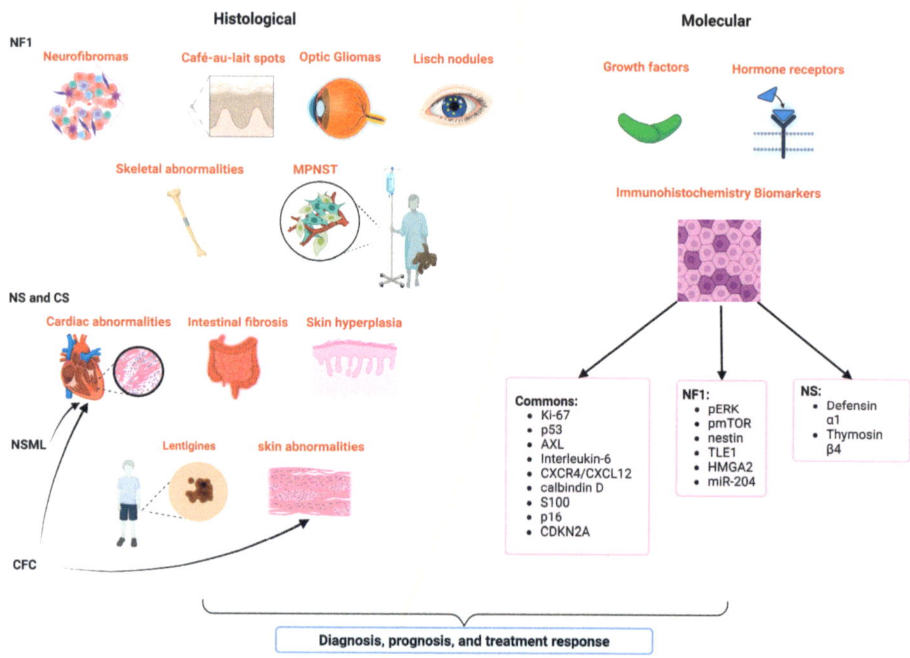

Figure 3. Histological and molecular characterization of RASopathies. Schematic representation of histological (**left panel**) and molecular (**right panel**) biomarkers commonly used in RASopathies for diagnosis, prognosis, and treatment response. Created with BioRender.com (accessed on 3 August 2024).

In NF1, specific biomarkers identify clinical and histological features such as neurofibromas [40], café-au-lait spots [125], optic gliomas [126], and Lisch nodules [127], highlighting distinct cellular compositions and tissue characteristics associated with the disease (Figure 3). Additionally, specific biomarkers such as hypercellularity and nuclear atypia in MPNSTs [128]; growth factors, hormone receptors, and signaling pathways in cNFs [120]; and insulin-like growth factor-1 and growth hormone receptors in neoplastic Schwann cells [120], along with ERK phosphorylation; mTOR pathway targets [129]; and histological markers such as calbindin D, S100, and nestin, are crucial for diagnosis, prognosis, patient selection, and treatment assessment [130]. Additional markers such as TLE1, HMGA2, p53, p16, CDKN2A, and miR-204 aid in differentiating tumor types and predicting patient outcomes [131–133] (Figure 3). LS shares clinical features with NF1, such as café-au-lait macules and freckling, but lacks neurofibromas [134], and while histological similarities include increased melanin levels [135], specific histological biomarkers unique to LS have not been identified.

Research on biomarkers in NS, NSML, and CS has provided valuable insights for their diagnosis and management. In NS, salivary inflammatory biomarkers such as defensin α1 and thymosin β4 are elevated [136], and males may show specific biomarkers indicating primary testicular insufficiency in Sertoli cells [137]. Histological findings in NS include cellular hyperplasia in skin and connective tissues, with cardiac manifestations suggesting potential histological changes in cardiac tissue [122,138,139]. NSML is characterized by cardiac features such as pulmonary valve stenosis and hypertrophic cardiomyopathy, along with lentigines on the skin [121]. CS presents with dysmorphic facial features, dermatological manifestations, and cardiac anomalies, with histological biomarkers such as abnormal collagen and elastic fibers contributing to its diagnosis and management [1]. Additionally, CS is associated with oncogenic predisposition, featuring both benign and malignant neoplasms such as cutaneous papillomas and rhabdomyosarcoma [123]. Histologically, specific biomarkers for CFC syndrome are not well defined. Clinically, CFC is characterized by cardiac anomalies such as pulmonic stenosis, septal defects, and hypertrophic cardiomyopathy; distinct facial features; and various skin abnormalities include dryness, hyperkeratosis, ichthyosis, keratosis pilaris, and hemangiomas [140]. Thus, histological biomarkers are essential and more reliable for clinical practice in the accurate identification and management of these diseases.

5. Physiologic Biomarkers

In delineating the physiologic markers of RASopathies, distinctions can be made among cardiac, bone, and embryonic indicators (Figure 4).

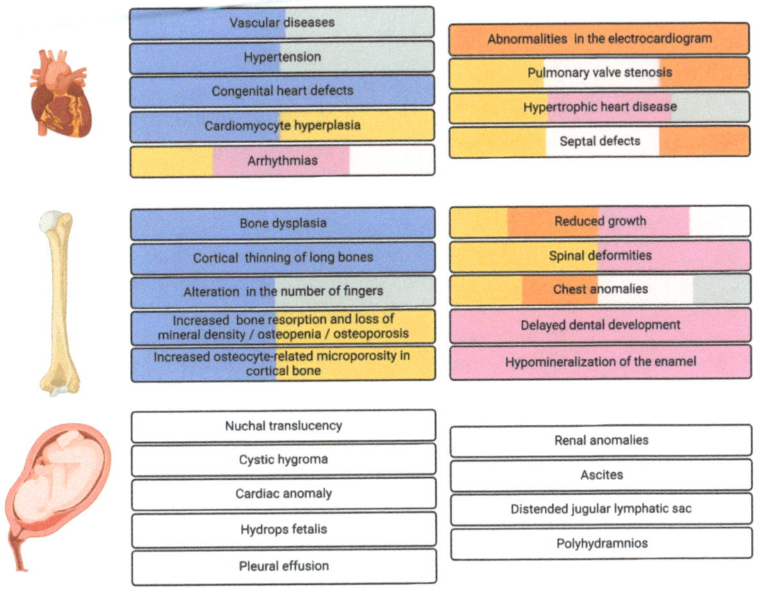

Figure 4. Cardiac, bone, and embryonic RASopathy physiological biomarkers. Each color represents a specific RASopathy. NF1: blue; LS: gray; NS: brown; CS: violet; NSML: orange; CFC: pink. Biomarkers common to all RASopathies are indicated in white. Created with BioRender.com (accessed on 3 August 2024).

Biomarkers pertaining to bone physiology, mainly increased bone resorption and loss of mineral density, alterations in the skeleton, reduced growth, alterations in the pectus, and alterations in the number of fingers, have been used in NF1, NS, NSML, CS, CFCS, and LS [22,121,141–144]. Regarding detection, most bone physiological markers present in RASopathies can be detected through X-rays. On the other hand, the loss of mineral density in bones is detectable by dual-energy X-ray absorptiometry (DXA) or by detecting urine pyridinium levels, since an increase leads to a decrease in mineral density [143].

Biomarkers of cardiac physiology, mainly arrhythmias and electrocardiogram abnormalities, pulmonary valve stenosis, septal defects, and hypertrophic cardiac diseases, are used in the same RASopathies as bone markers [22,38,121,141,143,144]. The detection of these biomarkers can be carried out during gestation through gestational ultrasonography. In adulthood, it is possible to use both electrocardiograms, which detect defects in the heart rhythm, and cardiac magnetic resonance, which can generate images to determine ventricular volume and analyze function as well as ejection fraction and myocardial mass, providing non-invasive mechanisms that make it possible to identify alterations in cardiac physiology [38,142,145]. During pregnancy, suspected prenatal RASopathies can be detected by ultrasound, which allows for the detection of increased nuchal translucency (>95th percentile), cystic hygroma, cardiac anomalies, hydrops fetalis, pleural effusion, renal anomalies, ascites, distended jugular lymphatic sac, and polyhydramnios [146].

Neurodevelopmental and endocrine biomarkers are crucial for diagnosis and management of RASopathies. Common symptoms include developmental delays, cognitive deficits, and structural brain malformations [147]. Neurodevelopmental biomarkers include cognitive and developmental assessments; neuroimaging findings; and variations in brain structure, such as changes in cortical surface area and thickness, as well as subcortical volume effects specific to each syndrome [148]. Conditions such as NF1, NS, and CS are associated with delays in speech and motor skills, learning disabilities, intellectual disabilities, and behavioral issues such as ADHD and autism spectrum disorders [1,147]. Endocrine complications frequently observed in RASopathies include short stature, reduced bone mineral density, and thyroid autoimmunity, with higher anti-TPO antibody levels noted in NS and CFC syndrome [149] (Figure 4). NF1 also presents endocrine challenges such as central precocious puberty, growth hormone deficiency, and hypersecretion, often associated with OPG affecting the hypothalamic–pituitary region [150].

6. Conclusions

Future research on biomarkers for RASopathies should focus on expanding metabolomic profiles, conducting longitudinal studies, and employing multi-omic approaches. Key areas include standardizing biomarker measurements, identifying biomarkers across diverse populations, and utilizing advanced imaging techniques. Additionally, exploring epigenetic biomarkers, validating therapeutic ones, and implementing artificial intelligence to analyze complex data and predict disease progression are essential. These efforts will enhance the understanding of disease progression, enable personalized treatment, and support effective clinical implementation, addressing current gaps and improving the management of RASopathies. Additionally, the transition from biomarker discovery to clinical application is a complex process that involves several critical steps and challenges. These include rigorous validation to ensure the biomarker's reliability and accuracy, as well as extensive clinical trials to establish its predictive value and utility in real-world settings. Furthermore, integrating insights from disciplines such as bioinformatics and systems biology can provide a more comprehensive understanding of the biomarker landscape. These fields contribute valuable tools and methods for analyzing large datasets, identifying patterns, and understanding the molecular mechanisms underlying disease processes. This multidisciplinary approach is essential for overcoming the challenges of biomarker implementation in routine clinical practice, ultimately enhancing patient diagnosis, monitoring, and treatment.

Biomarkers can provide a more objective way to diagnose RASopathies, which can lead to earlier and more effective treatments. They can also help researchers better understand how RASopathies progress, thereby aiding in the development of targeted therapies that are more effective and have fewer side effects. In the study of RASopathies, genetic markers remain the most crucial and widely used biomarkers due to their direct correlation with disease-causing mutations. Non-coding RNAs, such as miRNAs and lncRNAs, are emerging as important biomarkers but are still primarily in the research phase. Protein expression levels are directly linked to the presence and severity of RASopathies but are less frequently used in clinical practice. Phosphorylated proteins, including pAKT, pS6RP, and pERK, show significant promise for monitoring disease progression and treatment efficacy, although further clinical validation is needed. Metabolic markers, while less commonly used, hold potential for monitoring disease severity and complications. Histological and molecular biomarkers also play a vital role in the diagnosis and clinical management of RASopathies, offering critical insights into tissue structure and pathological states. Specific histopathological features, such as the presence of neurofibromas in NF1 or characteristic cardiac malformations in NS and NMSL, as well as dermatological features in CS, provide critical diagnostic clues. The reliability of these biomarkers supports accurate diagnosis, prognosis, and patient selection for targeted treatments. IHC, utilizing antibodies to detect biomarkers such as Ki-67, p53, and S100, plays a crucial role on this process, enabling precise clinical decision making and personalized therapeutic approaches. Physiological biomarkers, including cardiac function assessments and neurodevelopmental evaluations, are important for ongoing monitoring of patients with RASopathies. These non-invasive assessments can help track disease progression and response to treatment, offering valuable information for clinical management. One of the main concerns of current medicine is to offer a personalized or precision approach, which consists of adapting medical treatment according to the patient's genomic variations, biochemical profile, environment, and lifestyle. There is a significant research gap in translating these biomarkers from research to clinical practice. Limited clinical validation, the need for standardized protocols, and a lack of longitudinal studies are major challenges that need to be addressed. In conclusion, while significant progress has been made in identifying and utilizing biomarkers for RASopathies, much work remains to fully translate these findings into clinical practice. Continued efforts in basic and clinical research will pave the way for more effective and personalized approaches to managing these complex disorders, ultimately improving the health and life expectancy of patients.

Supplementary Materials: The following supporting information can be downloaded at: https://www.mdpi.com/article/10.3390/ijms25168563/s1.

Author Contributions: Conceptualization, investigation, and writing—original draft preparation: J.L., N.F., J.B.-F., M.R.-M., J.S.-R. and A.C.; writing—review and editing: J.L., N.F., J.B.-F., M.R.-M., J.S.-R. and A.C.; supervision: J.L.; funding acquisition: P.P.-M., M.I.-G. and J.L. All authors have read and agreed to the published version of the manuscript.

Funding: This study received support from the Alicia Koplowitz Foundation through the Research Grants program (Code: FAK21/001). Noemi Ferrito was supported by a predoctoral fellowship from the University of Salamanca (USAL). Juan Báez Flores received a predoctoral fellowship from Banco Santander and the University of Salamanca (USAL). Mario Rodríguez Martín's salary was funded by the Programa Investigo under the European Union—NextGenerationEU. We sincerely thank the Alicia Koplowitz Foundation, SEPE, the European Union, Banco Santander, and USAL for their invaluable support, which was crucial for the successful completion of this study.

Data Availability Statement: No new data were created in this study. Data sharing is not applicable to this article as no datasets were generated.

Conflicts of Interest: The authors declare no conflicts of interest.

References

1. Rauen, K.A. The RASopathies. *Annu. Rev. Genom. Hum. Genet.* **2013**, *14*, 355–369. [CrossRef] [PubMed]
2. Tidyman, W.E.; Rauen, K.A. The RASopathies: Developmental Syndromes of Ras/MAPK Pathway Dysregulation. *Curr. Opin. Genet. Dev.* **2009**, *19*, 230–236. [CrossRef] [PubMed]
3. Rauen, K.A. Defining RASopathy. *Dis. Models Mech.* **2022**, *15*, dmm049344. [CrossRef]
4. Montero-Bullón, J.F.; González-Velasco, Ó.; Isidoro-García, M.; Lacal, J. Integrated in Silico MS-Based Phosphoproteomics and Network Enrichment Analysis of RASopathy Proteins. *Orphanet J. Rare Dis.* **2021**, *16*, 303. [CrossRef] [PubMed]
5. Saint-Laurent, C.; Mazeyrie, L.; Yart, A.; Edouard, T. Novel Therapeutic Perspectives in Noonan Syndrome and RASopathies. *Eur. J. Pediatr.* **2023**, *183*, 1011–1019. [CrossRef] [PubMed]
6. Kauffman, H.; Ahrens-Nicklas, R.C.; Calderon-Anyosa, R.J.C.; Ritter, A.L.; Lin, K.Y.; Rossano, J.W.; Quartermain, M.D.; Banerjee, A. Genotype–Phenotype Association by Echocardiography Offers Incremental Value in Patients with Noonan Syndrome with Multiple Lentigines. *Pediatr. Res.* **2021**, *90*, 444–451. [CrossRef] [PubMed]
7. Huckstadt, V.; Chinton, J.; Gomez, A.; Obregon, M.G.; Gravina, L.P. Noonan Syndrome with Loose Anagen Hair with Variants in the PPP1CB Gene: First Familial Case Reported. *Am. J. Med. Genet. Part A* **2021**, *185*, 1256–1260. [CrossRef] [PubMed]
8. Wang, Q.; Cheng, S.; Fu, Y.; Yuan, H. Case Report: A de Novo RASopathy-Causing SHOC2 Variant in a Chinese Girl with Noonan Syndrome-like with Loose Anagen Hair. *Front. Genet.* **2022**, *13*, 1040124. [CrossRef] [PubMed]

9. Bertola, D.R.; Castro, M.A.A.; Yamamoto, G.L.; Honjo, R.S.; Ceroni, J.R.; Buscarilli, M.M.; Freitas, A.B.; Malaquias, A.C.; Pereira, A.C.; Jorge, A.A.L.; et al. Phenotype–Genotype Analysis of 242 Individuals with RASopathies: 18-Year Experience of a Tertiary Center in Brazil. *Am. J. Med. Genet. Part C Semin. Med. Genet.* **2020**, *184*, 896–911. [CrossRef]
10. Pabst, L.; Carroll, J.; Lo, W.; Truxal, K.V. Moyamoya Syndrome in a Child with Legius Syndrome: Introducing a Cerebral Vasculopathy to the SPRED1 Phenotype? *Am. J. Med. Genet. Part A* **2021**, *185*, 223–227. [CrossRef]
11. Scorrano, G.; David, E.; Calì, E.; Chimenz, R.; La Bella, S.; Di Ludovico, A.; Di Rosa, G.; Gitto, E.; Mankad, K.; Nardello, R.; et al. The Cardiofaciocutaneous Syndrome: From Genetics to Prognostic–Therapeutic Implications. *Genes* **2023**, *14*, 2111. [CrossRef] [PubMed]
12. Leoni, C.; Paradiso, F.V.; Foschi, N.; Tedesco, M.; Pierconti, F.; Silvaroli, S.; Diego, M.D.; Birritella, L.; Pantaleoni, F.; Rendeli, C.; et al. Prevalence of Bladder Cancer in Costello Syndrome: New Insights to Drive Clinical Decision-Making. *Clin. Genet.* **2022**, *101*, 454–458. [CrossRef] [PubMed]
13. Nicholson, C.L.; Flanagan, S.; Murati, M.; Boull, C.; McGough, E.; Ameduri, R.; Weigel, B.; Maguiness, S. Successful Management of an Arteriovenous Malformation with Trametinib in a Patient with Capillary-Malformation Arteriovenous Malformation Syndrome and Cardiac Compromise. *Pediatr. Dermatol.* **2022**, *39*, 316–319. [CrossRef] [PubMed]
14. Báez-Flores, J.; Rodríguez-Martín, M.; Lacal, J. The Therapeutic Potential of Neurofibromin Signaling Pathways and Binding Partners. *Commun. Biol.* **2023**, *6*, 436. [CrossRef] [PubMed]
15. Chen, Z.; Mo, J.; Brosseau, J.P.; Shipman, T.; Wang, Y.; Liao, C.P.; Cooper, J.M.; Allaway, R.J.; Gosline, S.J.C.; Guinney, J.; et al. Spatiotemporal Loss of NF1 in Schwann Cell Lineage Leads to Different Types of Cutaneous Neurofibroma Susceptible to Modification by the Hippo Pathway. *Cancer Discov.* **2019**, *9*, 114–129. [CrossRef] [PubMed]
16. Lodi, M.; Boccuto, L.; Carai, A.; Cacchione, A.; Miele, E.; Colafati, G.S.; Camassei, F.D.; de Palma, L.; de Benedictis, A.; Ferretti, E.; et al. Low-Grade Gliomas in Patients with Noonan Syndrome: Case-Based Review of the Literature. *Diagnostics* **2020**, *10*, 582. [CrossRef] [PubMed]
17. Luscan, A.; Shackleford, G.G.; Masliah-Planchon, J.; Laurendeau, I.; Ortonne, N.; Varin, J.; Lallemand, F.; Leroy, K.; Dumaine, V.; Hivelin, M.; et al. The Activation of the WNT Signaling Pathway Is a Hallmark in Neurofibromatosis Type 1 Tumorigenesis. *Clin. Cancer Res.* **2014**, *20*, 358–371. [CrossRef] [PubMed]
18. Ou, F.S.; Michiels, S.; Shyr, Y.; Adjei, A.A.; Oberg, A.L. Biomarker Discovery and Validation: Statistical Considerations. *J. Thorac. Oncol.* **2021**, *16*, 537–545. [CrossRef] [PubMed]
19. Califf, R.M. Biomarker Definitions and Their Applications. *Exp. Biol. Med.* **2018**, *243*, 213–221. [CrossRef]
20. McCann, M.R.; De la Rosa, M.V.G.; Rosania, G.R.; Stringer, K.A. L-Carnitine and Acylcarnitines: Mitochondrial Biomarkers for Precision Medicine. *Metabolites* **2021**, *11*, 51. [CrossRef]
21. Gurusamy, N.; Rajasingh, S.; Sigamani, V.; Rajasingh, R.; Isai, D.G.; Czirok, A.; Bittel, D.; Rajasingh, J. Noonan Syndrome Patient-Specific Induced Cardiomyocyte Model Carrying SOS1 Gene Variant c.1654A>G. *Exp. Cell Res.* **2021**, *400*, 112508. [CrossRef] [PubMed]
22. Jafry, M.; Sidbury, R. RASopathies. *Clin. Dermatol.* **2020**, *38*, 455–461. [CrossRef]
23. Motta, M.; Pannone, L.; Pantaleoni, F.; Bocchinfuso, G.; Radio, F.C.; Cecchetti, S.; Ciolfi, A.; Di Rocco, M.; Elting, M.W.; Brilstra, E.H.; et al. Enhanced MAPK1 Function Causes a Neurodevelopmental Disorder within the RASopathy Clinical Spectrum. *Am. J. Hum. Genet.* **2020**, *107*, 499–513. [CrossRef] [PubMed]
24. Tamburrino, F.; Mazzanti, L.; Scarano, E.; Gibertoni, D.; Sirolli, M.; Zioutas, M.; Schiavariello, C.; Perri, A.; Mantovani, A.; Rossi, C.; et al. Lipid Profile in Noonan Syndrome and Related Disorders: Trend by Age, Sex and Genotype. *Front. Endocrinol.* **2023**, *14*, 1209339. [CrossRef] [PubMed]
25. Zenker, M. Clinical Overview on RASopathies. *Am. J. Med. Genet. Part C Semin. Med. Genet.* **2022**, *190*, 414–424. [CrossRef] [PubMed]
26. Linglart, L.; Gelb, B.D. Congenital Heart Defects in Noonan Syndrome: Diagnosis, Management, and Treatment. *Am. J. Med. Genet. Part C Semin. Med. Genet.* **2020**, *184*, 73–80. [CrossRef] [PubMed]
27. Tang, S.; Yuan, K.; Chen, L. Molecular Biomarkers, Network Biomarkers, and Dynamic Network Biomarkers for Diagnosis and Prediction of Rare Diseases. *Fundam. Res.* **2022**, *2*, 894–902. [CrossRef] [PubMed]
28. Legius, E.; Messiaen, L.; Wolkenstein, P.; Pancza, P.; Avery, R.A.; Berman, Y.; Blakeley, J.; Babovic-Vuksanovic, D.; Cunha, K.S.; Ferner, R.; et al. Revised Diagnostic Criteria for Neurofibromatosis Type 1 and Legius Syndrome: An International Consensus Recommendation. *Genet. Med.* **2021**, *23*, 1506–1513. [CrossRef] [PubMed]
29. Richards, S.; Aziz, N.; Bale, S.; Bick, D.; Das, S.; Gastier-Foster, J.; Grody, W.W.; Hegde, M.; Lyon, E.; Spector, E.; et al. Standards and Guidelines for the Interpretation of Sequence Variants: A Joint Consensus Recommendation of the American College of Medical Genetics and Genomics and the Association for Molecular Pathology. *Genet. Med.* **2015**, *17*, 405–424. [CrossRef]
30. Moog, U.; Felbor, U.; Has, C.; Zirn, B. Disorders Caused by Genetic Mosaicism. *Dtsch. Arztebl. Int.* **2020**, *117*, 119–125. [CrossRef]
31. Chang, C.A.; Perrier, R.; Kurek, K.C.; Estrada-Veras, J.; Lehman, A.; Yip, S.; Hendson, G.; Diamond, C.; Pinchot, J.W.; Tran, J.M.; et al. Novel Findings and Expansion of Phenotype in a Mosaic RASopathy Caused by Somatic KRAS Variants. *Am. J. Med. Genet. Part A* **2021**, *185*, 2829–2845. [CrossRef] [PubMed]

32. Amani, V.; Riemondy, K.A.; Fu, R.; Griesinger, A.M.; Grimaldo, E.; De Sousa, G.R.; Gilani, A.; Hemenway, M.; Foreman, N.K.; Donson, A.M.; et al. Integration of Single-Nuclei RNA-Sequencing, Spatial Transcriptomics and Histochemistry Defines the Complex Microenvironment of NF1-Associated Plexiform Neurofibromas. *Acta Neuropathol. Commun.* **2023**, *11*, 158. [CrossRef] [PubMed]
33. Douben, H.; Hoogeveen-Westerveld, M.; Nellist, M.; Louwen, J.; Haan, M.K.D.; Punt, M.; Van Ommeren, B.; Van Unen, L.; Elfferich, P.; Kasteleijn, E.; et al. Functional Assays Combined with Pre-mRNA-Splicing Analysis Improve Variant Classification and Diagnostics for Individuals with Neurofibromatosis Type 1 and Legius Syndrome. *Hum. Mutat.* **2023**, *2023*, 9628049. [CrossRef]
34. Koster, R.; Brandão, R.D.; Tserpelis, D.; van Roozendaal, C.E.P.; van Oosterhoud, C.N.; Claes, K.B.M.; Paulussen, A.D.C.; Sinnema, M.; Vreeburg, M.; van der Schoot, V.; et al. Pathogenic Neurofibromatosis Type 1 (NF1) RNA Splicing Resolved by Targeted RNAseq. *npj Genom. Med.* **2021**, *6*, 95. [CrossRef] [PubMed]
35. Hartung, A.M.; Swensen, J.; Uriz, I.E.; Lapin, M.; Kristjansdottir, K.; Petersen, U.S.S.; Bang, J.M.V.; Guerra, B.; Andersen, H.S.; Dobrowolski, S.F.; et al. The Splicing Efficiency of Activating HRAS Mutations Can Determine Costello Syndrome Phenotype and Frequency in Cancer. *PLoS Genet.* **2016**, *12*, e1006039. [CrossRef] [PubMed]
36. Devonshire, A.S.; Sanders, R.; Wilkes, T.M.; Taylor, M.S.; Foy, C.A.; Huggett, J.F. Application of next Generation qPCR and Sequencing Platforms to mRNA Biomarker Analysis. *Methods* **2013**, *59*, 89–100. [CrossRef] [PubMed]
37. Hanses, U.; Kleinsorge, M.; Roos, L.; Yigit, G.; Li, Y.; Barbarics, B.; El-Battrawy, I.; Lan, H.; Tiburcy, M.; Hindmarsh, R.; et al. Intronic CRISPR Repair in a Preclinical Model of Noonan Syndrome-Associated Cardiomyopathy. *Circulation* **2020**, *142*, 1059–1076. [CrossRef] [PubMed]
38. Drenckhahn, J.D.; Nicin, L.; Akhouaji, S.; Krück, S.; Blank, A.E.; Schänzer, A.; Yörüker, U.; Jux, C.; Tombor, L.; Abplanalp, W.; et al. Cardiomyocyte Hyperplasia and Immaturity but Not Hypertrophy Are Characteristic Features of Patients with RASopathies. *J. Mol. Cell. Cardiol.* **2023**, *178*, 22–35. [CrossRef] [PubMed]
39. Coccia, E.; Valeri, L.; Zuntini, R.; Caraffi, S.G.; Peluso, F.; Pagliai, L.; Vezzani, A.; Pietrangiolillo, Z.; Leo, F.; Melli, N.; et al. Prenatal Clinical Findings in RASA1-Related Capillary Malformation-Arteriovenous Malformation Syndrome. *Genes* **2023**, *14*, 549. [CrossRef]
40. Brosseau, J.P.; Sathe, A.A.; Wang, Y.; Nguyen, T.; Glass, D.A.; Xing, C.; Le, L.Q. Human Cutaneous Neurofibroma Matrisome Revealed by Single-Cell RNA Sequencing. *Acta Neuropathol. Commun.* **2021**, *9*, 11. [CrossRef]
41. Imada, E.L.; Strianese, D.; Edward, D.P.; alThaqib, R.; Price, A.; Arnold, A.; Al-Hussain, H.; Marchionni, L.; Rodriguez, F.J. RNA-Sequencing Highlights Differential Regulated Pathways Involved in Cell Cycle and Inflammation in Orbitofacial Neurofibromas. *Brain Pathol.* **2022**, *32*, e13007. [CrossRef]
42. Yan, H.; Bu, P. Non-Coding RNA in Cancer. *Essays Biochem.* **2021**, *65*, 625–639. [CrossRef]
43. de Carvalho, J.B.; de Morais, G.L.; Vieira, T.C.d.S.; Rabelo, N.C.; Llerena, J.C.; de Carvalho Gonzalez, S.M.; de Vasconcelos, A.T.R. miRNA Genetic Variants Alter Their Secondary Structure and Expression in Patients with RASopathies Syndromes. *Front. Genet.* **2019**, *10*, 1144. [CrossRef]
44. Yu, T.T.; Xu, Q.F.; Li, S.Y.; Huang, H.J.; Dugan, S.; Shao, L.; Roggenbuck, J.A.; Liu, X.T.; Liu, H.Z.; Hirsch, B.A.; et al. Deletion at an 1q24 Locus Reveals a Critical Role of Long Noncoding RNA DNM3OS in Skeletal Development. *Cell Biosci.* **2021**, *11*, 47. [CrossRef]
45. Amirnasr, A.; Verdijk, R.M.; van Kuijk, P.F.; Kartal, P.; Vriends, A.L.M.; French, P.J.; van Royen, M.E.; Taal, W.; Sleijfer, S.; Wiemer, E.A.C. Deregulated microRNAs in Neurofibromatosis Type 1 Derived Malignant Peripheral Nerve Sheath Tumors. *Sci. Rep.* **2020**, *10*, 2927. [CrossRef]
46. Khosravi, T.; Oladnabi, M. The Role of miRNAs and lncRNAs in Neurofibromatosis Type 1. *J. Cell. Biochem.* **2023**, *124*, 17–30. [CrossRef]
47. Nix, J.S.; Yuan, M.; Imada, E.L.; Ames, H.; Marchionni, L.; Gutmann, D.H.; Rodriguez, F.J. Global microRNA Profiling Identified miR-10b-5p as a Regulator of Neurofibromatosis 1 (NF1)-Glioma Migration. *Neuropathol. Appl. Neurobiol.* **2021**, *47*, 96–107. [CrossRef]
48. Mulero-Navarro, S.; Sevilla, A.; Roman, A.C.; Lee, D.F.; D'Souza, S.L.; Pardo, S.; Riess, I.; Su, J.; Cohen, N.; Schaniel, C.; et al. Myeloid Dysregulation in a Human Induced Pluripotent Stem Cell Model of PTPN11-Associated Juvenile Myelomonocytic Leukemia. *Cell Rep.* **2015**, *13*, 504–515. [CrossRef] [PubMed]
49. Herman, A.B.; Tsitsipatis, D.; Gorospe, M. Integrated lncRNA Function upon Genomic and Epigenomic Regulation. *Mol. Cell* **2022**, *82*, 2252–2266. [CrossRef] [PubMed]
50. Tritto, V.; Ferrari, L.; Esposito, S.; Zuccotti, P.; Bianchessi, D.; Natacci, F.; Saletti, V.; Eoli, M.; Riva, P. Non-Coding RNA and Tumor Development in Neurofibromatosis Type 1: ANRIL Rs2151280 Is Associated with Optic Glioma Development and a Mild Phenotype in Neurofibromatosis Type 1 Patients. *Genes* **2019**, *10*, 892. [CrossRef] [PubMed]
51. Gyorffy, B.; Schafer, R. Biomarkers Downstream of RAS: A Search for Robust Transcriptional Targets. *Curr. Cancer Drug Targets* **2010**, *10*, 858–868. [CrossRef] [PubMed]
52. Campbell, S.L.; Philips, M.R. Post-Translational Modification of RAS Proteins. *Curr. Opin. Struct. Biol.* **2021**, *71*, 180–192. [CrossRef]
53. Jang, H.H. Regulation of Protein Degradation by Proteasomes in Cancer. *J. Cancer Prev.* **2018**, *23*, 153–161. [CrossRef] [PubMed]

54. Dohlman, H.G.; Campbell, S.L. Regulation of Large and Small G Proteins by Ubiquitination. *J. Biol. Chem.* **2019**, *294*, 18613–18623. [CrossRef] [PubMed]
55. Kim, S.E.; Yoon, J.Y.; Jeong, W.J.; Jeon, S.H.; Park, Y.; Yoon, J.B.; Park, Y.N.; Kim, H.; Choi, K.Y. H-Ras Is Degraded by Wnt/β-Catenin Signaling via β-TrCP-Mediated Polyubiquitylation. *J. Cell Sci.* **2009**, *122*, 842–848. [CrossRef] [PubMed]
56. Yoshino, H.; Yin, G.; Kawaguchi, R.; Popov, K.I.; Temple, B.; Sasaki, M.; Kofuji, S.; Wolfe, K.; Kofuji, K.; Okumura, K.; et al. Identification of Lysine Methylation in the Core GTPase Domain by GoMADScan. *PLoS ONE* **2019**, *14*, e0219436. [CrossRef] [PubMed]
57. Castellano, E.; Santos, E. Functional Specificity of Ras Isoforms: So Similar but so Different. *Genes Cancer* **2011**, *2*, 216–231. [CrossRef] [PubMed]
58. Ratner, N.; Miller, S.J. A RASopathy Gene Commonly Mutated in Cancer: The Neurofibromatosis Type 1 Tumour Suppressor. *Nat. Rev. Cancer* **2015**, *15*, 290–301. [CrossRef] [PubMed]
59. Dasgupta, B.; Yi, Y.; Chen, D.Y.; Weber, J.D.; Gutmann, D.H. Proteomic Analysis Reveals Hyperactivation of the Mammalian Target of Rapamycin Pathway in Neurofibromatosis 1-Associated Human and Mouse Brain Tumors. *Cancer Res.* **2005**, *65*, 2755–2760. [CrossRef]
60. Bergoug, M.; Doudeau, M.; Godin, F.; Mosrin, C.; Vallée, B.; Bénédetti, H. Neurofibromin Structure, Functions and Regulation. *Cells* **2020**, *9*, 2365. [CrossRef]
61. McFall, T.; Stites, E.C. Identification of RAS Mutant Biomarkers for EGFR Inhibitor Sensitivity Using a Systems Biochemical Approach. *Cell Rep.* **2021**, *37*, 110096. [CrossRef]
62. Koliou, X.; Fedonidis, C.; Kalpachidou, T.; Mangoura, D. Nuclear Import Mechanism of Neurofibromin for Localization on the Spindle and Function in Chromosome Congression. *J. Neurochem.* **2016**, *136*, 78–91. [CrossRef]
63. Feng, L.; Yunoue, S.; Tokuo, H.; Ozawa, T.; Zhang, D.; Patrakitkomjorn, S.; Ichimura, T.; Saya, H.; Araki, N. PKA Phosphorylation and 14-3-3 Interaction Regulate the Function of Neurofibromatosis Type I Tumor Suppressor, Neurofibromin. *FEBS Lett.* **2004**, *557*, 275–282. [CrossRef]
64. Harder, A.; Rosche, M.; Reuß, D.E.; Holtkamp, N.; Uhlmann, K.; Friedrich, R.; Mautner, V.F.; Von Deimling, A. Methylation Analysis of the Neurofibromatosis Type 1 (NF1) Promoter in Peripheral Nerve Sheath Tumours. *Eur. J. Cancer* **2004**, *40*, 2820–2828. [CrossRef]
65. Beauclair, G.; Bridier-Nahmias, A.; Zagury, J.F.; Säb, A.; Zamborlini, A. JASSA: A Comprehensive Tool for Prediction of SUMOylation Sites and SIMs. *Bioinformatics* **2015**, *31*, 3483–3491. [CrossRef]
66. Yan, W.; Markegard, E.; Dharmaiah, S.; Urisman, A.; Drew, M.; Esposito, D.; Scheffzek, K.; Nissley, D.V.; McCormick, F.; Simanshu, D.K. Structural Insights into the SPRED1-Neurofibromin-KRAS Complex and Disruption of SPRED1-Neurofibromin Interaction by Oncogenic EGFR. *Cell Rep.* **2020**, *32*, 107909. [CrossRef]
67. Pandit, B.; Sarkozy, A.; Pennacchio, L.A.; Carta, C.; Oishi, K.; Martinelli, S.; Pogna, E.A.; Schackwitz, W.; Ustaszewska, A.; Landstrom, A.; et al. Gain-of-Function RAF1 Mutations Cause Noonan and LEOPARD Syndromes with Hypertrophic Cardiomyopathy. *Nat. Genet.* **2007**, *39*, 1007–1012. [CrossRef]
68. Razzaque, M.A.; Nishizawa, T.; Komoike, Y.; Yagi, H.; Furutani, M.; Amo, R.; Kamisago, M.; Momma, K.; Katayama, H.; Nakagawa, M.; et al. Germline Gain-of-Function Mutations in RAF1 Cause Noonan Syndrome. *Nat. Genet.* **2007**, *39*, 1013–1017. [CrossRef]
69. Heidorn, S.J.; Milagre, C.; Whittaker, S.; Nourry, A.; Niculescu-Duvas, I.; Dhomen, N.; Hussain, J.; Reis-Filho, J.S.; Springer, C.J.; Pritchard, C.; et al. Kinase-Dead BRAF and Oncogenic RAS Cooperate to Drive Tumor Progression through CRAF. *Cell* **2010**, *140*, 209–221. [CrossRef] [PubMed]
70. Roberts, A.E.; Allanson, J.E.; Tartaglia, M.; Gelb, B.D. Noonan Syndrome. *Lancet* **2013**, *381*, 333–342. [CrossRef] [PubMed]
71. Galperin, E.; Wilson, P.; Abdelmoti, L.; Norcross, R.; Palayam, M. Proteins of the Ubiquitin System in the Shoc2—ERK1/2 Signaling Axis and Noonan-like Syndrome with Loose Anagen Hair (NSLAH) RASopathy. *FASEB J.* **2022**, *36*, r1999. [CrossRef]
72. Bivona, T.G.; Quatela, S.E.; Bodemann, B.O.; Ahearn, I.M.; Soskis, M.J.; Mor, A.; Miura, J.; Wiener, H.H.; Wright, L.; Saba, S.G.; et al. PKC Regulates a Farnesyl-Electrostatic Switch on K-Ras That Promotes Its Association with Bcl-XL on Mitochondria and Induces Apoptosis. *Mol. Cell* **2006**, *21*, 481–493. [CrossRef]
73. Choi, B.H.; Philips, M.R.; Chen, Y.; Lu, L.; Dai, W. K-Ras Lys-42 Is Crucial for Its Signaling, Cell Migration, and Invasion. *J. Biol. Chem.* **2018**, *293*, 17574–17581. [CrossRef] [PubMed]
74. Ottaiano, A.; Normanno, N.; Facchini, S.; Cassata, A.; Nappi, A.; Romano, C.; Silvestro, L.; Stefano, A.D.; Rachiglio, A.M.; Roma, C.; et al. Study of Ras Mutations' Prognostic Value in Metastatic Colorectal Cancer: Storia Analysis. *Cancers* **2020**, *12*, 1919. [CrossRef] [PubMed]
75. Simão, S.; Agostinho, R.R.; Martínez-Ruiz, A.; Araújo, I.M. Regulation of Ras Signaling by S-Nitrosylation. *Antioxidants* **2023**, *12*, 1562. [CrossRef] [PubMed]
76. Martin-Vega, A.; Cobb, M.H. Navigating the ERK1/2 MAPK Cascade. *Biomolecules* **2023**, *13*, 1555. [CrossRef] [PubMed]
77. Yang, X.; You, J.; Luo, W.; Yue, J.; Ma, L.; Xiao, W.; Zhu, D.; Wu, Z.; Wang, D.; Nadiminty, N.; et al. The N-Terminal Kinase Suppressor of Ras Complex Has a Weak Nucleoside Diphosphate Kinase Activity. *Thorac. Cancer* **2010**, *1*, 109–115. [CrossRef] [PubMed]

78. Hahn, A.; Lauriol, J.; Thul, J.; Behnke-Hall, K.; Logeswaran, T.; Schänzer, A.; Böğürcü, N.; Garvalov, B.K.; Zenker, M.; Gelb, B.D.; et al. Rapidly Progressive Hypertrophic Cardiomyopathy in an Infant with Noonan Syndrome with Multiple Lentigines: Palliative Treatment with a Rapamycin Analog. *Am. J. Med. Genet. Part A* **2015**, *167*, 744–751. [CrossRef] [PubMed]
79. Ranza, E.; Guimier, A.; Verloes, A.; Capri, Y.; Marques, C.; Auclair, M.; Mathieu-Dramard, M.; Morin, G.; Thevenon, J.; Faivre, L.; et al. Overlapping Phenotypes between SHORT and Noonan Syndromes in Patients with PTPN11 Pathogenic Variants. *Clin. Genet.* **2020**, *98*, 10–18. [CrossRef]
80. Endo, M.; Yamamoto, H.; Setsu, N.; Kohashi, K.; Takahashi, Y.; Ishii, T.; Iida, K.I.; Matsumoto, Y.; Hakozaki, M.; Aoki, M.; et al. Prognostic Significance of AKT/mTOR and MAPK Pathways and Antitumor Effect of mTOR Inhibitor in NF1-Related and Sporadic Malignant Peripheral Nerve Sheath Tumors. *Clin. Cancer Res.* **2013**, *19*, 450–461. [CrossRef]
81. Solares, I.; Viñal, D.; Morales-Conejo, M.; Rodriguez-Salas, N.; Feliu, J. Novel Molecular Targeted Therapies for Patients with Neurofibromatosis Type 1 with Inoperable Plexiform Neurofibromas: A Comprehensive Review. *ESMO Open* **2021**, *6*, 100223. [CrossRef] [PubMed]
82. Ugwu, N.; Cheraghlou, S.; Ko, C.J.; Cohen, J.M. Incidence, Survival, and Prognostic Factors Associated with Malignant Nodular Hidradenoma in the United States. *J. Am. Acad. Dermatol.* **2023**, *88*, 875–877. [CrossRef] [PubMed]
83. Ullrich, N.J.; Prabhu, S.P.; Reddy, A.T.; Fisher, M.J.; Packer, R.; Goldman, S.; Robison, N.J.; Gutmann, D.H.; Viskochil, D.H.; Allen, J.C.; et al. A Phase II Study of Continuous Oral mTOR Inhibitor Everolimus for Recurrent, Radiographic-Progressive Neurofibromatosis Type 1-Associated Pediatric Low-Grade Glioma: A Neurofibromatosis Clinical Trials Consortium Study. *Neuro Oncol.* **2020**, *22*, 1527–1535. [CrossRef] [PubMed]
84. Miyoshi, K.; Wakioka, T.; Nishinakamura, H.; Kamio, M.; Yang, L.; Inoue, M.; Hasegawa, M.; Yonemitsu, Y.; Komiya, S.; Yoshimura, A. The Sprouty-Related Protein, Spred, Inhibits Cell Motility, Metastasis, and Rho-Mediated Actin Reorganization. *Oncogene* **2004**, *23*, 5567–5576. [CrossRef] [PubMed]
85. Brown, J.A.; Diggs-Andrews, K.A.; Gianino, S.M.; Gutmann, D.H. Neurofibromatosis-1 Heterozygosity Impairs CNS Neuronal Morphology in a cAMP/PKA/ROCK-Dependent Manner. *Mol. Cell. Neurosci.* **2012**, *49*, 13–22. [CrossRef] [PubMed]
86. Ozawa, T.; Araki, N.; Yunoue, S.; Tokuo, H.; Feng, L.; Patrakitkomjorn, S.; Hara, T.; Ichikawa, Y.; Matsumoto, K.; Fuji, K.; et al. The Neurofibromatosis Type 1 Gene Product Neurofibromin Enhances Cell Motility by Regulating Actin Filament Dynamics via the Rho-ROCK-LIMK2-Cofilin Pathway. *J. Biol. Chem.* **2005**, *280*, 39524–39533. [CrossRef] [PubMed]
87. Langdon, Y.; Tandon, P.; Paden, E.; Duddy, J.; Taylor, J.M.; Conlon, F.L. SHP-2 Acts via ROCK to Regulate the Cardiac Actin Cytoskeleton. *Development* **2012**, *139*, 948–957. [CrossRef]
88. Chen, M.; Lu, L.; Cheng, D.; Zhang, J.; Liu, X.; Zhang, J.; Zhang, T. Icariin Promotes Osteogenic Differentiation in a Cell Model with NF1 Gene Knockout by Activating the cAMP/PKA/CREB Pathway. *Molecules* **2023**, *28*, 5128. [CrossRef] [PubMed]
89. Mazuelas, H.; Magallón-Lorenz, M.; Uriarte-Arrazola, I.; Negro, A.; Rosas, I.; Blanco, I.; Castellanos, E.; Lázaro, C.; Gel, B.; Carrió, M.; et al. Unbalancing cAMP and Ras/MAPK Pathways as a Therapeutic Strategy for Cutaneous Neurofibromas. *JCI Insight* **2024**, *9*, 168826. [CrossRef]
90. Biayna, J.; Mazuelas, H.; Gel, B.; Terribas, E.; Dumbovic, G.; Rosas, I.; Fernández-Rodriguez, J.; Blanco, I.; Castellanos, E.; Carrió, M.; et al. Using Antisense Oligonucleotides for the Physiological Modulation of the Alternative Splicing of NF1 Exon 23a during PC12 Neuronal Differentiation. *Sci. Rep.* **2021**, *11*, 3661. [CrossRef]
91. Machado Almeida, P.; Lago Solis, B.; Stickley, L.; Feidler, A.; Nagoshi, E. Neurofibromin 1 in Mushroom Body Neurons Mediates Circadian Wake Drive through Activating cAMP–PKA Signaling. *Nat. Commun.* **2021**, *12*, 5758. [CrossRef]
92. Xue, C.; Yao, Q.; Gu, X.; Shi, Q.; Yuan, X.; Chu, Q.; Bao, Z.; Lu, J.; Li, L. Evolving Cognition of the JAK-STAT Signaling Pathway: Autoimmune Disorders and Cancer. *Signal Transduct. Target. Ther.* **2023**, *8*, 204. [CrossRef] [PubMed]
93. Miot, H.A.; Criado, P.R.; de Castro, C.C.S.; Ianhez, M.; Talhari, C.; Ramos, P.M. JAK-STAT Pathway Inhibitors in Dermatology. *An. Bras. Dermatol.* **2023**, *98*, 656–677. [CrossRef] [PubMed]
94. Erdogan, F.; Radu, T.B.; Orlova, A.; Qadree, A.K.; de Araujo, E.D.; Israelian, J.; Valent, P.; Mustjoki, S.M.; Herling, M.; Moriggl, R.; et al. JAK-STAT Core Cancer Pathway: An Integrative Cancer Interactome Analysis. *J. Cell. Mol. Med.* **2022**, *26*, 2049–2062. [CrossRef] [PubMed]
95. Chan, G.; Kalaitzidis, D.; Neel, B.G. The Tyrosine Phosphatase Shp2 (PTPN11) in Cancer. *Cancer Metastasis Rev.* **2008**, *27*, 179–192. [CrossRef] [PubMed]
96. Seif, F.; Khoshmirsafa, M.; Aazami, H.; Mohsenzadegan, M.; Sedighi, G.; Bahar, M. The Role of JAK-STAT Signaling Pathway and Its Regulators in the Fate of T Helper Cells. *Cell Commun. Signal.* **2017**, *15*, 23. [CrossRef] [PubMed]
97. Wu, J.; Keng, V.W.; Patmore, D.M.; Kendall, J.J.; Patel, A.V.; Jousma, E.; Jessen, W.J.; Choi, K.; Tschida, B.R.; Silverstein, K.A.T.; et al. Insertional Mutagenesis Identifies a STAT3/Arid1b/β-Catenin Pathway Driving Neurofibroma Initiation. *Cell Rep.* **2016**, *14*, 1979–1990. [CrossRef] [PubMed]
98. Ma, S.; Meng, Z.; Chen, R.; Guan, K.-L. The Hippo Pathway: Biology and Pathophysiology. *Annu. Rev. Biochem.* **2019**, *88*, 577–604. [CrossRef] [PubMed]
99. O'Neill, E. Ras and the Hippo Pathway in Cancer. In *Conquering RAS: From Biology to Cancer Therapy*; Elsevier Inc.: Amsterdam, The Netherlands, 2017; pp. 25–39, ISBN 978-0-12-803505-4.
100. Faden, D.L.; Asthana, S.; Tihan, T.; De Risi, J.; Kliot, M. Whole Exome Sequencing of Growing and Non-Growing Cutaneous Neurofibromas from a Single Patient with Neurofibromatosis Type 1. *PLoS ONE* **2017**, *12*, e0170348. [CrossRef]

101. Vélez-Reyes, G.L.; Koes, N.; Ryu, J.H.; Kaufmann, G.; Berner, M.; Weg, M.T.; Wolf, N.K.; Rathe, S.K.; Ratner, N.; Moriarity, B.S.; et al. Transposon Mutagenesis-Guided Crispr/Cas9 Screening Strongly Implicates Dysregulation of Hippo/Yap Signaling in Malignant Peripheral Nerve Sheath Tumor Development. *Cancers* **2021**, *13*, 1584. [CrossRef]
102. Zhao, H.; Ming, T.; Tang, S.; Ren, S.; Yang, H.; Liu, M.; Tao, Q.; Xu, H. Wnt Signaling in Colorectal Cancer: Pathogenic Role and Therapeutic Target. *Mol. Cancer* **2022**, *21*, 144. [CrossRef] [PubMed]
103. Jeong, W.J.; Ro, E.J.; Choi, K.Y. Interaction between Wnt/β-Catenin and RAS-ERK Pathways and an Anti-Cancer Strategy via Degradations of β-Catenin and RAS by Targeting the Wnt/β-Catenin Pathway. *npj Precis. Oncol.* **2018**, *2*, 5. [CrossRef] [PubMed]
104. He, K.; Gan, W.J. Wnt/β-Catenin Signaling Pathway in the Development and Progression of Colorectal Cancer. *Cancer Manag. Res.* **2023**, *15*, 435–448. [CrossRef] [PubMed]
105. Tompa, M.; Nagy, A.; Komoly, S.; Kalman, B. Wnt Pathway Markers in Molecular Subgroups of Glioblastoma. *Brain Res.* **2019**, *1718*, 114–125. [CrossRef]
106. Noda, S.; Takahashi, A.; Hayashi, T.; Tanuma, S.I.; Hatakeyama, M. Determination of the Catalytic Activity of LEOPARD Syndrome-Associated SHP2 Mutants toward Parafibromin, a Bona Fide SHP2 Substrate Involved in Wnt Signaling. *Biochem. Biophys. Res. Commun.* **2016**, *469*, 1133–1139. [CrossRef] [PubMed]
107. Derynck, R.; Budi, E.H. Specificity, Versatility, and Control of TGF-b Family Signaling. *Sci. Signal.* **2019**, *12*, aav5183. [CrossRef] [PubMed]
108. Tzavlaki, K.; Moustakas, A. TGF-B Signaling. *Biomolecules* **2020**, *10*, 487. [CrossRef] [PubMed]
109. Torres, K.C.L.; Lima, G.; Simões e Silva, A.C.; Lubambo, I.; Rodrigues, L.O.; Rodrigues, L.; Silveira, K.D.; Vieira, É.L.M.; Romano-Silva, M.A.; Miranda, D.M. Immune Markers in the RASopathy Neurofibromatosis Type 1. *J. Neuroimmunol.* **2016**, *295–296*, 122–129. [CrossRef] [PubMed]
110. Ju, Y.; Park, J.S.; Kim, D.; Kim, B.; Lee, J.H.; Nam, Y.; Yoo, H.W.; Lee, B.H.; Han, Y.M. SHP2 Mutations Induce Precocious Gliogenesis of Noonan Syndrome-Derived iPSCs during Neural Development in Vitro. *Stem Cell Res. Ther.* **2020**, *11*, 209. [CrossRef]
111. Jin, Q.; Ma, R.C.W. Metabolomics in Diabetes and Diabetic Complications: Insights from Epidemiological Studies. *Cells* **2021**, *10*, 2832. [CrossRef]
112. Occelli, C.; Levraut, J.; Pourcher, T. Metabolomics, the Future of Biomarkers? *Eur. J. Emerg. Med.* **2024**, *31*, 7–8. [CrossRef] [PubMed]
113. Martins, A.S.; Jansen, A.K.; Rodrigues, L.O.C.; Matos, C.M.; Souza, M.L.R.; Miranda, D.M.; de Rezende, N.A. Increased Insulin Sensitivity in Individuals with Neurofibromatosis Type 1. *Arch. Endocrinol. Metab.* **2018**, *62*, 34–39. [CrossRef] [PubMed]
114. Martins, A.S.; Jansen, A.K.; Rodrigues, L.O.C.; Matos, C.M.; Souza, M.L.R.; de Souza, J.F.; de Fátima Haueisen Sander Diniz, M.; Barreto, S.M.; Diniz, L.M.; de Rezende, N.A.; et al. Lower Fasting Blood Glucose in Neurofibromatosis Type 1. *Endocr. Connect.* **2016**, *5*, 28–33. [CrossRef] [PubMed]
115. Tritz, R.; Benson, T.; Harris, V.; Hudson, F.Z.; Mintz, J.; Zhang, H.; Kennard, S.; Chen, W.; Stepp, D.W.; Csanyi, G.; et al. Nf1 Heterozygous Mice Recapitulate the Anthropometric and Metabolic Features of Human Neurofibromatosis Type 1. *Transl. Res.* **2021**, *228*, 52–63. [CrossRef] [PubMed]
116. Noronha, R.M.; Villares, S.M.F.; Torres, N.; Quedas, E.P.S.; Homma, T.K.; Albuquerque, E.V.A.; Moraes, M.B.; Funari, M.F.A.; Bertola, D.R.; Jorge, A.A.L.; et al. Noonan Syndrome Patients beyond the Obvious Phenotype: A Potential Unfavorable Metabolic Profile. *Am. J. Med. Genet. Part A* **2021**, *185*, 774–780. [CrossRef] [PubMed]
117. Fahrner, J.A.; Frazier, A.; Bachir, S.; Walsh, M.F.; Applegate, C.D.; Thompson, R.; Halushka, M.K.; Murphy, A.M.; Gunay-Aygun, M. A Rasopathy Phenotype with Severe Congenital Hypertrophic Obstructive Cardiomyopathy Associated with a PTPN11 Mutation and a Novel Variant in SOS1. *Am. J. Med. Genet. A* **2012**, *158A*, 1414–1421. [CrossRef] [PubMed]
118. Gripp, K.W.; Kawame, H.; Viskochil, D.H.; Nicholson, L. Elevated Catecholamine Metabolites in Patients with Costello Syndrome. *Am. J. Med. Genet.* **2004**, *128A*, 48–51. [CrossRef] [PubMed]
119. Marciano, D.P.; Snyder, M.P. Personalized Metabolomics. In *High-Throughput Metabolomics: Methods and Protocols*; D'Alessandro, A., Ed.; Springer: New York, NY, USA, 2019; pp. 447–456, ISBN 978-1-4939-9236-2.
120. Wallis, D.; Stemmer-Rachamimov, A.; Adsit, S.; Korf, B.; Pichard, D.; Blakeley, J.; Sarin, K.Y. Status and Recommendations for Incorporating Biomarkers for Cutaneous Neurofibromas Into Clinical Research. *Neurology* **2021**, *97*, S42–S49. [CrossRef] [PubMed]
121. Hilal, N.; Chen, Z.; Chen, M.H.; Choudhury, S. RASopathies and Cardiac Manifestations. *Front. Cardiovasc. Med.* **2023**, *10*, 1176828. [CrossRef]
122. Meier, A.B.; Raj Murthi, S.; Rawat, H.; Toepfer, C.N.; Santamaria, G.; Schmid, M.; Mastantuono, E.; Schwarzmayr, T.; Berutti, R.; Cleuziou, J.; et al. Cell Cycle Defects Underlie Childhood-Onset Cardiomyopathy Associated with Noonan Syndrome. *iScience* **2022**, *25*, 103596. [CrossRef]
123. Siegel, D.H.; Mann, J.A.; Krol, A.L.; Rauen, K.A. Dermatological Phenotype in Costello Syndrome: Consequences of Ras Dysregulation in Development. *Br. J. Dermatol.* **2012**, *166*, 601–607. [CrossRef] [PubMed]
124. Magaki, S.; Hojat, S.A.; Wei, B.; So, A.; Yong, W.H. An Introduction to the Performance of Immunohistochemistry. In *Methods in Molecular Biology*; NIH Public Access: Rockville Pike Bethesda, MD, USA, 2019, Volume 1897, pp. 289–298.
125. Peltonen, S.; Kallionpää, R.A.; Peltonen, J. Neurofibromatosis Type 1 (NF1) Gene: Beyond Café Au Lait Spots and Dermal Neurofibromas. *Exp. Dermatol.* **2017**, *26*, 645–648. [CrossRef] [PubMed]

126. Helfferich, J.; Nijmeijer, R.; Brouwer, O.F.; Boon, M.; Fock, A.; Hoving, E.W.; Meijer, L.; den Dunnen, W.F.A.; de Bont, E.S.J.M. Neurofibromatosis Type 1 Associated Low Grade Gliomas: A Comparison with Sporadic Low Grade Gliomas. *Crit. Rev. Oncol. Hematol.* **2016**, *104*, 30–41. [CrossRef] [PubMed]
127. Ozarslan, B.; Russo, T.; Argenziano, G.; Santoro, C.; Piccolo, V. Cutaneous Findings in Neurofibromatosis Type 1. *Cancers* **2021**, *13*, 463. [CrossRef] [PubMed]
128. Miettinen, M.M.; Antonescu, C.R.; Fletcher, C.D.M.; Kim, A.; Lazar, A.J.; Quezado, M.M.; Reilly, K.M.; Stemmer-Rachamimov, A.; Stewart, D.R.; Viskochil, D.; et al. Histopathologic Evaluation of Atypical Neurofibromatous Tumors and Their Transformation into Malignant Peripheral Nerve Sheath Tumor in Patients with Neurofibromatosis 1—A Consensus Overview. *Hum. Pathol.* **2017**, *67*, 1–10. [CrossRef] [PubMed]
129. Kim, A.; Dombi, E.; Tepas, K.; Fox, E.; Martin, S.; Wolters, P.; Balis, F.M.; Jayaprakash, N.; Turkbey, B.; Muradyan, N.; et al. Phase I Trial and Pharmacokinetic Study of Sorafenib in Children with Neurofibromatosis Type I and Plexiform Neurofibromas. *Pediatr. Blood Cancer* **2013**, *60*, 396–401. [CrossRef] [PubMed]
130. Meyerholz, D.K.; Ofori-Amanfo, G.K.; Leidinger, M.R.; Goeken, J.A.; Khanna, R.; Sieren, J.C.; Darbro, B.W.; Quelle, D.E.; Weimer, J.M. Immunohistochemical Markers for Prospective Studies in Neurofibromatosis-1 Porcine Models. *J. Histochem. Cytochem.* **2017**, *65*, 607–618. [CrossRef] [PubMed]
131. Guedes-Corrêa, J.; Cardoso, R. Immunohistochemical Markers for Schwannomas, Neurofibromas and Malignant Peripheral Nerve Sheath Tumors—What Can the Recent Literature Tell Us? *Arq. Bras. Neurocir. Braz. Neurosurg.* **2018**, *37*, 105–112. [CrossRef]
132. Martin, E.; Acem, I.; Grünhagen, D.J.; Bovée, J.V.M.G.; Verhoef, C. Prognostic Significance of Immunohistochemical Markers and Genetic Alterations in Malignant Peripheral Nerve Sheath Tumors: A Systematic Review. *Front. Oncol.* **2020**, *10*, 594069. [CrossRef]
133. Johansson, G.; Peng, P.C.; Huang, P.Y.; Chien, H.F.; Hua, K.T.; Kuo, M.L.; Chen, C.T.; Lee, M.J. Soluble AXL: A Possible Circulating Biomarker for Neurofibromatosis Type 1 Related Tumor Burden. *PLoS ONE* **2014**, *9*, e115916. [CrossRef]
134. Brems, H.; Pasmant, E.; Van Minkelen, R.; Wimmer, K.; Upadhyaya, M.; Legius, E.; Messiaen, L. Review and Update of SPRED1 Mutations Causing Legius Syndrome. *Hum. Mutat.* **2012**, *33*, 1538–1546. [CrossRef] [PubMed]
135. Zhang, J.; Li, M.; Yao, Z. Molecular Screening Strategies for NF1-like Syndromes with Café-Au-Lait Macules (Review). *Mol. Med. Rep.* **2016**, *14*, 4023–4029. [CrossRef] [PubMed]
136. Guglielmi, F.; Kirschner, F.; Staderini, E.; Iavarone, F.; Fiorino, A.; Gallenzi, P. Proteomic Analysis of Salivary Inflammatory Biomarkers of Developmental Gingival Enlargements in Patients with West and Noonan Syndromes: A Preliminary Pilot Single-Center Retrospective Study. *Eur. Rev. Med. Pharmacol. Sci.* **2023**, *27*, 11093–11102. [CrossRef] [PubMed]
137. Moniez, S.; Pienkowski, C.; Lepage, B.; Hamdi, S.; Daudin, M.; Oliver, I.; Jouret, B.; Cartault, A.; Diene, G.; Verloes, A.; et al. Noonan Syndrome Males Display Sertoli Cell-Specific Primary Testicular Insufficiency. *Eur. J. Endocrinol.* **2018**, *179*, 409–418. [CrossRef] [PubMed]
138. Sun, L.; Xie, Y.M.; Wang, S.S.; Zhang, Z.W. Cardiovascular Abnormalities and Gene Mutations in Children with Noonan Syndrome. *Front. Genet.* **2022**, *13*, 915129. [CrossRef] [PubMed]
139. Pierpont, E.I.; Hudock, R.L.; Foy, A.M.; Semrud-Clikeman, M.; Pierpont, M.E.; Berry, S.A.; Shanley, R.; Rubin, N.; Sommer, K.; Moertel, C.L. Social Skills in Children with RASopathies: A Comparison of Noonan Syndrome and Neurofibromatosis Type 1. *J. Neurodev. Disord.* **2018**, *10*, 21. [CrossRef] [PubMed]
140. Pierpont, M.E.M.; Magoulas, P.L.; Adi, S.; Kavamura, M.I.; Neri, G.; Noonan, J.; Pierpont, E.I.; Reinker, K.; Roberts, A.E.; Shankar, S.; et al. Cardio-Facio-Cutaneous Syndrome: Clinical Features, Diagnosis, and Management Guidelines. *Pediatrics* **2014**, *134*, e1149. [CrossRef] [PubMed]
141. Hebron, K.E.; Hernandez, E.R.; Yohe, M.E. The RASopathies: From Pathogenetics to Therapeutics. *DMM Dis. Models Mech.* **2022**, *15*, dmm049107. [CrossRef] [PubMed]
142. Palit, A.; Inamadar, A.C. RASopathies: Dermatologists' Viewpoints. *Indian J. Dermatol. Venereol. Leprol.* **2022**, *88*, 452–463. [CrossRef]
143. Fowlkes, J.L.; Thrailkill, K.M.; Bunn, R.C. RASopathies: The Musculoskeletal Consequences and Their Etiology and Pathogenesis. *Bone* **2021**, *152*, 116060. [CrossRef]
144. Lioncino, M.; Monda, E.; Verrillo, F.; Moscarella, E.; Calcagni, G.; Drago, F.; Marino, B.; Digilio, M.C.; Putotto, C.; Calabrò, P.; et al. Hypertrophic Cardiomyopathy in RASopathies: Diagnosis, Clinical Characteristics, Prognostic Implications, and Management. *Heart Fail. Clin.* **2022**, *18*, 19–29. [CrossRef] [PubMed]
145. Delogu, A.B.; Blandino, R.; Leoni, C.; Tartaglia, M.; Zampino, G. RASopathies and Sigmoid-Shaped Ventricular Septum Morphology: Evidence of a Previously Unappreciated Cardiac Phenotype. *Pediatr. Res.* **2023**, *93*, 752–754. [CrossRef] [PubMed]
146. Mangels, R.; Blumenfeld, Y.J.; Homeyer, M.; Mrazek-Pugh, B.; Hintz, S.R.; Hudgins, L. RASopathies: A Significant Cause of Polyhydramnios? *Prenat. Diagn.* **2021**, *41*, 362–367. [CrossRef] [PubMed]
147. Kim, Y.E.; Baek, S.T. Neurodevelopmental Aspects of RASopathies. *Mol. Cells* **2019**, *42*, 441–447. [CrossRef] [PubMed]
148. McGhee, C.A.; Honari, H.; Siqueiros-Sanchez, M.; Serur, Y.; van Staalduinen, E.K.; Stevenson, D.; Bruno, J.L.; Raman, M.M.; Green, T. Influences of RASopathies on Neuroanatomical Variation in Children. *Biol. Psychiatry Cogn. Neurosci. Neuroimaging* 2024, *in press*. [CrossRef] [PubMed]

149. Siano, M.A.; Pivonello, R.; Salerno, M.; Falco, M.; Mauro, C.; De Brasi, D.; Klain, A.; Sestito, S.; De Luca, A.; Pinna, V.; et al. Endocrine System Involvement in Patients with RASopathies: A Case Series. *Front. Endocrinol.* **2022**, *13*, 1030398. [CrossRef]
150. Bizzarri, C.; Bottaro, G. Endocrine Implications of Neurofibromatosis 1 in Childhood. *Horm. Res. Paediatr.* **2015**, *83*, 232–241. [CrossRef]

International Journal of *Molecular Sciences*

Review

Obesity-Related Ciliopathies: Focus on Advances of Biomarkers

Qianwen Zhang [1,†], Yiguo Huang [1,†], Shiyang Gao [1], Yu Ding [1], Hao Zhang [2], Guoying Chang [1,*] and Xiumin Wang [1,*]

1 Department of Endocrinology and Metabolism, Shanghai Children's Medical Center, School of Medicine, Shanghai Jiao Tong University, Shanghai 200127, China; zhangqianwen@scmc.com.cn (Q.Z.); huangyg29@sjtu.edu.cn (Y.H.); gaoshiyang@scmc.com.cn (S.G.); dingyu@scmc.com.cn (Y.D.)

2 Heart Center and Shanghai Institute of Pediatric Congenital Heart Disease, Shanghai Children's Medical Center, National Children's Medical Center, School of Medicine, Shanghai Jiao Tong University, Shanghai 200127, China; zhanghao@scmc.com.cn

* Correspondence: changguoying@scmc.com.cn (G.C.); wangxiumin@scmc.com.cn (X.W.)

† These authors contributed equally to this work.

Abstract: Obesity-related ciliopathies, as a group of ciliopathies including Alström Syndrome and Bardet–Biedl Syndrome, exhibit distinct genetic and phenotypic variability. The understanding of these diseases is highly significant for understanding the functions of primary cilia in the human body, particularly regarding the relationship between obesity and primary cilia. The diagnosis of these diseases primarily relies on clinical presentation and genetic testing. However, there is a significant lack of research on biomarkers to elucidate the variability in clinical manifestations, disease progression, prognosis, and treatment responses. Through an extensive literature review, the paper focuses on obesity-related ciliopathies, reviewing the advancements in the field and highlighting the potential roles of biomarkers in the clinical presentation, diagnosis, and prognosis of these diseases.

Keywords: obesity; ciliopathy; biomarker; rare disease

Citation: Zhang, Q.; Huang, Y.; Gao, S.; Ding, Y.; Zhang, H.; Chang, G.; Wang, X. Obesity-Related Ciliopathies: Focus on Advances of Biomarkers. *Int. J. Mol. Sci.* **2024**, *25*, 8484. https://doi.org/10.3390/ijms25158484

Academic Editor: Mariarosaria Santillo

Received: 30 June 2024
Revised: 27 July 2024
Accepted: 1 August 2024
Published: 3 August 2024

1. Introduction

1.1. Primary Cilia and Obesity

Primary cilia, also known as sensory cilia, are highly conserved hair-like organelles [1]. Cilia consist of a microtubule core called the axoneme, which extends from a modified centriole known as the basal body. A variety of receptors and ion channels are embedded within the ciliary membranes, facilitating the detection and transmission of stimuli from the extracellular environment, while also being capable of dispatching signals outward [2].

Obesity prevention and control is a global public health challenge. The 2023 World Obesity Map predicts that the worldwide prevalence of obesity will rise to 24% in 2035, with a total of nearly two billion people, and the obese population is also showing a trend in younger groups [3,4].

There is a strong link between primary cilia and obesity. A series of obesity-associated loci identified by genome-wide association studies (GWAS) are confirmed to be related to hypothalamic cilia, including adenylate cyclase 3 (ADCY3) and the melanocortin-4 receptor (MC4R). ADCY3 has been seen the marker of cilia in neurons, and MC4R is found to be specifically located on primary cilia in the hypothalamic paraventricular nucleus of mice [5].

Meanwhile, the causative genes associated with morbid obesity in humans were later verified to be associated with primary cilia, suggesting that primary cilia may have an important role in obesity and energy metabolism [6].

1.2. Obesity-Related Ciliopathies

Ciliopathies are a class of genetic disorders whose etiology is related to ciliary dysfunction [7]. The proteins these genes encode are localized in the cilia–centromere complex, affecting the assembly, maintenance or function of the centrioles or cilia [8–10]. Cilia are expressed and play roles in most mature cells of the human body; thus, ciliary defects affect nearly all organs and tissues, leading to complex symptoms. Common clinical manifestations include obesity, kidney abnormalities, vision and hearing impairments, heart defects, and insulin resistance [8,10–12]. In total, five ciliopathies are characterized with obesity, including Bardet–Biedl syndrome (BBS, OMIM #209900), Alström syndrome (ALMS, OMIM #203800), Carpenter syndrome (CRPT, OMIM #201000), mental retardation, truncal obesity, retinal dystrophy, and micropenis syndrome (MORMS, OMIM #610156), and morbid obesity and spermatogenic failure (MOSPGF, OMIM #615703) [8,13–17]. Most of these diseases are inherited in an autosomal recessive pattern and the pathogenic mutations usually lead to the loss of the protein or the production of a non-functional or truncated protein. BBS has also been described to have a triallelic pattern of inheritance [18].

ALMS is caused by mutations in the *ALMS1* gene, with an incidence of 1–9 per 1,000,000 [13]. Defect *ALMS1* can lead to multi-organ system damage, with primary features including early childhood obesity, insulin resistance and hyperinsulinemia, retinal cone dystrophy, and hearing impairment [19]. The clinical phenotype, onset time, and severity vary significantly among patients [8,13,20–22].

BBS is also a ciliopathy affecting multiple organ systems, with the estimated incidence of 1 per 160,000 in northern European populations [14]. Unlike ALMS, it could be caused by mutations in many genes, including *BBS1-22*, *IFT74*, *SCLT1*, *SCAPER*, and *NPHP* genes [14,23]. Major clinical manifestations include retinal dystrophy, polydactyly, early-onset obesity, hypogonadism, intellectual disability, and renal abnormalities [23,24]. Secondary clinical manifestations include congenital heart disease, liver involvement (such as liver fibrosis and liver cysts), endocrine disorders, like diabetes, hypothyroidism, and hypercholesterolemia, enamel hypoplasia, ataxia, delayed speech and growth, craniofacial anomalies, and olfactory abnormalities [23,24].

CRPT, with an incidence estimated at 1 per 1 million births [25], is characterized by craniosynostosis, polydactyly, cardiac defects, and obesity [26–28]. It can be divided into two types. Carpenter syndrome-1 (CRPT1) is caused by homozygous mutations in the *RAB23* gene [28]. According to the Human Gene Mutation Database (HGMD), only 17 pathogenic *RAB23* variants have been described in patients with CRPT1 [29]. Carpenter syndrome-2 (CRPT2) is caused by mutations in the *MEGF8* gene [15].

MORMS is caused by homozygous mutations in the *INPP5E* gene [30]. The primary clinical features of MORMS include intellectual disability, truncal obesity, retinal dystrophy, and a micropenis [16,30]. To date, only one family has been reported with this syndrome [16]. Obesity development is one of the hallmark features of MORMS. The case described by Torkar et al. [31] highlighted prominent phenotypic characteristics, including early-onset and severe obesity, accompanied by the development of metabolic syndrome.

MOSPGF is caused by mutations in the *CEP19* gene [17]. The clinical symptom of this disease is much less complicated compared to other ciliopathies. The primary features of MOSPGF include morbid obesity and spermatogenic failure [17].

In addition, Joubert syndrome (JBTS, OMIM #213300), as a ciliopathy, is not associated with obesity generally, but, recently, Sophie et al. [32] found that a patient with JBTS caused by mutations in the *ARL13B* gene exhibited an obesity phenotype, expanding the spectrum of JBTS.

1.2.1. Localization and Functions in Cilia

These proteins encoded by the genes of obesity-related ciliopathies are located in the different accessory substructures of the primary cilia and have unique functions (Figure 1). ALMS1 and CEP19 are located in the basal body [33,34]. CEP19 participates in the process of triggering the entry of intraflagellar transport (IFT) into the cilium [34]. RAB23 localizes to the basal region of the cilium adjacent to one of the centrioles, promoting cilium formation [35]. Eight of the BBS proteins (BBS1, BBS2, BBS4, BBS5, BBS7, BBS8/TTC8, BBS9, and BBS18/BBIP1) form a transport complex called BBSome [36] while proteins

BBS6/MKKS, BBS10, and BBS12 are known as chaperone complexes, facilitating the BBSome assembly [37]. In addition, BBS3/ARL6 is a GTPase regulating the BBSome entry to (and exit from) the cilium [38]. ARL13B is also a GTPase ciliary affecting transmembrane protein localizations and anterograde IFT assembly stability though Sonic Hedgehog (Shh) signaling [39]. Proteins encoded by other genes of BBS are believed to be located on the basal body or region, playing roles on recruitment of BBSome [38]. *INPP5E* encodes a 72 kDa phosphatase localized in the axoneme of primary cilia and plays a crucial role in regulating the PI3K signaling pathway within cilia [40]. Megf8 is now believed to take part in forming a membrane-tethered ubiquitin complex that can fine-tune the strength of Hedgehog (Hh) signaling [41].

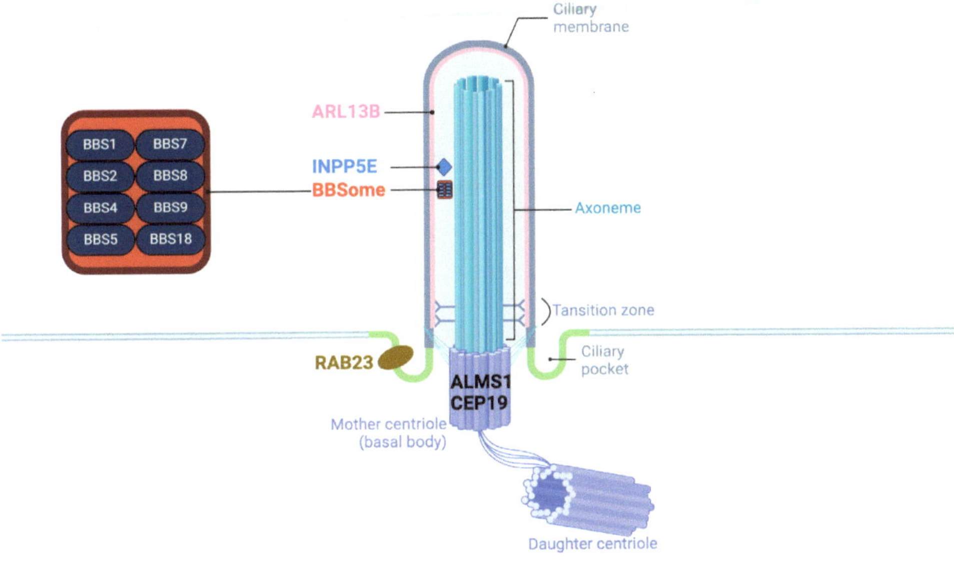

Figure 1. The structure of primary cilia and the localization of proteins of obesity-related ciliopathies.

1.2.2. Clinical Presentations, Diagnosis, and Treatment

There are similarities as well as differences in the clinical presentations between obesity-related ciliopathies due to the different roles of cilia-related genes in various organs. Table 1 categorizes and lists the confirmed clinical phenotypes of five obesity-related ciliopathies, grouped by systems and organs [8,13–17,19–30,38,42,43].

Table 1. All associated phenotypes of the ciliopathies characterized with obesity.

System or Organ	BBS [8,14,23,24,38,43]	ALMS [8,13,19–22,42]	CRPT [8,15,25–29]	MORMS [8,16,30]	MOSPGF [8,17]
Nervous system	Macrocephaly, intellectual disability, global developmental delay, cognitive impairment, poor coordination, hydrocephalus, brachycephaly, specific learning disability, neurological speech impairment, gait imbalance, delayed speech and language development, ataxia	Seizures, obsessive-compulsive behaviour, autism, aplasia/hypoplasia of the cerebellum	Trigonocephaly, sagittal craniosynostosis, oxycephaly, large foramen magnum, lambdoidal craniosynostosis, frontal bossing, brachycephaly, intellectual disability, cerebral atrophy, aplasia/hypoplasia of the corpus callosum, coronal craniosynostosis	Delayed speech and language development, intellectual disability, moderate	Intellectual disability
Endocrine system	Obesity, diabetes mellitus, nephrogenic diabetes insipidus, polycystic ovaries, hypertrichosis	Type II diabetes mellitus, insulin-resistant diabetes mellitus, hyperuricemia, hypertriglyceridemia, hyperinsulinemia, gynecomastia, precocious puberty, multinodular goiter, menstrual irregularities, hypothyroidism, hypergonadotropic hypogonadism, growth hormone deficiency, diabetes insipidus, polycystic ovaries, hypertrichosis, alopecia, acanthosis nigricans, global developmental delay, accelerated skeletal maturation, truncal obesity, short stature	Precocious puberty, short stature, obesity	Childhood-onset truncal obesity	Hyperbetalipoproteinemia, hypercholesterolemia, hypertriglyceridemia, insulin resistance, type II diabetes mellitus, obesity

Table 1. *Cont.*

System or Organ	BBS [8,14,23,24,38,43]	ALMS [8,13,19–22,42]	CRPT [8,15,25–29]	MORMS [8,16,30]	MOSPGF [8,17]
Motion system	Syndactyly, postaxial hand polydactyly, foot polydactyly, mesoaxial polydactyly, finger syndactyly, radial deviation of finger, polydactyly, short foot, broad foot, brachydactyly syndrome	Scoliosis, kyphosis	Joint contracture of the hand, camptodactyly, webbed neck, short neck, toe syndactyly, talipes equinovarus, shallow acetabular fossae, pseudoepiphyses of the proximal phalanges of the hand, preaxial polydactyly, preaxial foot polydactyly, postaxial polydactyly, postaxial hand polydactyly, metatarsus adductus, lateral displacement of patellae, genu varum, genu valgum, duplication of the proximal phalanx of the hallux, cutaneous finger syndactyly, coxa vara, coxa valga, complete duplication of proximal phalanx of the thumb, clinodactyly of the 5th finger, broad thumb, brachydactyly syndrome, aplasia/hypoplasia of the middle phalanges of the toes, aplasia/hypoplasia of the middle phalanges of the hand, aplasia of the middle phalanx of the hand, scoliosis, pectus excavatum, pectus carinatum, flared iliac wings, sacral dimple, spina bifida occulta	NM	NM
Urinary system	Multicystic kidney dysplasia, renal insufficiency, renal hypoplasia, renal cyst, Stage 5 chronic kidney disease, renal agenesis, nephrotic syndrome, abnormality of the kidney, recurrent respiratory infections	Vesicoureteral reflux, tubulointerstitial nephritis, renal insufficiency, abnormality of the urethra, nephrocalcinosis, nephritis, glomerulopathy	Hydroureter, hydronephrosis		

Table 1. *Cont.*

System or Organ	BBS [8,14,23,24,38,43]	ALMS [8,13,19–22,42]	CRPT [8,15,25–29]	MORMS [8,16,30]	MOSPGF [8,17]
Reproductive system	Cryptorchidism, hypospadias, micropenis, hypoplasia of penis, hypogonadism, vaginal atresia, external genital hypoplasia, decreased testicular size, abnormality of the ovary	Decreased fertility, abnormality of the testis, abnormality of female external genitalia	Shawl scrotum, micropenis, external genital hypoplasia, cryptorchidism	Micropenis	Azoospermia, infertility, oligospermia
Cardiovascular system	Hypertension, left ventricular hypertrophy, bicuspid aortic valve, atria septal defect	Renovascular hypertension, pulmonary hypertension, hypertrophic cardiomyopathy, dilated cardiomyopathy, congestive heart failure, atherosclerosis	Ventricular septal defect, transposition of the great arteries, tetralogy of fallot, patent ductus arteriosus, atria septal defect, pulmonic stenosis	NM	Congestive heart failure, hypertension, myocardial infarction, premature coronary artery disease
Digestive system	Hepatic fibrosis, hepatic failure, biliary tract abnormality, aganglionic megacolon	Splenomegaly, portal hypertension, hepatomegaly, hepatic steatosis, elevated hepatic transaminases, cirrhosis, chronic hepatic failure, chronic active hepatitis	NM	NM	Hepatic steatosis
Respiratory system	Respiratory distress, bronchiolitis, recurrent respiratory infections, asthma	Recurrent pneumonia, asthma, respiratory insufficiency, pulmonary fibrosis	NM	NM	NM
Face	High palate, low-set, posteriorly rotated ears, prominent nasal bridge, short neck, downslanted palpebral fissures, medial flaring of the eyebrow, hypodontia, dental crowding, abnormalities of the teeth	Round face, hyperostosis frontalis interna, gingivitis, abnormalities of the teeth	Wide nasal bridge, upslanted palpebral fissure, underdeveloped supraorbital ridges, sparse eyebrow, retrognathia, preauricular pit, persistence of primary teeth, narrow palate, micrognathia, malar flattening, hypoplasia of the maxilla, hypoplasia of midface, highly arched eyebrow, high palate, depressed nasal bridge, anteverted nares, agenesis of permanent teeth	NM	NM

Table 1. *Cont.*

System or Organ	BBS [8,14,23,24,38,43]	ALMS [8,13,19–22,42]	CRPT [8,15,25–29]	MORMS [8,16,30]	MOSPGF [8,17]
Eyes	Strabismus, retinopathy, rod-cone dystrophy, abnormal electroretinogram, cataract, myopia, retinal degeneration, cone/cone–rod dystrophy, abnormality of retinal pigmentation, nyctalopia, macular dystrophy, glaucoma, congenital primary aphakia, astigmatism, nystagmus	Deeply set eye, subcapsular cataract, pigmentary retinopathy, photophobia, nystagmus, cone/cone–rod dystrophy, chorioretinal abnormality, blindness	Hypertelorism, epicanthus, telecanthus, optic atrophy, opacification of the corneal stroma, microcornea	Cataract, retinal dystrophy, visual impairment	NM
Ear	Hearing impairment, recurrent otitis media	Progressive sensorineural hearing impairment, otitis media	Sensorineural hearing impairment, protruding ear, low-set ears, conductive hearing impairment, abnormality of the pinna	NM	NM
Other	Hyposmia, situs inversus totalis	Pes planus, hypoalphalipoproteinemia	Umbilical hernia, omphalocele, polysplenia, wide intermamillary distance, supernumerary nipple, hypoplastic nipples, situs inversus totalis, cutis laxa	NM	Hypoalphalipoproteinemia

BBS, Bardet–Biedl syndrome; ALMS, Alström syndrome; CRPT, Carpenter syndrome; MORMS, mental retardation, truncal obesity, retinal dystrophy, and micropenis syndrome; MOSPGF, morbid obesity and spermatogenic failure; NM, not mentioned. From Table 1, it can be deduced that the five genetic ciliopathies exhibit similar phenotypes of the neurologic, endocrine and reproductive systems. In the neurological system, common abnormalities include developmental delays, cognitive impairments, seizures, and motor coordination disorders. In the endocrine system, obesity, diabetes, and insulin resistance could usually be observed. Patients with obesity-related ciliopathies also complain about problems with their reproductive systems, such as genital development issues (e.g., micropenis, vaginal atresia, underdeveloped external genitalia) and reduced fertility (e.g., azoospermia, infertility) [8,13–17,19–30,38,42,43].

Apart from the above three systems, the clinical manifestations presented in different obesity-related ciliopathy are different. In the musculoskeletal system, polydactyly is typical for BBS [8,14,23,24,38], while spinal abnormalities, such as scoliosis and kyphosis, are common in ALMS [13,19–22,42]. The symptoms vary a lot in CRPT [15,25–29].

Patients with BBS, ALMS, and CRPT exhibit urinary system abnormalities, with specific differences observed among these conditions [8,13–15,19–29,38,42,44]. Chronic kidney disease is a dominant feature in BBS and could lead to increased morbidity and premature death [45]. The symptom in other diseases is much less severe in comparison with that in BBS [24,45]. Vesicoureteral reflux and nephritis could be observed in patients with ALMS [20], while patients with CRPT usually have hydronephrosis or pyelonephritis [8,26]. In the cardiovascular system, cardiomyopathy is common in ALMS while being rare in other diseases [8,19]. Patients with CRPT are more likely to have congenital heart defects, such as tetralogy of Fallot, patent ductus arteriosus, and atrial septal defect [8]. Abnormalities in the digestive and respiratory systems are usually in BBS and ALMS. Abnormal liver function is common, while non-alcoholic fatty liver disease shows higher incidence in patients with ALMS and BBS. They also exhibit issues related to recurrent respiratory infections [19,23,24]. No abnormalities in these systems were identified in patients with CRPT and MORMS [8,15,16,25–30].

The diagnosis of obesity-related ciliopathies primarily consists of two parts: clinical diagnosis and molecular diagnosis [19,24]. There are specific criterial for diagnosis for patients with ALMS and BBS. For example, the diagnosis of BBS can be confirmed if a patient meets four out of six major clinical symptoms or three major symptoms plus two secondary criteria [46]. The clinical diagnosis is made based on the clinical findings (signs and symptoms), medical history, and family history. Genetic sequencing is a powerful tool to confirm a molecular diagnosis. Other strategies, such as PCR and hybridization-based tests, can also be employed, particularly for cascade testing or in populations where there are recurrent hotspot mutations. The number of patients with CRPT, MORMS, and MOSPGF is relatively small, and research on clinical and molecular diagnosis is limited. However, the diagnostic strategy is similar to that for ALMS and BBS.

Currently, symptomatic treatment remains the primary approach for managing obesity-related ciliopathies [47]. For the obesity seen in ALMS and BBS, the MC4R agonist setmelanotide [48] and GLP-1 receptor agonists (GLP-1 RAs) [49] are promising therapeutic options. Notably, the U.S. FDA approved setmelanotide for the chronic management of weight in adults and pediatric patients aged 6 and older with deficiencies in POMC, LEPR, or PCSK1 in 2020 [50]. In June 2022, the indication for setmelanotide was expanded to include patients with BBS [51]. However, the variability in treatment response among patients requires further research to better understand the underlying mechanisms. These new drug targets also have the potential to serve as biomarkers for monitoring treatment efficacy and prognosis.

1.2.3. Characteristics of Obesity

The characteristics of obesity associated with the five ciliopathies are summarized in Table 2. The Table describes the age of onset, prevalence, BMI, and common comorbidities associated with obesity for each condition [14,16,17,19,24,26,28,30]. The onset age and prevalence of obesity vary among the ciliopathies, with BBS, ALMS, and MOSPGF showing early onset obesity [17,19,24], while MORMS typically manifests obesity when patients grow older (5 to 15 years) [16]. The prevalence of obesity is high in ALMS, MOSPGF (91%), and BBS (89%) [17,19,24]. The features of obesity also differ, with central obesity in BBS and truncal obesity in MORMS, whereas MOSPGF is characterized by morbid obesity with BMI > 40.0 kg/m^2 [17]. CRPT is notable for high birth weight and obesity present at birth, with a prevalence of 90%, but specific onset range and BMI are not reported [26,28].

Table 2. Obesity features between BBS, ALMS, CRPT, MORMS, and MOSPGF.

Diseases	Gene	Age of Onset Range	Incidence	BMI	Common Co-Morbidities	Comment
BBS [14,24]	*BBS1-22, IFT74, SCLT1, SCAPER, NPHP*	Birth—3 years	89%	35.7 ± 8.0 kg/m^2	Retinal cone–rod dystrophy, postaxial polydactyly, cognitive impairment, hypogonadism and genitourinary abnormalities, kidney disease	Central obesity; birth weight typically normal.
ALMS [19]	*ALMS1*	Birth—5 years	70–98%	NM	Cone–rod dystrophy, progressive sensorineural hearing loss, short stature, hypogonadism (central or primary), progressive renal disease, insulin resistance/type 2 diabetes mellitus	Birth weight typically normal; hyperphagia and excessive weight gain begin during the first years, resulting in childhood obesity.
CRPT [26,28]	*MEGF8, RAB23*	NM	90%	NM	Craniosynostosis, polydactyly, cardiac defects	High birth weight and obesity were prevalent.
MORMS [16,30]	*INPP5E*	5–15 years	NM	NM	Impaired intellectual development, retinal dystrophy, micropenis	Truncal obesity
MOSPGF [17]	*CEP19*	Birth—3 years	91%	>40.0 kg/m^2	Spermatogenic failure, hypertension, type 2 diabetes mellitus	Morbid obesity

BBS, Bardet–Biedl syndrome; ALMS, Alström syndrome; CRPT, Carpenter syndrome; MORMS, mental retardation, truncal obesity, retinal dystrophy, and micropenis syndrome; MOSPGF, morbid obesity and spermatogenic failure; BMI, body mass index; NM, not mentioned.

1.2.4. Mechanism of Obesity

Obesity in the human body is actually the result of an imbalance between energy intake and energy expenditure (Figure 2). Most cells, including neurons and glial cells throughout the central nervous system, possess primary cilia. The hypothalamus is especially critical for the regulation of energy homeostasis. The agouti-related protein (AgRP) neurons and the pro-opiomelanocortin (POMC) neurons in the arcuate nucleus of the hypothalamic tuberal region positively and negatively regulate feeding behavior, respectively [52]. They are also regulated by the periphery, mainly by insulin secreted by the pancreas, leptin secreted by adipocytes, ghrelin secreted by stomach, and cholecystokinin secreted by small intestine. MC4R neurons in the paraventricular nucleus, as their downstream neurons, directly participate in the regulation of energy metabolism [52].

The mechanisms associated with ciliary dysfunction and obesity in these syndromes are complex and only partially understood. It is currently believed that obesity in these diseases is associated through the leptin–melanocortin pathway influencing the hypothalamus [7]. Defect function of cilia in these diseases probably affects signaling between neurons/glial cells in this pathway while the causative mechanisms vary in different diseases. Obesity of BBS is caused by defects in the neurological control of the appetite, although it is unclear whether defective leptin signaling [53] or signaling by anorexigenic GPCR neuropeptide Y family receptors is the primary cause [54]. Similar to the situation in BBS, an increase in energy intake was observed both in human and mice with ALMS [55]. The percentage of ciliated hypothalamic neurons is significantly reduced in *Alms1* mutant mice [56]. Apart from that, a significant reduction in energy expenditure also accounts for the process of obesity, though the specific mechanism is still unknown [55]. The obesity in CRPT may result from damaged Hh signaling or *RAB23* itself potentially regulating adipogenesis [28,29].

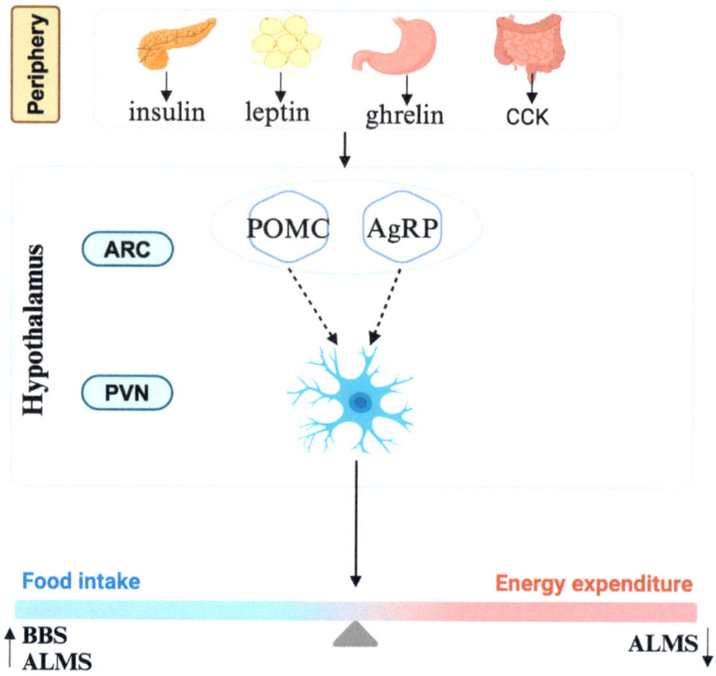

Figure 2. Mechanism of obesity. CCK, cholecystokinin; ARC, arcuate nucleus; PVN, paraventricular nucleus; POMC, pro-opiomelanocortin; AgRP, agouti-related protein; BBS, Bardet–Biedl syndrome; ALMS, Alström syndrome.

2. Advance of Biomarkers

Due to rarity of patients with obesity-related ciliopathies, the progress in identifying related biomarkers is based predominantly on ALMS and BBS studies [6]. These two syndromes have been the focus of more extensive research, leading to a better understanding of their underlying mechanisms and potential biomarkers for diagnosis, variability in clinical presentation, progress, and prognosis and treatment.

2.1. Diagnosis and Differential Diagnosis

The diagnosis of obesity-related ciliopathies primarily involves two components: clinical diagnosis and molecular diagnosis [19,24]. For ALMS and BBS, clinical diagnosis is mainly based on characteristic clinical features, while molecular diagnosis, through genetic testing, is considered the gold standard for definitive diagnosis [19,24]. With the advancement of gene sequencing technologies, next-generation sequencing (NGS) has become the primary method for diagnosing monogenic diseases, including obesity-related ciliopathies [57]. These diseases are autosomal recessive disorders, meaning that the diagnosis can be confirmed by detecting two pathogenic variants in the associated genes. However, there are clinical scenarios where the diagnosis may be challenging. In some cases, patients may exhibit high clinical suspicion for a ciliopathy, while the result of genetic testing may be negative, or only one pathogenic variant is detected, which is insufficient to confirm a diagnosis [58].

Gene functional studies can provide new evidence for the ACMG classification of gene variants. Chen et al. [59] conducted a study on patients with a clinical suspicion of ALMS who carried one pathogenic variant and one variant of uncertain significance (VUS). They performed ALMS1 protein expression staining on skin fibroblasts and discovered that ALMS1 expression at the centrosome was severely impaired. This finding provided evidence to prove the pathogenicity of the missense variant of *ALMS1* and confirmed the diagnosis of ALMS. In contrast, for patients carrying only VUS, staining results indicated that ALMS1 expression was not impaired, excluding the diagnosis of ALMS; however, further genetic testing confirmed the diagnosis of BBS. Thus, this study suggests that ALMS1 expression detection could be a promising differential diagnostic marker for ALMS.

2.2. Variability in Clinical Presentation, Progress, and Prognosis

Obesity-related ciliopathies exhibit highly complex and variable clinical manifestations. As genetic disorders, genotype–phenotype analyses can partly explain the variability in clinical presentations, disease progression, and prognosis.

2.2.1. Genotype–Phenotype Correlation

The clinical manifestations of obesity-related ciliopathies are highly complex, with the same disease potentially caused by different genes. The clinical presentation of BBS can vary depending on the specific gene involved. In a meta-analysis by Niederlova et al. [38], differences in clinical symptoms caused by various BBS genes were compared (Table 3). Patients with *BBS3* mutations tend to have fewer clinical symptoms. Retinal dystrophy is usually milder in patients with *BBS1* gene mutations while being severe in patients with *BBS2*, *BBS3*, and *BBS4* Mutations. Patients with *BBS2* mutations have a higher incidence of polydactyly, and *BBS10* gene mutations are linked to more pronounced obesity and insulin resistance [38].

Table 3. Genotype–phenotype correlation in ALMS and BBS.

Genotype	Distinguishing Clinical Features/Comments
ALMS1 variants in exon 16 [60]	Early onset (before one year old) of retinal degeneration, urinary system dysfunction, dilated cardiomyopathy, and diabetes
ALMS1 variants in exon 8 [60]	Milder symptoms or later onset of kidney disease
ALMS1 truncated around exon 10 [22]	A higher prevalence of liver dysfunction and experience worse disease progression
ALMS1 c.7911dupC [61]	Severe cardiomyopathy
BBS1 [38]	Relatively less "syndromic" penetrance of renal anomalies
BBS2 [38]	Relatively more "syndromic" penetrance of renal anomalies "Leanest" of obesity phenotype
BBS3 [38] (*ARL6*)	Lowest "syndromic" penetrance of cognitive impairment and renal anomalies
BBS4 [38]	Low penetrance of renal anomalies Early-onset morbid obesity
BBS5 [38]	Relatively more "syndromic"
BBS6 [38] (*MKKS*)	More likely to have CHD and genitourinary malformations
BBS7 [38]	Relatively more "syndromic" penetrance of renal anomalies
BBS8 [38] (*TTC8*)	Relatively less "syndromic" penetrance of renal anomalies
BBS9 [38]	High penetrance of renal anomalies
BBS10 [38]	Most severe renal impairment Significant adiposity
BBS12 [38]	Significant adiposity
BBS21 [38] (*C8rorf37*)	High penetrance of polydactyly

Unlike BBS, which is associated with many genes, ALMS is a monogenic disease. However, the *ALMS1* gene is very large, and different mutation sites within this gene can result in varying severity, progression, and prognosis of clinical manifestations [58]. Patients with variants in exon 16 usually present with early onset (before one-year old) of retinal degeneration, urinary system dysfunction, dilated cardiomyopathy, and diabetes, while those with variants in exon 8 are prone to have milder symptoms or later onset of kidney disease [60]. A recent meta-analysis conducted by Brais Bea-Mascato et al. [22] revealed important insights into the genotype–phenotype, based on data from 227 ALMS patients. Patients with the longest allele of the *ALMS1* gene truncated around exon 10 (E10) exhibit a higher prevalence of liver dysfunction and experience worse disease progression. However, no significant differences in the prevalence of dilated cardiomyopathy (DCM), hypertrophic cardiomyopathy (HCM), and type 2 diabetes mellitus (T2DM) are observed among patients grouped by longevity of allele. Savas et al. found that the c.7911dupC (p. Asn2638Glnfs*24) mutation can be related to severe cardiomyopathy in ALMS [61].

2.2.2. Other Biomarkers from Multi-Omics Data

Additionally, advances in multi-omics data have also made great contributions to the discovery of new biomarkers (Table 4).

Fatty liver is a common issue in these diseases and can progress to liver fibrosis in advanced stages. This condition is related to obesity but may also be linked to pathogenic genes. ALMS is currently considered a classic model for non-alcoholic fatty liver disease (NAFLD) [62].

For monitoring fatty liver disease, liver biopsy is considered as the gold standard. However, there is a limit of widespread use in clinic considering its invasiveness and complexity. Non-invasive indexes, such as the alanine aminotransferase (ALT)/aspartate aminotransferase (AST) ratio, the AST-to-platelet ratio index (APRI), and the Fibrosis-4 Index (FIB-4) could predict liver fibrosis in liver disease to some extent [63,64]. Recently, imaging methods, such as transient elastography (TE), FibroScan, and shear-wave elastography (SWE), have advanced. Silvia et al. [62] found that SWE plays an important predictive role in the progression of fatty liver to fibrosis in patients with ALMS.

Cardiomyopathy is another significant clinical manifestation of ALMS. It can be identified and monitored using biomarkers, such as NT-proBNP, high-sensitivity troponin, and T-wave inversion on a 12-lead electrocardiogram [65]. However, Nicola et al. [66] suggested that evidence for using NT-proBNP to monitor cardiomyopathy in ALMS may be insufficient. They thought that extracellular volume (ECV) expansion might be a more robust predictor of adverse cardiovascular outcomes. Additionally, they also proposed that elevated triglycerides could be used as a potential marker for cardiac fibrosis, but more prospective studies are needed to validate this.

Agnieszka et al. [67] conducted a study to detect and analyze microRNA (miRNA) expression in the serum of patients with ALMS, BBS, obese controls, and normal controls. They found that miR-301a-3p expression was significantly reduced in both ALMS and BBS patients. Meanwhile, miR-92b-3p expression was decreased in ALMS but increased in BBS. Additionally, they identified eight miRNAs (miR-30a-5p, miR-92b-3p, miR-99a-5p, miR-122-5p, miR-192-5p, miR-193a-5p, miR-199a-3p, and miR-205-5p) that showed significant correlations with clinical parameters, including lipid profiles, serum creatinine, cystatin C, fasting glucose, insulin, C-peptide levels, HbA1c values, and insulin resistance (HOMA-IR). These findings suggest that miRNAs could serve as valuable biomarkers of disease progression in patients with ALMS and BBS syndromes. However, further research is needed to validate these results and to fully understand the potential of miRNAs as biomarkers for these conditions.

Krzysztof et al. [68] conducted a non-targeted metabolomics analysis on patients with ALMS, BBS, obese controls, and normal controls. They found that metabolic changes in ALMS/BBS patients were similar to those observed in non-syndrome obesity, such as the higher levels of acylcarnitines, and tetrahydroaldosterone-3-glucuronide, which have been correlated with insulin resistance and hypertension, respectively, in numerous studies [69–71]. They also found that ALMS/BBS patients were characterized by elevated levels of oxidized phosphatidylcholines (PCs), which are recognized as markers of oxidative stress. These diseases exhibit clinical manifestations that often progress with age. Through further age-related analysis, they further found that lipids, primarily lysophosphatidylethanolamines (LPE), are major markers of disease progression. Interestingly, only one metabolite, long-chain fatty acid (FA 26:1; O_2), showed a negative correlation with age in ALMS and BBS patients, while it correlated positively in the other groups, suggesting it may have a unique indicative biomarker role [67].

In another study, Krzysztof et al. conducted serum bone metabolism marker testing on patients with ALMS and BBS. They found serum osteocalcin (OC) and urinary deoxypyridinoline (DPD) levels were negatively correlated with the HOMA-IR index. Additionally, serum receptor activator of nuclear factor kappa-B ligand (s-RANKL) levels were negatively correlated with fasting blood glucose concentrations [72].

Fibrotic changes have been reported in multiple organs in ALMS. Merlin et al. [73] analyzed miRNA expression on peripheral lymphocytes from six ALMS patients and found that fibrosis-related miR-324-5p was upregulated by 2.1-fold in ALMS males. Additionally, passenger strand members of the same mature miRNA category (miR-27a vs. miR-27a star), along with other miRNAs previously reported to be associated with fibrosis (miR-27a, miR-27b, miR-29b, and miR-25), showed disturbances. These findings suggest that miRNAs could potentially serve as biomarkers for fibrosis in ALMS.

Chronic kidney disease (CKD) is the most common cause of death among BBS patients, with significant variability in severity. Early identification and intervention are crucial for managing CKD in these patients [24]. Magnetic resonance diffusion tensor imaging (DTI) can be used to assess the microstructural integrity of the kidneys. Compared to the control group, BBS patients exhibited lower cortical fractional anisotropy (FA) and axial diffusivity, and higher mean diffusivity and radial diffusivity in the kidneys [74]. The urine protein profile of BBS patients can provide information on the risk and predictive factors for adverse renal outcomes and urinary markers of renal insufficiency. Notably, the abundance of urinary fibronectin (u-FN), CD44 antigen, and lysosomal α-glucosidase significantly correlated with glomerular filtration rate [75]. By comparing the urinary metabolomics of BBS patients and control subjects, they discovered that the excretion of several monocarboxylates, including lactate, was increased in the early and late stages of CKD [76]. Emanuela et al. conducted a targeted serum metabolomics study comparing BBS patients to control subjects. They found that renal insufficiency in BBS patients was associated with abnormal levels of plasma phosphatidylcholines and acylcarnitines. Miriam et al. assessed the renal function of 54 BBS patients. They discovered that maximum urine osmolality (max-Uosm) was correlated with the annual decline in estimated glomerular filtration rate (ΔeGFR), suggesting that a defect in urine concentration may predict disease progression [77].

Table 4. Other biomarkers from multi-omics data.

Clinical Features	Correlation Detection	Biomarkers/Specific Methods	Comment	References
Fatty liver	Imaging analysis	SWE, TE	SWE plays an important predictive role in the progression of fatty liver to fibrosis in patients with ALMS	Silvia et al. [62]
	Metabolomics	ALT/AST ratio, APRI, FIB-4	Predicts liver fibrosis in liver disease to some extent	Guangqin et al. [63], Erin et al. [64]

Table 4. *Cont.*

Clinical Features	Correlation Detection	Biomarkers/Specific Methods	Comment	References
Cardiomyopathy	Imaging analysis	ECV	ECV expansion might be a more robust predictor of adverse cardiovascular outcomes	Nicola et al. [66]
	Metabolomics	NT-proBNP, high-sensitivity troponin, elevated triglycerides	Elevated triglycerides are proposed as a potential marker for cardiac fibrosis	Ashwin et al. [65], Nicola et al. [66]
Kidney	Proteomics	u-FN, CD44 antigen, lysosomal α-glucosidase	u-FN, CD44 antigen, and lysosomal α-glucosidase significantly correlated with glomerular filtration rate	Marianna et al. [75]
Fibrosis	MiRNA sequencing	miRNAs (miR-324-5p, miR-27a, miR-27b, miR-29b, and miR-25)	MiRNAs could potentially serve as biomarkers for fibrosis in ALMS	Merlin et al. [73]
Obesity and related metabolic syndromes	MiRNA sequencing	miRNAs (miR-30a-5p, miR-92b-3p, miR-99a-5p, miR-122-5p, miR-192-5p, miR-193a-5p, miR-199a-3p, and miR-205-5p)	MiR-301a-3p expression was significantly reduced in both ALMS and BBS patients; miR-92b-3p expression was decreased in ALMS but increased in BBS; miRNAs could serve as valuable biomarkers of disease progression in patients with ALMS and BBS	Agnieszka et al. [67]
	Metabolomics: Non-targeted metabolomics analysis	PC, LPE, long-chain fatty acid (FA 26:1; O_2), acylcarnitines, tetrahydroaldosterone-3-glucuronide	LPEs are major markers of disease progression; only long-chain fatty acid (FA 26:1; O_2) showed a negative correlation with age in ALMS and BBS patients	Krzysztof et al. [68]
	Metabolomics: Serum bone metabolism marker testing	OC, DPD, s-RANKL	OC and DPD levels were negatively correlated with the HOMA-IR index	Krzysztof et al. [72]

SWE, shear-wave elastography; TE, transient elastography; ECV, extracellular volume; ALMS, Alström syndrome; BBS, Bardet–Biedl syndrome; ALT, alanine aminotransferase; AST, aspartate aminotransferase; APRI, AST-to-platelet ratio index; FIB-4, Fibrosis-4 Index; miRNA, microRNA; PC, oxidized phosphatidylcholine; LPE, lysophosphatidylethanolamines; OC, osteocalcin; DPD, deoxypyridinoline; HOMA-IR, Homeostasis Model Assessment of Insulin Resistance; s-RANKL, serum receptor activator of nuclear factor kappa-B ligand; u-FN, urinary fibronectin.

3. Discussion

Primary cilia, present in most mature human cells, play crucial roles in various organ systems, and defects in cilia can lead to widespread clinical manifestations. Obesity-related ciliopathies constitute a small part of ciliopathies; however, they provide important clues to the study of the relationship between primary cilia and obesity in humans.

However, obesity-related ciliopathies are very rare, making the discovery of biomarkers much more difficult than other diseases. Regarding the importance of biomarkers in diagnosis, monitoring, and follow-up of these diseases, there is an urgent need for relevant biomarker studies and relevant reviews. Therefore, this paper focuses on the topic of biomarkers for obesity-related ciliopathies and reviews the advances of biomarkers in the clinical presentation, diagnosis, and prognosis of these diseases.

For a long time, the underlying causes of these diseases remained unknown, leading to unclear or delayed diagnoses for many patients. The development of genetic diagnostic technologies, especially NGS, has provided solutions, enabling accurate diagnosis [78]. Trio exome sequencing would be a great option considering the autosomal recessive inheritance pattern. Proband-only medical exome sequencing is considered as a more cost-effective test, especially in China [79], and WES has also proved to be a useful and cost-effective tool in diagnosing ciliopathy [80]. Some countries prefer using targeted gene panels, which focus on a specific set of genes known to be associated with ciliopathies [81]. However, the high cost of genetic testing is a significant barrier to widespread implementation, particularly in remote and low-income areas. The disparity in access to genetic testing in China may contribute to the uneven geographic distribution of ALMS diagnosed cases, with patients in remote and impoverished regions being less likely to receive a diagnosis [42]. RNA sequencing [82] and multiplex PCR [83] could be an optional diagnostic technique for complicated heterogeneous ciliopathies. For physicians with limited literature information of these disease, identifying pathogenic mutations with WES is a promising tool to provide clues to diagnose, although it could lead to overuse and the waste of resources.

For a subset of patients with unclear diagnoses, protein functional studies offer a pathway to resolve uncertainties, thereby providing additional evidence for the classification of variants. This approach not only improves diagnostic rates but also potentially provides new insights into disease severity assessment. Chen et al. [59] conducted immunofluorescence staining for *ALMS1* protein expression on skin fibroblasts from 16 ALMS patients. They found that patients with significant *ALMS1* expression fluorescence exhibited later onset and milder vision impairments.

The development of multi-omics technologies and analytical methods, including metabolomics, proteomics, imaging analysis, and miRNA sequencing, provide new possibilities for discovering novel biomarkers in these diseases. However, these advancements are still largely focused on explaining the variability in clinical manifestations. Additionally, the majority of research concentrated on AS and BBS, with limited data available for other obesity-related ciliopathies. There is currently a lack of biomarkers that can effectively indicate treatment efficacy. Up to now, only the genotype–phenotype correlation result is powerful enough to be used in clinical practice. For the biomarkers identified from multi-omics data, more evidence should be provided in the future to confirm the correlation and to validate their effectiveness and accuracy in clinical use.

A significant challenge in discovering biomarkers for obesity-related ciliopathies is the limited number of patients, which restricts the statistical power for identifying genotype–phenotype correlations and other omics biomarkers. For example, due to the limited number of patients, Krzysztof et al. did not separately conduct non-targeted metabolomics analysis for BBS and ALMS, which is a noted limitation [68].

Understanding these diseases is not only beneficial for the diagnosis and treatment of the disease itself, but also for the discovery of the role of cilia-related genes in obesity, which is expected to lead to the discovery of new therapeutic targets for obesity. For instance, *ALMS1* is considered a biomarker for the diagnosis and prognosis of acute myocardial infarction (AMI) [84], with specific variants, such as G/A variant (rs674804) and glutamic acid repeat polymorphism, being markers for early-onset myocardial infarction [85,86]. Furthermore, Edwige et al. [87] identified a crucial protein–protein interaction between ALMS1 and protein kinase C-α (PKCα), which could be a new pharmacological class for treating IR and its numerous comorbidities, such as type 2 diabetes and CVD, in large populations.

4. Conclusions

Obesity-related ciliopathy is rare. Identifying the pathogenic gene with genetic sequencing is the main method to diagnose these diseases, making it a dominant biomarker. Genotype–phenotype correlation could also explain some variability in clinical presentation. Biomarkers identified with multi-omics are restricted to ALMS and BBS patients. There is currently a lack of biomarkers that can effectively indicate treatment efficacy.

5. Future Directions

More cost-effective and convenient genetic testing is crucial for enhancing the early diagnosis and management of obesity-related ciliopathies.

It is crucial to conduct international multi-center studies, pooling patient resources to carry out comprehensive analyses and particularly conducting lipidomics analysis in ALMS [68] and genotype–phenotype analyses focusing on different pathogenic variants within the same BBS gene.

Building on these efforts, integrating multi-omics data with longitudinal cohort studies is essential. Prospective studies that follow patients over time can provide valuable insights into the progression of diseases, prognosis, and treatment efficacy. By combining genomic, transcriptomic, proteomic, and metabolomic data, researchers can uncover new biomarkers and better understand their roles in disease mechanisms.

Author Contributions: Conceptualization, Q.Z., Y.H., S.G. and X.W.; writing—original draft preparation, Q.Z., Y.H. and S.G.; writing—review and editing, Q.Z., Y.H., S.G., Y.D., H.Z., G.C. and X.W. All authors have read and agreed to the published version of the manuscript.

Funding: This work was supported by the National Key Research and Development Program of China (2022YFC2703102) and National Nature Science Foundation of China (82170910).

Institutional Review Board Statement: This study did not require ethical approval because it reports other studies.

Data Availability Statement: Not applicable.

Conflicts of Interest: The authors declare no conflicts of interest.

Abbreviations

GWAS	Genome-wide association studies
ADCY3	Adenylate cyclase 3
MC4R	Melanocortin 4 Receptor
GLP-1	Glucagon-like Peptide-1
BBS	Bardet–Biedl syndrome
ALMS	Alström syndrome
CRPT	Carpenter syndrome
MORMS	Mental retardation, truncal obesity, retinal dystrophy, and micropenis syndrome
MOSPGF	Morbid obesity and spermatogenic failure
JBTS	Joubert syndrome
HGMD	Human Gene Mutation Database
IFT	Intraflagellar transport
Shh	Sonic Hedgehog
Hh	Hedgehog
BMI	Body mass index
AgRP	Agouti-related protein
POMC	Pro-opiomelanocortin
NGS	Next-generation sequencing
ACMG	American College of Medical Genetics and Genomics
VUS	Variant of uncertain significance
DCM	Dilated cardiomyopathy
HCM	Hypertrophic cardiomyopathy
T2DM	Type 2 diabetes mellitus
NAFLD	Non-alcoholic fatty liver disease
ALT	Alanine aminotransferase
AST	Aspartate aminotransferase
APRI	AST-to-platelet ratio index
FIB-4	Fibrosis-4 Index
TE	Transient elastography
SWE	Shear-wave elastography

NT-proBNP	N-terminal pro-brain natriuretic peptide
ECV	Extracellular volume
miRNA	MicroRNA
HOMA-IR	Homeostasis Model Assessment of Insulin Resistance
HbA1c	Glycated Hemoglobin A1c
PC	Phosphatidylcholine
LPE	Lysophosphatidylethanolamine
FA	Fatty acid
OC	Osteocalcin
DPD	Deoxypyridinoline
s-RANKL	Serum receptor activator of nuclear factor kappa-B ligand
CKD	Chronic kidney disease
DTI	Diffusion tensor imaging
u-FN	Urinary fibronectin
GFR	Glomerular filtration rate
max-Uosm	Maximum urine osmolality
ΔeGFR	Estimated glomerular filtration rate
WES	Whole-exome sequencing
PCR	Polymerase chain reaction
AMI	Acute myocardial infarction
CCK	Cholecystokinin
ARC	Arcuate nucleus
PVN	Paraventricular nucleus

References

1. Wallmeier, J.; Nielsen, K.G.; Kuehni, C.E.; Lucas, J.S.; Leigh, M.W.; Zariwala, M.A.; Omran, H. Motile ciliopathies. *Nat. Rev. Dis. Primers* **2020**, *6*, 77. [CrossRef] [PubMed]
2. Wang, L.; Wen, X.; Wang, Z.; Lin, Z.; Li, C.; Zhou, H.; Yu, H.; Li, Y.; Cheng, Y.; Chen, Y.; et al. Ciliary transition zone proteins coordinate ciliary protein composition and ectosome shedding. *Nat. Commun.* **2022**, *13*, 3997. [CrossRef] [PubMed]
3. Hong, Y.; Ullah, R.; Wang, J.B.; Fu, J.F. Trends of obesity and overweight among children and adolescents in China. *World J. Pediatr.* **2023**, *19*, 1115–1126. [CrossRef] [PubMed]
4. World Obesity Atlas 2023. Available online: https://www.worldobesityday.org/assets/downloads/ (accessed on 31 July 2024).
5. Siljee, J.E.; Wang, Y.; Bernard, A.A.; Ersoy, B.A.; Zhang, S.; Marley, A.; Von Zastrow, M.; Reiter, J.F.; Vaisse, C. Subcellular localization of MC4R with ADCY3 at neuronal primary cilia underlies a common pathway for genetic predisposition to obesity. *Nat. Genet.* **2018**, *50*, 180–185. [CrossRef] [PubMed]
6. Vaisse, C.; Reiter, J.F.; Berbari, N.F. Cilia and Obesity. *Cold Spring Harb. Perspect. Biol.* **2017**, *9*, a028217. [CrossRef] [PubMed]
7. Engle, S.E.; Bansal, R.; Antonellis, P.J.; Berbari, N.F. Cilia signaling and obesity. *Semin. Cell Dev. Biol.* **2021**, *110*, 43–50. [CrossRef] [PubMed]
8. Lovera, M.; Lüders, J. The ciliary impact of nonciliary gene mutations. *Trends Cell Biol.* **2021**, *31*, 876–887. [CrossRef] [PubMed]
9. Reiter, J.F.; Leroux, M.R. Genes and molecular pathways underpinning ciliopathies. *Nat. Rev. Mol. Cell Biol.* **2017**, *18*, 533–547. [CrossRef] [PubMed]
10. Hildebrandt, F.; Benzing, T.; Katsanis, N. Ciliopathies. *N. Engl. J. Med.* **2011**, *364*, 1533–1543. [CrossRef]
11. Tobin, J.L.; Beales, P.L. The nonmotile ciliopathies. *Genet. Med.* **2009**, *11*, 386–402. [CrossRef]
12. Legendre, M.; Zaragosi, L.-E.; Mitchison, H.M. Motile cilia and airway disease. *Semin. Cell Dev. Biol.* **2021**, *110*, 19–33. [CrossRef] [PubMed]
13. Marshall, J.D.; Beck, S.; Maffei, P.; Naggert, J.K. Alström Syndrome. *Eur. J. Hum. Genet.* **2007**, *15*, 1193–1202. [CrossRef] [PubMed]
14. Forsythe, E.; Beales, P.L. Bardet-Biedl syndrome. *Eur. J. Hum. Genet.* **2013**, *21*, 8–13. [CrossRef] [PubMed]
15. Twigg, S.R.F.; Lloyd, D.; Jenkins, D.; Elçioglu, N.E.; Cooper, C.D.O.; Al-Sannaa, N.; Annagür, A.; Gillessen-Kaesbach, G.; Hüning, I.; Knight, S.J.L.; et al. Mutations in multidomain protein MEGF8 identify a Carpenter syndrome subtype associated with defective lateralization. *Am. J. Hum. Genet.* **2012**, *91*, 897–905. [CrossRef] [PubMed]
16. Hampshire, D.J.; Ayub, M.; Springell, K.; Roberts, E.; Jafri, H.; Rashid, Y.; Bond, J.; Riley, J.H.; Woods, C.G. MORM syndrome (mental retardation, truncal obesity, retinal dystrophy and micropenis), a new autosomal recessive disorder, links to 9q34. *Eur. J. Hum. Genet.* **2006**, *14*, 543–548. [CrossRef] [PubMed]
17. Shalata, A.; Ramirez, M.C.; Desnick, R.J.; Priedigkeit, N.; Buettner, C.; Lindtner, C.; Mahroum, M.; Abdul-Ghani, M.; Dong, F.; Arar, N.; et al. Morbid obesity resulting from inactivation of the ciliary protein CEP19 in humans and mice. *Am. J. Hum. Genet.* **2013**, *93*, 1061–1071. [CrossRef] [PubMed]
18. Katsanis, N.; Ansley, S.J.; Badano, J.L.; Eichers, E.R.; Lewis, R.A.; Hoskins, B.E.; Scambler, P.J.; Davidson, W.S.; Beales, P.L.; Lupski, J.R. Triallelic inheritance in Bardet-Biedl syndrome, a Mendelian recessive disorder. *Science* **2001**, *293*, 2256–2259. [CrossRef] [PubMed]

19. Paisey, R.B.; Steeds, R.; Barrett, T.; Williams, D.; Geberhiwot, T.; Gunay-Aygun, M. Alström Syndrome. In *GeneReviews*®; Adam, M.P., Feldman, J., Mirzaa, G.M., Pagon, R.A., Wallace, S.E., Bean, L.J.H., Gripp, K.W., Amemiya, A., Eds.; University of Washington: Seattle, WA, USA, 1993. Available online: http://www.ncbi.nlm.nih.gov/books/NBK1267/ (accessed on 25 April 2024).
20. Marshall, J.D.; Maffei, P.; Collin, G.B.; Naggert, J.K. Alström Syndrome: Genetics and Clinical Overview. *Curr. Genom.* **2011**, *12*, 225–235. [CrossRef]
21. Marshall, J.D.; Bronson, R.T.; Collin, G.B.; Nordstrom, A.D.; Maffei, P.; Paisey, R.B.; Carey, C.; MacDermott, S.; Russell-Eggitt, I.; Shea, S.E.; et al. New Alström Syndrome Phenotypes Based on the Evaluation of 182 Cases. *Arch. Intern. Med.* **2005**, *165*, 675–683. [CrossRef]
22. Bea-Mascato, B.; Valverde, D. Genotype–phenotype associations in Alström syndrome: A systematic review and meta-analysis. *J. Med. Genet.* **2024**, *61*, 18–26. [CrossRef]
23. Melluso, A.; Secondulfo, F.; Capolongo, G.; Capasso, G.; Zacchia, M. Bardet-Biedl Syndrome: Current Perspectives and Clinical Outlook. *Ther. Clin. Risk Manag.* **2023**, *19*, 115–132. [CrossRef] [PubMed]
24. Forsyth, R.; Gunay-Aygun, M. Bardet-Biedl Syndrome Overview. In *GeneReviews*®; Adam, M.P., Feldman, J., Mirzaa, G.M., Pagon, R.A., Wallace, S.E., Bean, L.J.H., Gripp, K.W., Amemiya, A., Eds.; University of Washington: Seattle, WA, USA, 1993. Available online: http://www.ncbi.nlm.nih.gov/books/NBK1363/ (accessed on 27 April 2024).
25. Bouaré, F.; Noureldine, M.H.A.; Hajhouji, F.; Ghannane, H.; Jallo, G.I.; Benali, S.A. Complex craniosynostosis in the context of Carpenter's syndrome. *Childs Nerv. Syst. ChNS Off. J. Int. Soc. Pediatr. Neurosurg.* **2022**, *38*, 831–835. [CrossRef] [PubMed]
26. Hidestrand, P.; Vasconez, H.; Cottrill, C. Carpenter syndrome. *J. Craniofac. Surg.* **2009**, *20*, 254–256. [CrossRef] [PubMed]
27. Temtamy, S.A. Carpenter's syndrome: Acrocephalopolysyndactyly. An autosomal recessive syndrome. *J. Pediatr.* **1966**, *69*, 111–120. [CrossRef] [PubMed]
28. Jenkins, D.; Seelow, D.; Jehee, F.S.; Perlyn, C.A.; Alonso, L.G.; Bueno, D.F.; Donnai, D.; Josifova, D.; Mathijssen, I.M.J.; Morton, J.E.V.; et al. RAB23 Mutations in Carpenter Syndrome Imply an Unexpected Role for Hedgehog Signaling in Cranial-Suture Development and Obesity. *Am. J. Hum. Genet.* **2007**, *80*, 1162–1170. [CrossRef] [PubMed]
29. Khairat, R.; Elhossini, R.; Sobreira, N.; Wohler, E.; Otaify, G.; Mohamed, A.M.; Raouf, E.R.A.; Sayed, I.; Aglan, M.; Ismail, S.; et al. Expansion of the phenotypic and mutational spectrum of Carpenter syndrome. *Eur. J. Med. Genet.* **2022**, *65*, 104377. [CrossRef]
30. Jacoby, M.; Cox, J.J.; Gayral, S.; Hampshire, D.J.; Ayub, M.; Blockmans, M.; Pernot, E.; Kisseleva, M.V.; Compère, P.; Schiffmann, S.N.; et al. INPP5E mutations cause primary cilium signaling defects, ciliary instability and ciliopathies in human and mouse. *Nat. Genet.* **2009**, *41*, 1027–1031. [CrossRef] [PubMed]
31. Drole Torkar, A.; Avbelj Stefanija, M.; Bertok, S.; Podkrajšek, K.T.; Debeljak, M.; Kranjc, B.S.; Battelino, T.; Kotnik, P. Novel Insights Into Monogenic Obesity Syndrome Due to INPP5E Gene Variant: A Case Report of a Female Patient. *Front. Endocrinol.* **2021**, *12*, 581134. [CrossRef]
32. Thomas, S.; Cantagrel, V.; Mariani, L.; Serre, V.; Lee, J.-E.; Elkhartoufi, N.; de Lonlay, P.; Desguerre, I.; Munnich, A.; Boddaert, N.; et al. Identification of a novel ARL13B variant in a Joubert syndrome-affected patient with retinal impairment and obesity. *Eur. J. Hum. Genet.* **2015**, *23*, 621–627. [CrossRef]
33. Hearn, T.; Spalluto, C.; Phillips, V.J.; Renforth, G.L.; Copin, N.; Hanley, N.A.; Wilson, D.I. Subcellular localization of ALMS1 supports involvement of centrosome and basal body dysfunction in the pathogenesis of obesity, insulin resistance, and type 2 diabetes. *Diabetes* **2005**, *54*, 1581–1587. [CrossRef]
34. The CEP19-RABL2 GTPase Complex Binds IFT-B to Initiate Intraflagellar Transport at the Ciliary Base—PubMed. Available online: https://pubmed.ncbi.nlm.nih.gov/28625565/ (accessed on 29 June 2024).
35. Gerondopoulos, A.; Strutt, H.; Stevenson, N.L.; Sobajima, T.; Levine, T.P.; Stephens, D.J.; Strutt, D.; Barr, F.A. Planar Cell Polarity Effector Proteins Inturned and Fuzzy Form a Rab23 GEF Complex. *Curr. Biol.* **2019**, *29*, 3323–3330.e8. [CrossRef] [PubMed]
36. Bardet-Biedl Syndrome: Genetics, Molecular Pathophysiology, and Disease Management—PubMed. Available online: https://pubmed.ncbi.nlm.nih.gov/27853007/ (accessed on 21 July 2024).
37. Gupta, N.; D'Acierno, M.; Zona, E.; Capasso, G.; Zacchia, M. Bardet–Biedl syndrome: The pleiotropic role of the chaperonin-like BBS6, 10, and 12. *Proteins* **2022**, *190*, 9–19. [CrossRef] [PubMed]
38. Niederlova, V.; Modrak, M.; Tsyklauri, O.; Huranova, M.; Stepanek, O. Meta-analysis of genotype-phenotype associations in Bardet-Biedl syndrome uncovers differences among causative genes. *Hum. Mutat.* **2019**, *40*, 2068–2087. [CrossRef] [PubMed]
39. Ferent, J.; Constable, S.; Gigante, E.D.; Yam, P.T.; Mariani, L.E.; Legué, E.; Liem, K.F., Jr.; Caspary, T.; Charron, F. The Ciliary Protein Arl13b Functions Outside of the Primary Cilium in Shh-Mediated Axon Guidance. *Cell Rep.* **2019**, *29*, 3356–3366.e3. [CrossRef] [PubMed]
40. Whiting, K.R.; Haer-Wigman, L.; Florijn, R.J.; van Beek, R.; Oud, M.M.; Plomp, A.S.; Boon, C.J.F.; Kroes, H.Y.; Roepman, R. Utilization of automated cilia analysis to characterize novel INPP5E variants in patients with non-syndromic retinitis pigmentosa. *Eur. J. Hum. Genet.* **2024**. [CrossRef] [PubMed]
41. Kopinke, D.; Norris, A.M.; Mukhopadhyay, S. Developmental and regenerative paradigms of cilia regulated hedgehog signaling. *Semin. Cell Dev. Biol.* **2021**, *110*, 89–103. [CrossRef] [PubMed]
42. Zhang, Q.; Ding, Y.; Feng, B.; Tang, Y.; Chen, Y.; Wang, Y.; Chang, G.; Liu, S.; Wang, J.; Li, Q.; et al. Molecular and Phenotypic Expansion of Alström Syndrome in Chinese Patients. *Front. Genet.* **2022**, *13*, 808919. [CrossRef]

43. Mujahid, S.; Hunt, K.F.; Cheah, Y.S.; Forsythe, E.; Hazlehurst, J.M.; Sparks, K.; Mohammed, S.; Tomlinson, J.W.; Amiel, S.A.; Carroll, P.V.; et al. The Endocrine and Metabolic Characteristics of a Large Bardet-Biedl Syndrome Clinic Population. *J. Clin. Endocrinol. Metab.* **2018**, *103*, 1834–1841. [CrossRef] [PubMed]
44. Yang, H.W. The characterization and comorbidities of heterozygous Bardet-Biedl syndrome carriers. *Int. J. Med. Sci.* **2024**, *21*, 784–794. [CrossRef]
45. Meyer, J.R.; Krentz, A.D.; Berg, R.L.; Richardson, J.G.; Pomeroy, J.; Hebbring, S.J.; Haws, R.M. Kidney failure in Bardet–Biedl syndrome. *Clin. Genet.* **2022**, *101*, 429–441. [CrossRef]
46. Forsythe, E.; Kenny, J.; Bacchelli, C.; Beales, P.L. Managing Bardet-Biedl Syndrome-Now and in the Future. *Front. Pediatr.* **2018**, *6*, 23. [CrossRef] [PubMed]
47. Faccioli, N.; Poitou, C.; Clément, K.; Dubern, B. Current Treatments for Patients with Genetic Obesity. *J. Clin. Res. Pediatr. Endocrinol.* **2023**, *15*, 108–119. [CrossRef] [PubMed]
48. Haws, R.M.; Gordon, G.; Han, J.C.; Yanovski, J.A.; Yuan, G.; Stewart, M.W. The efficacy and safety of setmelanotide in individuals with Bardet-Biedl syndrome or Alström syndrome: Phase 3 trial design. *Contemp. Clin. Trials Commun.* **2021**, *22*, 100780. [CrossRef] [PubMed]
49. Ali, S.; Baig, S.; Wanninayake, S.; da Silva Xavier, G.; Dawson, C.; Paisey, R.; Geberhiwot, T. Glucagon-like peptide-1 analogues in monogenic syndromic obesity: Real-world data from a large cohort of Alström syndrome patients. *Diabetes Obes. Metab.* **2024**, *26*, 989–996. [CrossRef] [PubMed]
50. Markham, A. Setmelanotide: First Approval. *Drugs* **2021**, *81*, 397–403. [CrossRef] [PubMed]
51. Trapp, C.M.; Censani, M. Setmelanotide: A promising advancement for pediatric patients with rare forms of genetic obesity. *Curr. Opin. Endocrinol. Diabetes Obes.* **2023**, *30*, 136–140. [CrossRef]
52. Brüning, J.C.; Fenselau, H. Integrative neurocircuits that control metabolism and food intake. *Science* **2023**, *381*, eabl7398. [CrossRef] [PubMed]
53. Liu, P.; Lechtreck, K.F. The Bardet-Biedl syndrome protein complex is an adapter expanding the cargo range of intraflagellar transport trains for ciliary export. *Proc. Natl. Acad. Sci. USA* **2018**, *115*, E934–E943. [CrossRef] [PubMed]
54. Loktev, A.V.; Jackson, P.K. Neuropeptide Y family receptors traffic via the Bardet-Biedl syndrome pathway to signal in neuronal primary cilia. *Cell Rep.* **2013**, *5*, 1316–1329. [CrossRef]
55. Stephenson, E.J.; Kinney, C.E.; Stayton, A.S.; Han, J.C. Energy expenditure deficits drive obesity in a mouse model of Alström syndrome. *Obesity* **2023**, *31*, 2786–2798. [CrossRef]
56. Hearn, T. ALMS1 and Alström syndrome: A recessive form of metabolic, neurosensory and cardiac deficits. *J. Mol. Med.* **2019**, *97*, 1–17. [CrossRef] [PubMed]
57. Xuan, J.; Yu, Y.; Qing, T.; Guo, L.; Shi, L. Next-generation sequencing in the clinic: Promises and challenges. *Cancer Lett.* **2013**, *340*, 284–295. [CrossRef] [PubMed]
58. Marshall, J.D.; Muller, J.; Collin, G.B.; Milan, G.; Kingsmore, S.F.; Dinwiddie, D.; Farrow, E.G.; Miller, N.A.; Favaretto, F.; Maffei, P.; et al. Alström Syndrome: Mutation spectrum of ALMS1. *Hum. Mutat.* **2015**, *36*, 660–668. [CrossRef] [PubMed]
59. Chen, J.H.; Geberhiwot, T.; Barrett, T.G.; Paisey, R.; Semple, R.K. Refining genotype-phenotype correlation in Alström syndrome through study of primary human fibroblasts. *Mol. Genet. Genomic. Med.* **2017**, *5*, 390–404. [CrossRef] [PubMed]
60. Marshall, J.D.; Hinman, E.G.; Collin, G.B.; Beck, S.; Cerqueira, R.; Maffei, P.; Milan, G.; Zhang, W.; Wilson, D.I.; Hearn, T.; et al. Spectrum of ALMS1 variants and evaluation of genotype-phenotype correlations in Alström syndrome. *Hum. Mutat.* **2007**, *28*, 1114–1123. [CrossRef] [PubMed]
61. Dedeoglu, S.; Dede, E.; Oztunc, F.; Gedikbasi, A.; Yesil, G.; Dedeoglu, R. Mutation identification and prediction for severe cardiomyopathy in Alström syndrome, and review of the literature for cardiomyopathy. *Orphanet. J. Rare Dis.* **2022**, *17*, 359. [CrossRef] [PubMed]
62. Bettini, S.; Bombonato, G.; Dassie, F.; Favaretto, F.; Piffer, L.; Bizzotto, P.; Busetto, L.; Chemello, L.; Senzolo, M.; Merkel, C.; et al. Liver Fibrosis and Steatosis in Alström Syndrome: A Genetic Model for Metabolic Syndrome. *Diagnostics* **2021**, *11*, 797. [CrossRef] [PubMed]
63. Xiao, G.; Zhu, S.; Xiao, X.; Yan, L.; Yang, J.; Wu, G. Comparison of laboratory tests, ultrasound, or magnetic resonance elastography to detect fibrosis in patients with nonalcoholic fatty liver disease: A meta-analysis. *Hepatology* **2017**, *66*, 1486–1501. [CrossRef]
64. Cleveland, E.; Bandy, A.; VanWagner, L.B. Diagnostic challenges of nonalcoholic fatty liver disease/nonalcoholic steatohepatitis. *Clin. Liver. Dis.* **2018**, *11*, 98–104. [CrossRef]
65. Roy, A.; Patel, L.; Yuan, M.; O'Shea, C.; Alvior, A.M.B.; Charalambides, M.; Moxon, D.; Baig, S.; Bunting, K.V.; Gehmlich, K.; et al. Defining the cardiovascular phenotype of adults with Alström syndrome. *Int. J. Cardiol.* **2024**, *409*, 132212. [CrossRef]
66. Edwards, N.C.; Moody, W.E.; Yuan, M.; Warfield, A.T.; Cramb, R.; Paisey, R.B.; Geberhiwot, T.; Steeds, R.P. Diffuse Left Ventricular Interstitial Fibrosis Is Associated With Sub-Clinical Myocardial Dysfunction in Alström Syndrome: An Observational Study. *Orphanet J. Rare Dis.* **2015**, *10*, 83. [CrossRef] [PubMed]
67. Zmyslowska, A.; Smyczynska, U.; Stanczak, M.; Jeziorny, K.; Szadkowska, A.; Fendler, W.; Borowiec, M. Association of circulating miRNAS in patients with Alstrőm and Bardet-Biedl syndromes with clinical course parameters. *Front. Endocrinol.* **2022**, *13*, 1057056. [CrossRef] [PubMed]

68. Jeziorny, K.; Pietrowska, K.; Sieminska, J.; Zmysłowska-Polakowska, E.; Kretowski, A.; Ciborowski, M.; Zmyslowska, A. Serum metabolomics identified specific lipid compounds which may serve as markers of disease progression in patients with Alström and Bardet-Biedl syndromes. *Front. Mol. Biosci.* **2023**, *10*, 1251905. [CrossRef] [PubMed]
69. Shi, M.; Han, S.; Klier, K.; Fobo, G.; Montrone, C.; Yu, S.; Harada, M.; Henning, A.-K.; Friedrich, N.; Bahls, M.; et al. Identification of candidate metabolite biomarkers for metabolic syndrome and its five components in population-based human cohorts. *Cardiovasc. Diabetol.* **2023**, *22*, 141. [CrossRef] [PubMed]
70. Mihalik, S.J.; Goodpaster, B.H.; Kelley, D.E.; Chace, D.H.; Vockley, J.; Toledo, F.G.S.; DeLany, J.P. Increased levels of plasma acylcarnitines in obesity and type 2 diabetes and identification of a marker of glucolipotoxicity. *Obesity* **2010**, *18*, 1695–1700. [CrossRef] [PubMed]
71. Yu, Z.R.; Ning, Y.; Yu, H.; Tang, N.J. A HPLC-Q-TOF-MS-based urinary metabolomic approach to identification of potential biomarkers of metabolic syndrome. *J. Huazhong Univ. Sci Technolog. Med. Sci.* **2014**, *34*, 276–283. [CrossRef] [PubMed]
72. Jeziorny, K.; Zmyslowska-Polakowska, E.; Wyka, K.; Pyziak-Skupień, A.; Borowiec, M.; Szadkowska, A.; Zmysłowska, A. Identification of bone metabolism disorders in patients with Alström and Bardet-Biedl syndromes based on markers of bone turnover and mandibular atrophy. *Bone Rep.* **2022**, *17*, 101600. [CrossRef] [PubMed]
73. Butler, M.G.; Wang, K.; Marshall, J.D.; Rethmeyer, J.; Gunewardena, S.; Manzardo, A. Coding and noncoding expression patterns associated with rare obesity-related disorders: Prader—Willi and Alström syndromes. *AGG* **2015**, *5*, 53–75. [CrossRef] [PubMed]
74. Borrelli, P.; Zacchia, M.; Cavaliere, C.; Basso, L.; Salvatore, M.; Capasso, G.; Aiello, M. Diffusion tensor imaging for the study of early renal dysfunction in patients affected by bardet-biedl syndrome. *Sci. Rep.* **2021**, *11*, 20855. [CrossRef]
75. Caterino, M.; Zacchia, M.; Costanzo, M.; Bruno, G.; Arcaniolo, D.; Trepiccione, F.; Siciliano, R.A.; Mazzeo, M.F.; Ruoppolo, M.; Capasso, G. Urine Proteomics Revealed a Significant Correlation Between Urine-Fibronectin Abundance and Estimated-GFR Decline in Patients with Bardet-Biedl Syndrome. *Kidney Blood Press Res.* **2018**, *43*, 389–405. [CrossRef]
76. Marchese, E.; Caterino, M.; Fedele, R.; Pirozzi, F.; Cevenini, A.; Gupta, N.; Ingrosso, D.; Perna, A.; Capasso, G.; Ruoppolo, M.; et al. Multi-Omics Studies Unveil Extraciliary Functions of BBS10 and Show Metabolic Aberrations Underlying Renal Disease in Bardet-Biedl Syndrome. *Int. J. Mol. Sci.* **2022**, *23*, 9420. [CrossRef] [PubMed]
77. Zacchia, M.; Blanco, F.D.V.; Torella, A.; Raucci, R.; Blasio, G.; Onore, M.E.; Marchese, E.; Trepiccione, F.; Vitagliano, C.; Di Iorio, V.; et al. Urine concentrating defect as presenting sign of progressive renal failure in Bardet-Biedl syndrome patients. *Clin. Kidney J.* **2021**, *14*, 1545–1551. [CrossRef]
78. Fernandez-Marmiesse, A.; Gouveia, S.; Couce, M.L. NGS Technologies as a Turning Point in Rare Disease Research, Diagnosis and Treatment. *Curr. Med. Chem.* **2018**, *25*, 404–432. [CrossRef] [PubMed]
79. Hu, X.; Li, N.; Xu, Y.; Li, G.; Yu, T.; Yao, R.-E.; Fu, L.; Wang, J.; Yin, L.; Yin, Y.; et al. Proband-only medical exome sequencing as a cost-effective first-tier genetic diagnostic test for patients without prior molecular tests and clinical diagnosis in a developing country: The China experience. *Genet. Med.* **2018**, *20*, 1045–1053. [CrossRef] [PubMed]
80. Castro-Sánchez, S.; Álvarez-Satta, M.; Tohamy, M.A.; Beltran, S.; Derdak, S.; Valverde, D. Whole exome sequencing as a diagnostic tool for patients with ciliopathy-like phenotypes. *PLoS ONE* **2017**, *12*, e0183081. [CrossRef] [PubMed]
81. Shaheen, R.; Szymanska, K.; Basu, B.; Patel, N.; Ewida, N.; Faqeih, E.; Al Hashem, A.; Derar, N.; Alsharif, H.; Aldahmesh, M.A. Characterizing the morbid genome of ciliopathies. *Genome. Biol.* **2016**, *17*, 242. [CrossRef] [PubMed]
82. Frésard, L.; Smail, C.; Ferraro, N.M.; Teran, N.A.; Li, X.; Smith, K.S.; Bonner, D.; Kernohan, K.D.; Marwaha, S.; Zappala, Z.; et al. Identification of rare-disease genes using blood transcriptome sequencing and large control cohorts. *Nat. Med.* **2019**, *25*, 911–919. [CrossRef] [PubMed]
83. Modarage, K.; Malik, S.A.; Goggolidou, P. Molecular Diagnostics of Ciliopathies and Insights Into Novel Developments in Diagnosing Rare Diseases. *Br. J. Biomed. Sci.* **2022**, *79*, 10221. [CrossRef] [PubMed]
84. Su, J.; Gao, C.; Wang, R.; Xiao, C.; Yang, M. Genes associated with inflammation and the cell cycle may serve as biomarkers for the diagnosis and prognosis of acute myocardial infarction in a Chinese population. *Mol. Med. Rep.* **2018**, *18*, 1311–1322. [CrossRef]
85. Zhang, S.Y.; Xuan, C.; Wang, Y.; Zhang, S.-Q.; Li, H.; He, G.-W.; Tian, Q.-W. Association between ALMS 1 variants and early-onset coronary artery disease: A case-control study in Chinese population. *Biosci. Rep.* **2020**, *40*, BSR20193637. [CrossRef]
86. Xuan, C.; Li, H.; Tian, Q.W.; Guo, J.J.; He, G.W.; Lun, L.M.; Wang, Q. Quantitative Assessment of Serum Amino Acids and Association with Early-Onset Coronary Artery Disease. *Clin. Interv. Aging* **2021**, *16*, 465–474. [CrossRef] [PubMed]
87. Schreyer, E.; Obringer, C.; Messaddeq, N.; Kieffer, B.; Zimmet, P.; Fleming, A.; Geberhiwot, T.; Marion, V. PATAS, a First-in-Class Therapeutic Peptide Biologic, Improves Whole-Body Insulin Resistance and Associated Comorbidities In Vivo. *Diabetes* **2022**, *71*, 2034–2047. [CrossRef] [PubMed]

International Journal of
Molecular Sciences

Review

Biomarkers Involved in the Pathogenesis of Hemophilic Arthropathy

Oana Viola Badulescu [1], Dragos-Viorel Scripcariu [2,*], Minerva Codruta Badescu [3], Manuela Ciocoiu [1,†], Maria Cristina Vladeanu [1,†], Carmen Elena Plesoianu [3], Andrei Bojan [2,*], Dan Iliescu-Halitchi [3], Razvan Tudor [2], Bogdan Huzum [2], Otilia Elena Frasinariu [4] and Iris Bararu-Bojan [1]

1 Department of Pathophysiology, University of Medicine and Pharmacy Grigore T. Popa, 700115 Iasi, Romania; oana.badulescu@umfiasi.ro (O.V.B.); manuela.ciocoiu@umfiasi.ro (M.C.); maria.apavaloaie@umfiasi.ro (M.C.V.); iris.bararu@umfiasi.ro (I.B.-B.)
2 Department of Surgical Sciences, University of Medicine and Pharmacy Grigore T. Popa, 700115 Iasi, Romania; rc_tudor@yahoo.com (R.T.); bogdan.huzum@umfiasi.ro (B.H.)
3 Department of Internal Medicine, University of Medicine and Pharmacy Grigore T. Popa, 700115 Iasi, Romania; minerva.badescu@umfiasi.ro (M.C.B.); carmen-elena.plesoianu@umfiasi.ro (C.E.P.); halitchi.iliescud@umfiasi.ro (D.I.-H.)
4 Department of Pediatry, University of Medicine and Pharmacy Grigore T. Popa, 700115 Iasi, Romania; frasinariu.otilia@umfiasi.ro
* Correspondence: dragos-viorel.scripcariu@umfiasi.ro (D.-V.S.); andrei.bojan@umfiasi.ro (A.B.)
† These authors contributed equally to this work.

Abstract: Hemophilia, which is a rare disease, results from congenital deficiencies of coagulation factors VIII and IX, respectively, leading to spontaneous bleeding into joints, resulting in hemophilic arthropathy (HA). HA involves complex processes, including synovial proliferation, angiogenesis, and tissue remodeling. Despite ongoing research, factors contributing to HA progression, especially in adults with severe HA experiencing joint pain, remain unclear. Blood markers, particularly collagen-related ones, have been explored to assess joint health in hemophilia. For example, markers like CTX-I and CTX-II reflect bone and cartilage turnover, respectively. Studies indicate elevated levels of certain markers post-bleeding episodes, suggesting joint health changes. However, longitudinal studies on collagen turnover and basement membrane or endothelial cell markers in relation to joint outcomes, particularly during painful episodes, are scarce. Given the role of the CX3CL1/CX3XR1 axis in arthritis, other studies investigate its involvement in HA. The importance of different inflammatory and bone damage biomarkers should be assessed, alongside articular cartilage and synovial membrane morphology, aiming to enhance understanding of hemophilic arthropathy progression.

Keywords: hemophilic arthropathy; biomarkers in hemophilic arthropathy; joint lesions in hemophilia

Citation: Badulescu, O.V.; Scripcariu, D.-V.; Badescu, M.C.; Ciocoiu, M.; Vladeanu, M.C.; Plesoianu, C.E.; Bojan, A.; Iliescu-Halitchi, D.; Tudor, R.; Huzum, B.; et al. Biomarkers Involved in the Pathogenesis of Hemophilic Arthropathy. *Int. J. Mol. Sci.* **2024**, *25*, 9897. https://doi.org/10.3390/ijms25189897

Academic Editors: Alfonso Baldi and Andrea Bernini

Received: 23 July 2024
Revised: 28 August 2024
Accepted: 9 September 2024
Published: 13 September 2024

1. Introduction

Hemophilia is an X-linked, inherited coagulation disorder characterized by deficiencies in specific plasma glycoproteins that are crucial for blood clotting. The disorder manifests in two primary forms: hemophilia A, caused by reduced levels of coagulation factor VIII (FVIII), and hemophilia B (also known as Christmas disease), caused by reduced levels of coagulation factor IX (FIX) [1]. These deficiencies impede the blood clotting process, leading to a propensity for prolonged bleeding.

One of the most common and severe complications associated with hemophilia is hemarthrosis, which refers to bleeding into the joints. This condition is particularly prevalent in individuals with hemophilia [2]. It is estimated that hemarthrosis accounts for 70–80% of all bleeding episodes in patients with hemophilia [3]. Repeated joint bleeds can lead to chronic pain, inflammation, and long-term joint damage, significantly impacting the quality of life for those affected. Managing hemarthrosis often requires a combination of prophylactic treatment with clotting factor concentrates, physical therapy, and sometimes surgical intervention to mitigate joint damage and maintain mobility [4].

The syndrome of joint manifestations that occurs in the course of hemophilia is known as hemophilic arthropathy (HA). Research indicates that around 90% of patients with severe hemophilia A (defined as having FVIII levels below 0.01 IU/mL in serum) experience joint bleeding by the age of 4.4 years. This early onset of joint bleeding significantly contributes to the development of HA, which is characterized by chronic joint pain, swelling, and reduced mobility.

Despite the availability of effective prophylactic factor replacement therapy (FRT) in many parts of the world, including highly developed countries [3], advanced and fully developed HA remains a significant global health issue. Prophylactic FRT involves the regular infusion of clotting factor concentrates to prevent bleeding episodes and protect joint health. However, the progression of HA can still occur, especially in severe cases of hemophilia, due to the development of inhibitors. These inhibitors are antibodies that neutralize the infused clotting factors (anti-FVIII/FIX), rendering the treatment less effective or even ineffective.

The management of HA requires a multifaceted approach. In addition to FRT, patients may benefit from physical therapy to maintain joint function and mobility. In cases where inhibitors are present, immune tolerance induction (ITI) therapy may be used to reduce inhibitor levels. Advanced cases of HA might also necessitate orthopedic interventions, such as synovectomy or joint replacement surgery, to alleviate pain and improve joint function [5].

Globally, efforts are ongoing to improve access to comprehensive hemophilia care, including prophylactic FRT and multidisciplinary management strategies, to reduce the incidence and severity of HA and enhance the quality of life for patients with hemophilia.

Our current understanding of the factors contributing to the progression of HA remains incomplete. It is particularly noteworthy that adults with hemophilia, especially those with advanced HA, frequently experience episodes of joint pain. These episodes can occur with or without accompanying joint bleeding, adding complexity to the condition. The precise impact of these pain episodes on the molecular mechanisms driving the progression of HA is not yet fully understood.

Research is ongoing to elucidate the underlying mechanisms that cause joint damage in hemophilia. The interaction between recurrent bleeding and subsequent inflammation is believed to play a crucial role in the deterioration of joint health. Inflammation resulting from bleeding episodes leads to synovial hypertrophy, cartilage degradation, and bone erosion, all of which contribute to the development and progression of HA.

Moreover, the role of subclinical bleeds—minor bleeding episodes that do not cause noticeable symptoms—remains a significant area of investigation. These bleeds can contribute to joint damage over time, even in the absence of acute pain or swelling. Understanding how these subclinical events influence the molecular pathways involved in HA is critical for developing more effective treatment strategies [6].

There is also interest in the potential genetic and environmental factors that may influence the severity and progression of HA. Differences in genetic makeup may affect an individual's susceptibility to joint damage and response to treatment. Environmental factors, such as physical activity levels and access to comprehensive hemophilia care, also play a role in the disease's trajectory.

To address these gaps in knowledge, researchers are employing advanced imaging techniques, molecular biology, and clinical studies to gain a deeper understanding of HA. The goal is to identify biomarkers that can predict disease progression and to develop targeted therapies that can prevent or mitigate joint damage more effectively. As our understanding of HA improves, it holds the promise of better management and improved outcomes for individuals living with hemophilia [7].

Regular prophylactic intravenous infusion of factor VIII is the standard treatment for individuals with severe hemophilia A. However, due to the relatively short half-life of factor VIII, more than two infusions per week are required to maintain protective trough levels. This regimen places a significant treatment burden on patients and can result in inadequate care for those who struggle to adhere to it. Even with regular prophylaxis, both clinical and subclinical bleeding events may still occur. Therefore, treatments with greater efficacy and reduced burden are needed.

Emicizumab (Hemlibra, F. Hoffmann–La Roche) is a recombinant, humanized, bispecific monoclonal antibody that bridges activated factor IX and factor X to compensate for the missing activated factor VIII, thereby restoring hemostasis. Its efficacy has been demonstrated in individuals with hemophilia A who have developed neutralizing anti-factor VIII alloantibodies (inhibitors), and it is administered once a week. Therefore, administering emicizumab once weekly or every two weeks resulted in significantly lower bleeding rates compared to no prophylaxis. In an intraindividual comparison, once-weekly emicizumab prophylaxis was associated with a significantly reduced bleeding rate compared to previous factor VIII prophylaxis [8].

2. Literature Search

We conducted systematic research in order to identify the key biomarkers associated with hemophilia arthropathy and to elucidate the pathogenesis of this condition.

To achieve this, we performed a systematic search on PubMed to gather relevant studies concerning biomarkers in hemophilic arthropathy. We used specific keywords such as "hemophilic arthropathy", "hemophilia", "biomarkers in hemophilia", and "biomarkers in hemophilic arthropathy".

Our search strategy was designed to include a broad range of studies to ensure comprehensive coverage of the topic. We set the following inclusion criteria: studies must have been published in English to ensure accurate interpretation and analysis; studies must have involved human subjects to ensure clinical relevance. Single-case reports were excluded to focus on studies with broader applicability and robust data.

Through this methodical approach, we aimed to compile a comprehensive and accurate overview of the current understanding of hemophilic arthropathy and its biomarkers. Our review not only highlights the most critical biomarkers but also discusses their potential roles in the pathogenesis and progression of the disease. Additionally, we explore how these biomarkers can be used in clinical practice to improve the diagnosis, monitoring, and treatment of hemophilic arthropathy.

By synthesizing findings from multiple studies, our review provides valuable insights into the molecular mechanisms underlying hemophilic arthropathy and offers a foundation for future research aimed at developing targeted therapies and improving patient outcomes.

3. Pathophysiology

The pathophysiology of blood-induced joint disease, such as HA, remains not fully understood. Some evidence suggests that the pathobiology of HA may be similar to that of rheumatoid arthritis (RA), as both conditions feature chronic proliferative synovitis and cartilage destruction. Additionally, the presence of metalloprotease enzymes, inflammatory cells, and inflammatory cytokines (such as interleukin-1 (IL-1) and tumor necrosis factor alpha (TNF-α)) has been observed in the synovium of both HA and RA.

However, a key difference between these diseases lies in the origin of inflammation: RA is driven by autoimmunity, while HA is triggered by blood-induced injury. In HA, synovial iron deposition resulting from blood in the joint is considered a primary factor in initiating and sustaining the inflammatory response and cell proliferation within the synovial membrane. Synovial macrophages responding to blood injury promote the inflammatory pathway of hemarthrosis and HA.

Despite these similarities, these factors alone do not fully explain the pathophysiology of HA. For instance, in hemochromatosis—a different arthropathy characterized by synovial iron deposition—significant synovial hyperplasia is not observed, although inflammatory cells and joint destruction can occur.

The distinguishing factor in HA is the presence of blood in the joint, directly contacting the synovial membrane and particularly affecting fibroblast-like synoviocytes (FLS), the unique resident cells in joints. This direct interaction between blood and the synovial membrane is pivotal in HA's pathology, setting it apart from other forms of arthritis.

The presence of extravasated blood directly within the joint triggers an inflammatory response, initiating a sequence of immunopathological processes [6]. Within this blood are numerous active morphotic elements, including granulocytes, monocytes, and lymphocytes, which actively secrete inflammatory mediators [8].

Over time, macrophages and cells of the synovial membrane gradually absorb the blood and its associated metabolites. However, with recurrent episodes of hemarthrosis, the efficiency of immune cells and the synovial membrane in removing these metabolites diminishes. Consequently, toxic blood components such as iron (Fe^{2+}) and hemosiderin accumulate within the synovial membrane cells.

This accumulation sets off a chain of pathological events, including degeneration of the articular cartilage and subchondral bone, proliferation of fibroblasts, and the formation of new blood vessels (angiogenesis). These processes collectively lead to hypertrophy of the synovial membrane [8–10].

The combined data on the pathophysiology of hemarthrosis (HA) and related arthritis suggest a clear link between the degeneration of joint tissues in the musculoskeletal system and a persistent, chronic inflammatory state [6]. This interconnectedness underscores the complex nature of joint diseases where ongoing inflammation plays a central role in tissue damage and dysfunction [11] (Table 1).

Table 1. This table summarizes the distinct aspects of HA.

Aspect	Hemophilic Arthropathy (HA)
Basic Pathophysiology	Chronic proliferative synovitis and cartilage destruction
Synovial Characteristics	Presence of metalloprotease enzymes, inflammatory cells, and cytokines (IL-1, TNF-α)
Origin of Inflammation	Triggered by blood-induced injury
Role of Synovial Iron Deposition	Primary factor initiating and sustaining inflammation and cell proliferation
Impact on Synovial Macrophages	Promotes inflammatory pathways of hemarthrosis and HA
Unique Pathological Features	Direct interaction of blood with synovial membrane and fibroblast-like synoviocytes (FLS)
Challenges in Understanding Pathophysiology	Incomplete understanding; complexities beyond iron deposition
Consequence of Blood Presence in Joint	Triggers inflammatory response; secretion of inflammatory mediators
Long-term Effects on Joint Tissues	Degeneration of articular cartilage, subchondral bone; synovial membrane hypertrophy
Persistent Inflammatory State	Interconnected with joint tissue degeneration

4. Chemokines and Cytokines

While certain cytokines have been implicated in the development of synovitis in hemophilic arthropathy, drawing parallels with rheumatoid arthritis (RA), studies demonstrating their precise functional role in the disease's pathogenic cascade remain limited [12]. Initially described as a degenerative rather than inflammatory joint disorder, recent research suggests that hemophilic arthropathy shares similarities with both the degenerative joint damage seen in osteoarthritis (OA) and the chronic inflammatory processes of RA [13].

The extensive array of mediators implicated in HA pathogenesis includes cytokines, chemokines, and their associated receptors and signaling pathways. In a study conducted by Wojdasiewicz et al. on 40 patients, it was proven that the CX3CL1/CX3CR1 signaling axis plays a role in the development of degenerative joint lesions in patients with HA. They found a statistically significant increase in the average concentration of CX3CL1 in the serum of patients with HA compared to patients with osteoarthritis (OA). However, no statistically significant difference in CX3CL1 concentration was observed in other biofluids, such as synovial fluid, between patients with HA and patients with OA, despite a higher mean level in patients with HA. Immunohistochemical and histological analyses revealed CX3CR1 overexpression in the synovial membrane of patients with HA and OA, with no significant differences noted in articular cartilage slices between groups. However, there was an observable tendency towards increased extracellular matrix staining, although without significant intergroup differences. These findings suggested that inflammatory and degenerative processes mediated by CX3CR1 activation exhibit similar intensity in the synovial membrane of affected joints in both patient groups. Nonetheless, patients with HA demonstrated more pronounced activity of synovial membrane cells in CX3CL1 production and a systemic tendency towards higher serum concentrations compared to patients with OA. The similar level of CX3CR1 expression in synovial membrane slices between patients with HA and patients with OA may have reflected comparable clinical stages of joint disease development and the implementation of effective management strategies, including regular FRT administration and appropriate care programs aimed at minimizing hemarthrosis and joint-related exacerbations [14].

The "inflammasome", a critical regulator of pro-inflammatory interleukin (IL)-1β maturation and secretion, has garnered attention in joint pathology [15]. Iron also plays a pivotal role by inducing the expression of several pro-inflammatory cytokines, including IL-1α, IL-6, and tumor necrosis factor (TNF)-α. Additionally, iron appears to contribute to the initiation of synovial pannus growth by disrupting the expression of key genes such as c-myc and mdm2, which govern synoviocyte proliferation. Synovitis, characterized by synovial tissue inflammation, involves hypertrophy, inflammatory cell migration, and substantial neo-angiogenesis. IL-1α, IL-6, IL-1β, and TNF-α activate monocytes/macrophages, initiating a catabolic response that includes the production of nitric oxide (NO), matrix metalloproteinases (MMPs), tissue plasminogen activator, and other matrix components [16–18]. These factors, in turn, influence T cells, fibroblasts, and osteoclasts through various inflammatory mediators, ultimately leading to degradation of articular cartilage and subchondral bone.

Mignot et al. proved that the cytokine expression profile analyzed using ELISPOT between HA-FLS and non-HA-FLS, both activated or not by LPS, appears to align with the inflammatory theory. Specifically, cytokines associated with innate immunity showed notable differences in expression between non-HA and HA-FLS. Additionally, HA-FLS were observed to be involved in processes related to bone remodeling and iron recycling, underscoring the pivotal role of FLS in hemophilic arthropathy pathophysiology. Interestingly, they observed that HA-FLS and RA-FLS did not secrete IL-1α following LPS treatment, whereas IL-1α was detected in supernatants from non-HA-FLS, THP-1 cell lines, and HR-FLS. The reasons for this paradoxical down-regulation of IL-1α in LPS-induced HA-FLS and RA-FLS remain unknown, despite the established role of IL-1β in the pathogenesis of HA [11,19].

TNF-α plays a significant role in inflammation in hemophilic arthropathy (HA), as highlighted in recent research focusing on the iRhom2/ADAM17/TNF-α pathway. It is also recognized as a crucial mediator of proliferative synovitis in hemophilia A.

In contrast to TNF-α and IL-1α, Mignot et al. observed elevated levels of IL-6 expression after LPS treatment in HA-FLS, consistent with previous findings suggesting the therapeutic potential of targeting IL-6 in HA. Interestingly, the cytokine profile of

LPS-induced HA-FLS closely correlated with the potential post-transcriptional control of 30 selected miRNAs. Following LPS activation, HA-FLS up-regulated cytokines associated with macrophage polarization (also influenced by miR-1246) and modulated the TH17 pathway (miR-10b-5p), which is implicated in conditions like ankylosing spondylitis. Additionally, many of the other miRNAs may directly or indirectly affect inflammatory cytokine expression through pathways such as NFκB, PI3K, or JAK/STAT (notably miR-146 and miR-196b-5p, which exhibited the highest expression ratios) [20].

IL-1 triggers the activation of nuclear factor kappa-light-chain-enhancer of activated B cells (NF-κB) transcription factor, along with other factors like c-Jun N-terminal kinase (JNK) and p38 mitogen-activated protein kinases (p38MAPK). This cascade results in increased expression of various genes responsible for synthesizing enzymes, adhesion molecules, and inflammatory mediators such as cytokines and chemokines [4]. This mechanism parallels the role of NFκB in the development of synovitis and cartilage degeneration observed in osteoarthritis (OA) and rheumatoid arthritis (RA) [6]. Consistently underscoring the involvement of IL-1β in hemophilic arthropathy (HA) pathophysiology, several studies have documented significantly elevated IL-1β levels in histological sections of synovial membranes obtained during synovectomy or joint replacement from patients with HA compared to non-hemophilic individuals [3,4,11].

Furthermore, IL-1β can enhance transferrin-bound iron uptake by type B synoviocytes, leading to hemosiderin deposition and autocrine IL-1β secretion, thereby contributing to the development of chronic synovitis [21].

TNFα, a member of the tumor necrosis factor superfamily, plays a critical role in the pathophysiology of hemophilic arthropathy (HA). It induces catabolic processes in synovial joints and directly regulates intra-articular levels of FVIIIa, thereby modulating thrombomodulin (TM) expression [22]. Specifically, TNFα inhibits the synthesis of proteoglycans and collagen type II (COL2) by chondrocytes. It also promotes the expression of metalloproteases (MMP-1, MMP-3, MMP-13, and ADAMTS4), which play pivotal roles in joint catabolism [4,22]. Moreover, TNFα increases the risk of recurrent bleeding. It substantially reduces TM synthesis by synoviocytes, leading to elevated TM levels in synovial fluid due to neutrophil and cytokine activity on synovial cells. Recent studies indicate higher TM levels in patients with HA (56 ± 25 ng/mL) compared to healthy controls (39 ± 21 ng/mL). Normally, TM binds thrombin in a 1:1 ratio, activating protein C (PC), which inhibits coagulation by degrading factors FVa and FVIIIa [23,24]. This interplay between inflammatory mediators and hemostasis components helps explain persistent hemorrhagic processes despite FVIII replacement therapies.

Thus, IL-1β and TNFα, by triggering and exacerbating inflammatory damage and its consequences on joints, are pivotal in the pathophysiology of HA. Additionally, recent findings indicate elevated expression of the TNFα/TNF receptor (TNF-R) system in synovial tissue. Activation of this system may serve as a critical mediator of synovial proliferation, presenting a potential novel target for therapeutic intervention [25].

Moreover, a recent study demonstrated that similar to osteoarthritis (OA) and rheumatoid arthritis (RA), patients with hemophilic arthropathy (HA) exhibit increased levels of progranulin (PGRN), a molecule known for its protective role against the catabolic effects of TNFα [26]. This finding suggests potential future investigations into its role as a biomarker for monitoring disease activity (Table 2).

These cytokines and chemokines play diverse roles in the pathogenesis of HA, reflecting overlapping and unique mechanisms compared to RA and OA. Their dysregulation contributes to joint inflammation, cartilage degradation, and bone resorption, highlighting potential targets for therapeutic interventions specific to HA.

Table 2. Key cytokines and their roles in hemophilic arthropathy (HA).

Cytokine/Chemokine	Role in HA
TNFα	Induces catabolic processes in joints; regulates FVIIIa and TM expression (22)
IL-1β	Triggers NF-κB activation; enhances synovitis and cartilage degradation (4)
IL-6	Associated with macrophage polarization; implicated in inflammatory responses (11)
CX3CL1	Role in degenerative joint lesions; systemic elevation in patients with HA (4)
Progranulin (PGRN)	Protective against TNFα catabolic effects; potential biomarker (27)

5. Soluble Adhesion Molecules

Several studies have shown that circulating concentrations of soluble adhesion molecules (CAMs), sE-selectin, and sP-selectin are elevated in conditions such as coronary artery disease, myocardial infarction, atherosclerosis, or hypertension [27,28].

5.1. sVCAM-1 Levels

Among patients with hemophilia A, the sVCAM-1 levels did not differ significantly between those with and without hypertension. This finding does not necessarily indicate that sVCAM-1 levels are independent of hypertension, as the study only included three patients with both hemophilia A and hypertension. Additionally, the influence of hypertension on sVCAM-1 levels may be overshadowed by factors related to hemophilia A. Previous studies have demonstrated that sVCAM-1 levels are significantly elevated in patients with hemodialysis, indicating chronic inflammation. Chronic hepatitis C virus (HCV) infection, which is common among patients with hemodialysis, can also trigger chronic inflammation. Levels of sICAM-1, sVCAM-1, and sE-selectin are higher in patients with hemodialysis who are anti-HCV-positive [29–31]. HCV infection is also highly prevalent in patients with hemophilia A. However, in this study, the sVCAM-1 levels among these patients did not show significant differences when compared to healthy volunteers and asymptomatic chronic hepatitis C carriers with minimal inflammatory activity. Notably, sVCAM-1 levels only increased significantly during the advanced fibrosis stage of HCV infection. This indicates that while HCV infection is common in hemophilia A patients, its impact on sVCAM-1 levels is not pronounced until the infection progresses to a more severe stage characterized by extensive liver fibrosis. Thus, monitoring sVCAM-1 levels could potentially serve as an indicator of disease progression in HCV-infected individuals, particularly in those with underlying conditions such as hemophilia A [32].

5.2. sICAM-1 Levels

Research has shown that soluble intercellular adhesion molecule-1 (sICAM-1), sE-selectin, and sP-selectin levels are influenced by ABO blood group. To determine whether the expression of these cell surface adhesion molecules is affected by ABO blood group, an analysis was conducted on 27 patients with hemophilia A. The study found no significant differences in the levels of sVCAM-1, sE-Selectin, or sP-selectin among different ABO blood groups ($p > 0.05$). This result suggests that elevated sVCAM-1 levels in patients with hemophilia and hemophilic arthropathy are independent of blood group. However, caution is advised in interpreting these findings, as the patient sample may not be representative of the general population [33].

5.3. Other Soluble Biomarkers

Hemarthrosis results in cartilage and bone degradation, inflammation, and angiogenesis in patients with hemophilia A. Research has explored the relationship between various soluble biomarkers and these pathological symptoms. The biomarkers studied include C-terminal telopeptides of type I collagen, cartilage oligomeric matrix protein (COMP), tissue inhibitor of metalloproteinases-1 (TIMP-1), matrix metalloproteinases-3 and -9 (MMP-3 and MMP-9), vascular endothelial growth factor (VEGF), and chondroitin sulfate 846 epitope (CS846). Despite extensive investigations, no strong correlations between these biomarkers and MRI joint scores have been observed, except for a positive association between CS846 and MRI joint scores reported in the Oldenburg study [34].

Joint arthropathy in hemophilia A presents unique challenges because it is not a systemic disease, unlike rheumatoid arthritis (RA) and osteoarthritis (OA). This localized nature complicates the analysis of biomarkers, making it more intricate than in systemic joint diseases.

6. Bone Turnover Markers

Hemophilic arthropathy was originally characterized as a degenerative joint disease rather than an inflammatory one. An essential regulator of bone biology is the molecular triad consisting of osteoprotegerin (OPG), the receptor activator of nuclear factor κB (RANK), and RANK ligand (RANKL). This triad governs local bone turnover and is pivotal in triggering bone resorption induced by inflammation. RANKL, a transmembrane ligand, is primarily expressed on osteoblasts and stromal cells within the bone microenvironment. It is also produced by lymphocytes and synovial cells, contributing to osteoclastogenesis through a mechanism enhanced by several cytokines, such as TNF-α, IL-1, and IL-17, which collectively promote inflammation and bone resorption [6].

RANKL interacts with its receptor RANK, located on the surface of osteoclast precursors, leading to the differentiation and maturation of osteoclasts, which in turn fosters bone resorption. In this regulatory system, OPG acts as a decoy receptor for RANKL, competing with RANK for binding to RANKL. This competition effectively inhibits osteoclast differentiation, activity, and survival both in vivo and in vitro, thereby reducing bone resorption [35].

In hemophiliacs, there is a notably high prevalence of osteoporosis, which is closely linked to the severity of arthropathy and is further exacerbated by HIV infection. This condition highlights a significant imbalance, where increased bone resorption is not adequately compensated for by bone formation. Research suggests that the OPG/RANK/RANKL triad could be a key regulator of bone remodeling in the synovial tissue of adult hemophiliacs. Decreased levels of OPG and strong expressions of RANK and RANKL have been observed, indicating a shift towards enhanced bone resorption (Figure 1).

The severity of hemophilic arthropathy has been shown to correlate strongly with instrumental findings such as the World Federation of Hemophilia (WFH) orthopedic joint scale, Petterson scores, and ultrasound (US) evaluations. Molecular markers of bone turnover in the synovial tissue of hemophiliacs clearly indicate osteoclastic activation, which is not counterbalanced by OPG. Strong expressions of RANK and RANKL are found in the synovium, regardless of the type of treatment, whereas the expression of OPG is significantly reduced in patients with hemophilic arthropathy. This almost complete lack of OPG expression implies that the balance of bone turnover is skewed towards osteoclastic activity and consequently, bone resorption [36,37].

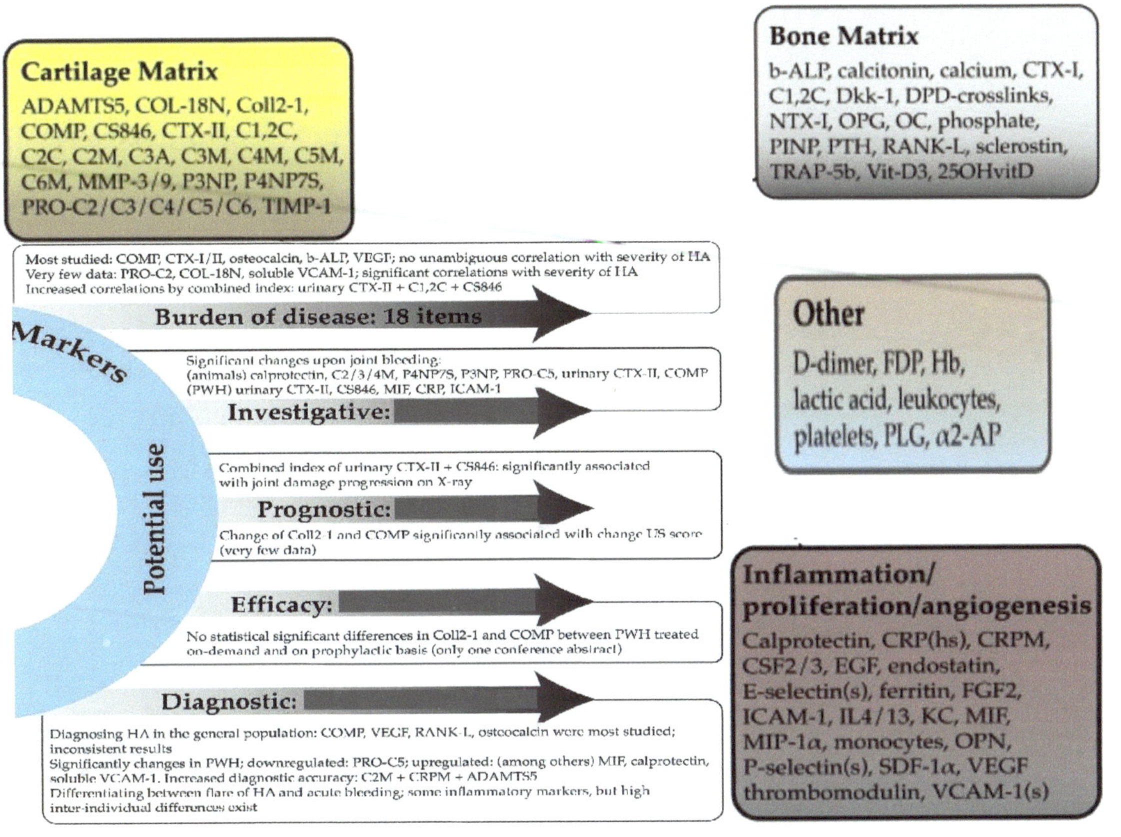

Figure 1. Biomarkers in HA [38].

Moreover, in vitro evidence suggests that reduced thrombin production leads to decreased thrombin-induced PAR-1 mediated proliferation of osteoblasts. These collective findings underscore the complexity of bone damage in hemophilic arthropathy, indicating that it is a multifactorial process. This complexity necessitates a comprehensive approach to understanding and treating bone damage in these patients, considering the intricate interplay of various molecular pathways involved in bone resorption and formation [39] (Table 3).

Table 3. Role of biomarkers in HA.

Markers	Role in HA	Details
Soluble Adhesion Molecules	**sVCAM-1 Levels**	- sVCAM-1 levels do not differ significantly between patients with hemophilia A with or without hypertension. - Previous studies show elevated sVCAM-1 in patients with hemodialysis, especially with HCV. - sVCAM-1 levels increase significantly only during advanced fibrosis in patients infected with HCV, indicating potential as a marker for disease progression.
	sICAM-1 Levels	- ABO blood group does not significantly affect levels of sICAM-1, sVCAM-1, sE-selectin, and sP-selectin in patients with hemophilia A. - Elevated sVCAM-1 levels are independent of blood group.

Table 3. *Cont.*

Markers	Role in HA	Details
	Other Soluble Biomarkers	- Various biomarkers studied (e.g., C-terminal telopeptides of type I collagen, COMP, TIMP-1, MMP-3, MMP-9, VEGF, and CS846). - No strong correlations with MRI joint scores, except a positive association between CS846 and MRI joint scores in the Oldenburg study. - Multifaceted biomarker approach recommended due to the localized nature of joint arthropathy in hemophilia A.
Bone Turnover Markers	**OPG/RANK/RANKL Triad**	- The OPG/RANK/RANKL system is crucial in bone remodeling and resorption. - RANKL promotes osteoclastogenesis, leading to bone resorption. - OPG acts as a decoy receptor, inhibiting osteoclast activity. - In patients with hemophilia A, reduced OPG levels and a strong RANK/RANKL expression shift the balance toward bone resorption, contributing to osteoporosis and arthropathy.
	Bone Resorption in Hemophilia	- Patients with hemophilia A have a high prevalence of osteoporosis, linked to arthropathy severity and exacerbated by HIV. - A strong correlation between hemophilic arthropathy severity and instrumental findings (WFH orthopedic joint scale, Petterson scores, and US evaluations). - Reduced thrombin production leads to decreased osteoblast proliferation, contributing to the complexity of bone damage in these patients.
Cytokine/Chemokine	**Role in HA**	
TNFα	Induces catabolic processes in joints; regulates FVIIIa and TM expression.	
IL-1β	Triggers NF-κB activation; enhances synovitis and cartilage degradation.	
IL-6	Associated with macrophage polarization; implicated in inflammatory responses.	
CX3CL1	Role in degenerative joint lesions; systemic elevation in patients with HA.	
Progranulin (PGRN)	Protective against TNFα catabolic effects; potential biomarker.	

7. Future Perspectives

The shift from episodic treatment to prophylaxis marked a significant advancement in hemophilia care. The next goal should be to ensure long-term protection, as preventing arthropathy relies on effectively preventing bleeding—something that current treatments have not yet been able to consistently achieve. A key unmet need in the management and monitoring of hemophilic arthropathy is the absence of serum and synovial biomarkers to gauge disease activity. Identifying and validating such biomarkers would enhance decision-making. Research on synovial tissue in patients with recent-onset RA has revealed significant correlations between histological findings, transcriptomic profiles, and clinical responses to treatments. Similarly, synovial biomarkers derived from transcriptomic analysis could help identify predictors of therapeutic responses and potentially uncover new targets for personalized management of hemophilic arthropathy. The increased use of joint ultrasound (US) as a routine clinical examination in comprehensive care centers is anticipated to streamline workflows, enable prompt diagnosis of acute hemarthrosis, and enhance early treatment. Additionally, implementing point-of-care US in clinical practice and at the bedside should facilitate regular monitoring of arthropathy progression, provided that US techniques are standardized [38,40,41].

7.1. The Importance of Physical Activity in HA Pathogenesis

Another key role is linked to the degree of physical activity and machine-controlled motions. The health benefits of physical activity for the general population are also applicable to individuals with hemophilia. Maintaining adequate muscle tone helps prevent injuries and reduces the risk of joint bleeding. Physical exercise is particularly important for people with hemophilia as it enhances their quality of life. Numerous studies highlight the significance of physical training for improving health, aligning with recommendations from both the World Health Organization and the World Federation of Hemophilia. Specifically, video games used for rehabilitation (exergaming) have positively influenced patients' attitudes towards exercise and have been shown to improve strength, coordination, and mobility. Additionally, motion capture (MoCap) sensors are increasingly utilized in medicine and physical therapy due to their availability and affordability compared to traditional 3D optical MoCap systems. The Kinect sensor, a popular MoCap option, has been successfully applied in various medical and rehabilitation contexts, including post-stroke limb rehabilitation, elderly exercise monitoring and fall prevention, range-of-motion evaluation in adhesive capsulitis, balance and postural control assessment and training, and virtual gyms for individuals with limited mobility [42].

7.2. The Importance of Animal Models in HA Pathogenesis

Factor VIII (FVIII) is among the most immunogenic biologics, and the development of inhibitors against FVIII poses a significant challenge to optimal care, leading to increased patient morbidity and mortality. Understanding the mechanisms that predispose individuals to or drive FVIII immune responses is crucial for clinical management. Early reports suggested that the α-CD20 monoclonal antibody rituximab could potentially eradicate FVIII inhibitors, but subsequent studies showing modest efficacy with rituximab monotherapy have tempered these expectations. This limited efficacy may be partly due to the persistence of rituximab-resistant $FVIII^+$ B cell subsets and the potential exacerbation of autoimmune disease due to increased BAFF (B cell activating factor) levels following B cell depletion. Doshi et al. indicate that a single dose of α-mBAFF monoclonal antibody in naive hemophilia A (HA) mice prevents the development of inhibitors, even after immune reconstitution. In mice with pre-existing inhibitors, combining α-mBAFF with α-mCD20 significantly reduces or eliminates FVIII inhibitors and maintains long-term suppression of FVIII-specific plasma cells (PCs). Recent evidence points to marginal zone (MZ) B cells as key mediators of the initial FVIII immune response, which can quickly differentiate into short-lived antibody-secreting cells (ASCs). High-affinity antibodies generally arise from germinal center (GC) reactions, where B cell receptor (BCR) rearrangement leads to differentiation into PCs or memory B cells. Within GCs, T follicular helper (Tfh) cells release BAFF to support the selection of high-affinity GC B cell clones. The substantial reduction in inhibitor levels with preemptive α-mBAFF monoclonal antibody treatment is likely due to MZ B cells' reliance on BAFF for survival and differentiation. We propose that these mice exhibit lower rates of GC B cell reactions and, consequently, fewer PCs and memory B cells, which promotes tolerance to FVIII. Importantly, these mice still mount a strong immune response to unrelated antigens, demonstrating the safety and specificity of this approach. These findings, along with data from enzyme replacement therapy studies, suggest that BAFF contributes to the immunogenicity of biotherapeutics. However, as some HA mice still developed high-titer anti-FVIII antibodies, further research is needed to determine if adjusting the dose or treatment duration could more effectively prevent inhibitor development [43].

Type II collagen degradation, indicated by C2M levels, may serve as a predictive marker for cartilage degradation and the development of arthropathy. In hemophilic rats, joint bleeding on day 0 did not initially affect serum C2M levels, but a significant increase was observed after a second bleed on day 14, which correlated with the severity of arthropathy on histology [44]. Other collagen markers, such as serum C4M and P4NP7S, significantly increased one day after the second joint bleed, while serum C3M levels significantly decreased. Serum PRO-C5 and P3NP increased significantly one week after both the first and second joint bleeds. In hemophilic mice, plasma C4M and PRO-C4 levels also significantly rose two weeks after induced hemarthrosis. In dogs with experimentally induced hemarthrosis, urinary CTX-II levels significantly increased from day two to seven (by 75% to 155%), and serum COMP levels rose by 46% from baseline to day two [45]. The idea that joint bleeds can trigger a systemic pro-inflammatory state and that inflammatory markers might be useful for detecting hemarthrosis was tested in hemophilic mice. It was found that serum calprotectin, a marker of residual inflammation, was significantly higher in mice with induced hemarthrosis at two and twelve weeks compared to control mice, where calprotectin was undetectable [46].

8. Plasminogen Activation and MMP-Mediated Joint Destruction as a Likely Mechanism for Joint Damage in HA

Although the existing literature supports the involvement of protease activation in the pathophysiology of HA, further investigation in experimental and clinical settings is needed. Nonetheless, several established characteristics of HA lend support to this hypothesis. Bleeding into the joint results in an influx of hemosiderin, which catalyzes the production of reactive oxygen species and triggers chondrocyte apoptosis. Type A synoviocytes absorb hemosiderin and release proinflammatory cytokines, such as IL-1α, IL-6, and TNF-α, creating a cytokine storm within the joint that exacerbates cartilage and bone damage. These cytokines also promote the recruitment and activation of monocytes and macrophages, which in turn stimulate the production of nitric oxide and proteases, including MMPs and uPA. An imbalance in hemostasis, characterized by low tissue factor (TF) and high thrombin-activatable fibrinolysis inhibitor (TAFI) levels, can lead to uncontrolled activity of uPA and tPA. This influx of blood exposes the synovium to high levels of plasminogen, setting the stage for uPA-mediated plasminogen activation and subsequent plasmin production. Fibrin deposition is frequently seen in the cavities of inflamed joints, and as a cofactor for tPA-mediated plasminogen activation, it facilitates substantial plasmin generation. Both uPA and plasmin can cleave pro-MMPs to produce active MMPs [44].

9. Diagnostic Markers Used in HA

Different studies compared biochemical markers between patients with HA and controls, focusing on serum COMP, VEGF, and bone markers like RANK-L and osteocalcin. Serum COMP findings were mixed: two studies reported higher levels in patients with HA, while one found lower levels. Cartilage and collagen markers also varied, with some (C2M, CTX-II, C4M) elevated and others (ADAMTS5, Coll2-1, C3A, C3M, C5M) decreased in hemophilia patients [47–49]. Most cartilage formation markers showed no difference between patients with HA and controls, except serum PRO-C5, which was lower in patients with hemophilia patients [50]. For bone metabolism, patients with HA had higher serum parathormone and lower 25-OH vitamin D in two studies. Serum osteocalcin and sclerostin levels showed conflicting results. Other markers, such as RANK-L, CTX-I, Dkk-1, and OPG, also had inconsistent findings [51,52]. Acharya et al. found a significant fourfold increase in plasma VEGF-A, SDF-1α, and MMP-9 levels in patients with hemophilia with joint disease. Other studies reported elevated levels of plasma MMP-9, SDF-1α, soluble VCAM-1, and several other markers in patients with hemophilia, though results for some markers were inconsistent. Three studies on acute joint bleeding in hemophilia found increased levels of serum D-dimer, ferritin, FDP, leukocytes, plasminogen, and VEGF, while plasma EGF, CSF2, IL4/13, FGF2, and MIP-1α were lower. Osteopontin was significantly higher in young patients with hemophilia with clinical synovitis [53,54].

10. Conclusions

In contrast to earlier views that depicted hemophilic arthropathy exclusively as a degenerative condition, recent and comprehensive research has uncovered a more intricate interplay of factors. These studies reveal that HA encompasses not only degenerative changes but also a significant inflammatory component, with synovitis emerging as a critical element in its pathophysiology.

This updated understanding of HA emphasizes that the condition is driven by a combination of joint degeneration and inflammatory processes rather than being solely degenerative. Synovitis, or inflammation of the synovial membrane, has been identified as a key player in the progression and severity of HA.

The recognition of these dual mechanisms opens up new avenues for therapeutic intervention. Targeting both the degenerative and inflammatory aspects of HA could lead to more effective treatment strategies.

Overall, this evolving perspective not only broadens the scope of potential therapeutic targets for HA but also underscores the importance of advanced diagnostic tools like imaging techniques in managing the condition more effectively. Currently, biomarkers hold scientific interest, but they offer no practical benefits to patients. As of now, they do not provide additional value. Our current understanding of biomarkers in hemophilia contributes to the scientific knowledge of pathophysiology, but it still has to prove its clinical relevance. Consequently, a multifaceted approach utilizing multiple biomarkers may provide a more effective and precise method for assessing the severity of joint arthropathy in patients with hemophilia A. By leveraging a panel of biomarkers, clinicians might better capture the complex and localized processes of joint damage and inflammation, leading to improved diagnostic and therapeutic strategies tailored to the specific needs of patients with hemophilia A.

Author Contributions: Conceptualization, A.B., I.B.-B., D.-V.S., O.E.F. and O.V.B.; methodology, A.B., M.C.V., C.E.P. and D.I.-H.; software, A.B.; validation, C.E.P., R.T., M.C., O.V.B. and M.C.B.; formal analysis, A.B., I.B.-B. and M.C.B.; investigation, A.B. and I.B.-B.; data curation, A.B., I.B.-B., C.E.P., M.C., O.V.B. and B.H.; writing—original draft preparation, A.B. and I.B.-B.; writing—review and editing, A.B., I.B.-B., D.-V.S., M.C., O.V.B. and O.E.F.; visualization, M.C.V., C.E.P. and M.C.B.; supervision, A.B., I.B.-B., M.C. and O.V.B. All authors have read and agreed to the published version of the manuscript.

Funding: This research received no external funding.

Conflicts of Interest: The authors declare no conflicts of interest.

References

1. Peyvandi, F.; Jayandharan, G.; Chandy, M.; Srivastava, A.; Nakaya, S.M.; Johnson, M.J.; Thompson, A.R.; Goodeve, A.; Garagiola, I.; Lavoretano, S.; et al. Genetic diagnosis of haemophilia and other inherited bleeding disorders. *Haemophilia* **2006**, *12*, 82–89. [CrossRef] [PubMed]
2. Valentino, L.A. Blood-induced joint disease: The pathophysiology of hemophilic arthropathy. *J. Thromb. Haemost.* **2010**, *8*, 1895–1902. [CrossRef] [PubMed]
3. Srivastava, A.; Brewer, A.K.; Mauser-Bunschoten, E.P.; Key, N.S.; Kitchen, S.; Llinas, A.; Ludlam, C.A.; Mahlangu, J.N.; Mulder, K.; Poon, M.C.; et al. Guidelines for the management of hemophilia. *Haemophilia* **2012**, *19*, e1–e47. [CrossRef] [PubMed]
4. Wojdasiewicz, P.; Poniatowski, Ł.A.; Kotela, A.; Skoda, M.; Pyzlak, M.; Stangret, A.; Kotela, I.; Szukiewicz, D. Comparative Analysis of the Occurrence and Role of CX3CL1 (Fractalkine) and Its Receptor CX3CR1 in Hemophilic Arthropathy and Osteoarthritis. *J. Immunol. Res.* **2020**, *2020*, 1–12. [CrossRef]
5. van Vulpen, L.F.D.; Roosendaal, G.; van Asbeck, B.S.; Mastbergen, S.C.; Lafeber, F.P.J.G.; Schutgens, R.E.G. The detrimental effects of iron on the joint: A comparison between haemochromatosis and haemophilia. *J. Clin. Pathol.* **2015**, *68*, 592–600. [CrossRef]
6. Melchiorre, D.; Manetti, M.; Matucci-Cerinic, M. Pathophysiology of Hemophilic Arthropathy. *J. Clin. Med.* **2017**, *6*, 63. [CrossRef]
7. Gopal, S.; Barnes, R.F.; Cooke, E.J.; Zhou, J.Y.; Levin, I.; Emery, P.; Hughes, T.H.; Karsdal, M.A.; Manon-Jensen, T.; von Drygalski, A. Systemic vascular basement membrane markers linked to synovial vascular remodeling are biomarkers of hemarthrosis in patients with hemophilia. *J. Thromb. Haemost.* **2021**, *19*, 1200–1211. [CrossRef]
8. Mahlangu, J.; Oldenburg, J.; Paz-Priel, I.; Negrier, C.; Niggli, M.; Mancuso, M.E.; Schmitt, C.; Jiménez-Yuste, V.; Kempton, C.; Dhalluin, C.; et al. Emicizumab Prophylaxis in Patients Who Have Hemophilia A without Inhibitors. *N. Engl. J. Med.* **2018**, *379*, 811–822. [CrossRef]
9. Wyseure, T.; Mosnier, L.O.; von Drygalski, A. Advances and challenges in hemophilic arthropathy. *Semin. Hematol.* **2016**, *53*, 10–19. [CrossRef]
10. Li, L.; Huang, L.; Sung, S.-S.J.; Vergis, A.L.; Rosin, D.L.; Rose, C.E., Jr.; Lobo, P.I.; Okusa, M.D. The chemokine receptors CCR2 and CX3CR1 mediate monocyte/macrophage trafficking in kidney ischemia–reperfusion injury. *Kidney Int.* **2008**, *74*, 1526–1537. [CrossRef]
11. Mignot, S.; Cagnard, N.; Albaud, B.; Bally, C.; Siavellis, J.; Hermine, O.; Frenzel, L. Unique inflammatory signature in haemophilic arthropathy: miRNA changes due to interaction between blood and fibroblast-like synoviocytes. *J. Cell. Mol. Med.* **2020**, *24*, 14453–14466. [CrossRef] [PubMed]

12. Koga, T.; Migita, K.; Sato, T.; Sato, S.; Umeda, M.; Nonaka, F.; Fukui, S.; Kawashiri, S.-Y.; Iwamoto, N.; Ichinose, K.; et al. MicroRNA-204-3p inhibits lipopolysaccharide-induced cytokines in familial Mediterranean fever via the phosphoinositide 3-kinase γ pathway. *Rheumatology* **2017**, *57*, 718–726. [CrossRef] [PubMed]
13. Haxaire, C.; Kalliolias, G.D.; DiCarlo, E.; Salmon, J.; Srivastava, A.; Blobel, C.P. Blood-Induced Arthropathy in Hemophilia: Mechanisms and Heterogeneity. *Semin. Thromb. Hemost.* **2015**, *41*, 832–837. [CrossRef] [PubMed]
14. Skinner, M.W. Haemophilia: Provision of factors and novel therapies: World Federation of Hemophilia goals and achievements. *Br. J. Haematol.* **2011**, *154*, 704–714. [CrossRef] [PubMed]
15. Oh, I.-S.; Suh, D.-W.; Park, S.-R.; Ha, K.-Y. Fractalkine receptor chemokine (CX3CR1) influences on cervical and lumbar disc herniation. *Indian J. Orthop.* **2015**, *49*, 239–244. [CrossRef] [PubMed]
16. Kellgren, J.H.; Lawrence, J.S. Radiological Assessment of Osteo-Arthrosis. *Ann. Rheum. Dis.* **1957**, *16*, 494–502. [CrossRef] [PubMed]
17. Jayadev, C.; Rout, R.; Price, A.; Hulley, P.; Mahoney, D. Hyaluronidase treatment of synovial fluid to improve assay precision for biomarker research using multiplex immunoassay platforms. *J. Immunol. Methods* **2012**, *386*, 22–30. [CrossRef]
18. Mathiessen, A.; Conaghan, P.G. Synovitis in osteoarthritis: Current understanding with therapeutic implications. *Arthritis Res. Ther.* **2017**, *19*, 1–9. [CrossRef]
19. van Vulpen, L.F.D.; Schutgens, R.E.G.; Coeleveld, K.; Alsema, E.C.; Roosendaal, G.; Mastbergen, S.C.; Lafeber, F.P.J.G. IL-1β, in contrast to TNFα, is pivotal in blood-induced cartilage damage and is a potential target for therapy. *Blood* **2015**, *126*, 2239–2246. [CrossRef]
20. Haxaire, C.; Hakobyan, N.; Pannellini, T.; Carballo, C.; McIlwain, D.; Mak, T.W.; Rodeo, S.; Acharya, S.; Li, D.; Szymonifka, J.; et al. Blood-induced bone loss in murine hemophilic arthropathy is prevented by blocking the iRhom2/ADAM17/TNF-α pathway. *Blood* **2018**, *132*, 1064–1074. [CrossRef]
21. Calcaterra, I.; Iannuzzo, G.; Dell'Aquila, F.; Di Minno MN, D. Pathophysiological Role of Synovitis in Hemophilic Arthropathy Development: A Two-Hit Hypothesis. *Front. Physiol.* **2020**, *11*, 541. [CrossRef] [PubMed]
22. Aggarwal, B.B.; Gupta, S.C.; Kim, J.H. Historical perspectives on tumor necrosis factor and its superfamily: 25 years later, a golden journey. *Blood* **2012**, *119*, 651–665. [CrossRef] [PubMed]
23. Anastasiou, G.; Gialeraki, A.; Merkouri, E.; Politou, M.; Travlou, A. Thrombomodulin as a regulator of the anticoagulant pathway. *Blood Coagul. Fibrinolysis* **2012**, *23*, 1–10. [CrossRef] [PubMed]
24. Dargaud, Y.; Simpson, H.; Chevalier, Y.; Scoazec, J.Y.; Hot, A.; Guyen, O.; Baglin, T.; Négrier, C. The potential role of synovial thrombomodulin in the pathophysiology of joint bleeds in haemophilia. *Haemophilia* **2012**, *18*, 818–823. [CrossRef]
25. Manetti, M.; Linari, S.; Romano, E.; Rosa, I.; Carulli, C.; Innocenti, M.; Matucci-Cerinic, M.; Ibba-Manneschi, L.; Castaman, G.; Melchiorre, D. TNF-α/TNF-R System May Represent a Crucial Mediator of Proliferative Synovitis in Hemophilia A. *J. Clin. Med.* **2019**, *8*, 939. [CrossRef]
26. Kotela, A.; Wojdasiewicz, P.; Łęgosz, P.; Sarzyńska, S.; Drela, K.; Pulik, Ł.; Kaleta, B.; Kniotek, M.; Borysowski, J.; Poniatowski, Ł.A.; et al. Increased serum levels of progranulin (PGRN) in patients with haemophilic arthropathy. *Clin. Exp. Pharmacol. Physiol.* **2018**, *46*, 373–379. [CrossRef]
27. Gualtierotti, R.; Solimeno, L.P.; Peyvandi, F. Hemophilic arthropathy: Current knowledge and future perspectives. *J. Thromb. Haemost.* **2021**, *19*, 2112–2121. [CrossRef]
28. Badulescu, O.V.; Filip, N.; Sirbu, P.D.; Bararu-Bojan, I.; Vladeanu, M.; Bojan, A.; Ciocoiu, M. Current practices in haemophilic patients undergoing orthopedic surgery—A systematic review. *Exp. Ther. Med.* **2020**, *20*, 207. [CrossRef]
29. Tseng, Y.-H.; Chiou, S.-S.; Zeng, Y.-S.; Tsai, S.-P.; Chen, C.-S.; Liao, Y.-M.; Lin, P.-C. Soluble vascular cell adhesion molecular-1 is a potential biological indicator of hemophilic arthropathy. *Medicine* **2016**, *95*, e5384. [CrossRef]
30. Tchalla, A.E.; Wellenius, G.A.; Travison, T.G.; Gagnon, M.; Iloputaife, I.; Dantoine, T.; Sorond, F.A.; Lipsitz, L.A. Circulating Vascular Cell Adhesion Molecule-1 Is Associated With Cerebral Blood Flow Dysregulation, Mobility Impairment, and Falls in Older Adults. *Hypertension* **2015**, *66*, 340–346. [CrossRef]
31. Okugawa, Y.; Miki, C.; Toiyama, Y.; Koike, Y.; Inoue, Y.; Kusunoki, M. Serum Level of Soluble Vascular Cell Adhesion Molecule 1 Is a Valuable Prognostic Marker in Colorectal Carcinoma. *Dis. Colon Rectum* **2009**, *52*, 1330–1336. [CrossRef] [PubMed]
32. Iacono, O.L.; Rincón, D.; Hernando, A.; Ripoll, C.; Catalina, M.V.; Salcedo, M.; Clemente, G.; Gomez, J.; Nuñez, O.; Matilla, A.; et al. Serum levels of soluble vascular cell adhesion molecule are related to hyperdynamic circulation in patients with liver cirrhosis. *Liver Int.* **2008**, *28*, 1129–1135. [CrossRef] [PubMed]
33. Kocijancic, M.; Cubranic, Z.; Vujicic, B.; Racki, S.; Dvornik, S.; Zaputovic, L. Soluble intracellular adhesion molecule-1 and omentin-1 as potential biomarkers of subclinical atherosclerosis in hemodialysis patients. *Int. Urol. Nephrol.* **2016**, *48*, 1145–1154. [CrossRef] [PubMed]
34. Oldenburg, J.; Zimmermann, R.; Katsarou, O.; Zanon, E.; Kellermann, E.; Lundin, B.; Ellinghaus, P. Potential biomarkers of haemophilic arthropathy: Correlations with compatible additive magnetic resonance imaging scores. *Haemophilia* **2016**, *22*, 760–764. [CrossRef] [PubMed]
35. Vandooren, B.; Cantaert, T.; Noordenbos, T.; Tak, P.P.; Baeten, D. The abundant synovial expression of the RANK/RANKL/Osteoprotegerin system in peripheral spondylarthritis is partially disconnected from inflammation. *Arthritis Rheum.* **2008**, *58*, 718–729. [CrossRef]
36. Lacey, D.L.; Timms, E.; Tan, H.L.; Kelley, M.J.; Dunstan, C.R.; Burgess, T.; Elliott, R.; Colombero, A.; Elliott, G.; Scully, S.; et al. Osteoprotegerin ligand is a cytokine that regulates osteoclast differentiation and activation. *Cell* **1998**, *93*, 165–176. [CrossRef]

37. Kamath, A.F.; Horneff, J.G.; Forsyth, A.; Nikci, V.; Nelson, C.L. Total Knee Arthroplasty in Hemophiliacs: Gains in Range of Motion Realized beyond Twelve Months Postoperatively. *Clin. Orthop. Surg.* **2012**, *4*, 121–128. [CrossRef]
38. de Oliveira, L.M.M.; Jardim, L.L.; Santana, M.A.P.; Cerqueira, M.H.; Lorenzato, C.S.; Franco, V.K.B.; Zuccherato, L.W.; Rezende, S.M.; Chaves, D.G. Effect of the First Factor VIII Infusions on Immunological Biomarkers in Previously Untreated Patients with Hemophilia A from the HEMFIL Study. *Thromb. Haemost.* **2021**, *121*, 891–899. [CrossRef] [PubMed]
39. Pasta, G.; Annunziata, S.; Polizzi, A.; Caliogna, L.; Jannelli, E.; Minen, A.; Mosconi, M.; Benazzo, F.; Di Minno, M.N.D. The Progression of Hemophilic Arthropathy: The Role of Biomarkers. *Int. J. Mol. Sci.* **2020**, *21*, 7292. [CrossRef]
40. Rodriguez-Merchan, E.C. Serological biomarkers in hemophilic arthropathy: Can they be used to monitor bleeding and ongoing progression of blood-induced joint disease in patients with hemophilia? *Blood Rev.* **2019**, *41*, 100642. [CrossRef]
41. van Bergen, E.; van Vulpen, L.; Schutgens, R.; Mastbergen, S.; Lafeber, F. Biochemical marker research in hemophilic arthropathy: A systematic review. *Blood Rev.* **2020**, *47*, 100781. [CrossRef] [PubMed]
42. Mateo, F.; Soria-Olivas, E.; Carrasco, J.J.; Bonanad, S.; Querol, F.; Pérez-Alenda, S.; Mateo, F.; Soria-Olivas, E.; Carrasco, J.J.; Bonanad, S.; et al. HemoKinect: A Microsoft Kinect V2 Based Exergaming Software to Supervise Physical Exercise of Patients with Hemophilia. *Sensors* **2018**, *18*, 2439. [CrossRef]
43. Doshi, B.S.; Rana, J.; Castaman, G.; Shaheen, M.A.; Kaczmarek, R.; Butterfield, J.S.; Meeks, S.L.; Leissinger, C.; Biswas, M.; Arruda, V.R. B cell–activating factor modulates the factor VIII immune response in hemophilia A. *J. Clin. Investig.* **2021**, *131*, e142906. [CrossRef]
44. Cooke, E.J.; Wyseure, T.; Zhou, J.Y.; Gopal, S.; Nasamran, C.A.; Fisch, K.M.; Manon-Jensen, T.; Karsdal, M.A.; Mosnier, L.O.; von Drygalski, A. Mechanisms of vascular permeability and remodeling associated with hemarthrosis in factor VIII-deficient mice. *J. Thromb. Haemost.* **2019**, *17*, 1815–1826. [CrossRef] [PubMed]
45. van Vulpen, L.; van Meegeren, M.; Roosendaal, G.; Jansen, N.; van Laar, J.; Schutgens, R.; Mastbergen, S.; Lafeber, F. Biochemical markers of joint tissue damage increase shortly after a joint bleed; an explorative human and canine in vivo study. *Osteoarthr. Cartil.* **2015**, *23*, 63–69. [CrossRef]
46. Cuesta-Barriuso, R.; Perez-Llanes, R.; Donoso-Ubeda, E.; Lopez-Pina, J.A.; Merono-Gallut, J. Effects of myofascial release on frequency of joint bleedings, joint status, and joint pain in patients with hemophilic elbow arthropathy. *Medicine* **2021**, *100*, e26025. [CrossRef]
47. Manon-Jensen, T.; Tangada, S.; Bager, C.; Chowdary, P.; Klamroth, R.; von Drygalski, A.; Windyga, J.; Escobar, M.; Frederiksen, P.; Engl, W.; et al. Evaluation of collagen turnover biomarkers as an objective measure for efficacy of treatment with rurioctocog alfa pegol in patients with hemophilia A: A secondary analysis of a randomized controlled trial. *J Thromb Haemost.* **2024**, *22*, 90–100. [CrossRef] [PubMed]
48. Hassab, H.M.; El-Gendy, W.M.; El-Noueam, K.I.; El Ghany, H.M.A.; Elwan, M.M. Serum cartilage oligomeric matrix protein reflects radiological damage and functional status in hemophilic arthropathy patients. *Egypt. Rheumatol.* **2016**, *38*, 241–245. [CrossRef]
49. Hua, B.; Olsen, E.H.N.; Sun, S.; Gudme, C.N.; Wang, L.; Vandahl, B.; Roepstorff, K.; Kjelgaard-Hansen, M.; Sørensen, B.B.; Zhao, Y.; et al. Serological biomarkers detect active joint destruction and inflammation in patients with haemophilic arthropathy. *Haemophilia* **2017**, *23*, e294–e300. [CrossRef]
50. Sun, X.; Zhou, X.; Zhuang, J.; Qiu, S.; Liu, Z.; Li, Q.; Zhang, Y.; Li, Y.; Du, D.; Sun, J. Relationship between serum cartilage turnover biomarkers and haemophilic arthropathy severity in adult patients with severe haemophilia A in China. *Haemophilia* **2022**, *29*, 362–364. [CrossRef]
51. Giordano, P.; Brunetti, G.; Lassandro, G.; Notarangelo, L.D.; Luciani, M.; Mura, R.M.; Lazzareschi, I.; Santagostino, E.; Piacente, L.; Ventura, A.; et al. High serum sclerostin levels in children with haemophilia A. *Br. J. Haematol.* **2015**, *172*, 293–295. [CrossRef] [PubMed]
52. Alioglu, B.; Selver, B.; Ozsoy, H.; Koca, G.; Ozdemir, M.; Dallar, Y. Evaluation of bone mineral density in Turkish children with severe haemophilia A: Ankara hospital experience. *Haemophilia* **2012**, *18*, 69–74. [CrossRef] [PubMed]
53. Acharya, S.S.; Kaplan, R.N.; Macdonald, D.; Fabiyi, O.T.; DiMichele, D.; Lyden, D. Neoangiogenesis contributes to the development of hemophilic synovitis. *Blood* **2011**, *117*, 2484–2493. [CrossRef]
54. Hauw, W.W.S.; Chia, J.S.J.; Nandurkar, H.H.; Sashindranath, M. The potential role of protease systems in hemophilic arthropathy. *Blood Adv.* **2022**, *6*, 5505–5515. [CrossRef] [PubMed]

International Journal of *Molecular Sciences*

MDPI

Article

Exploratory Untargeted Metabolomics of Dried Blood Spot Samples from Newborns with Maple Syrup Urine Disease

Abeer Z. Alotaibi [1], Reem H. AlMalki [2], Maha Al Mogren [2], Rajaa Sebaa [3], Mohammad Alanazi [1], Minnie Jacob [2], Ahamd Alodaib [2], Ahmad Alfares [2] and Anas M. Abdel Rahman [2,4,*]

1 Genome Research Chair, Department of Biochemistry, College of Science, King Saud University, P.O. Box 22452, Riyadh 11652, Saudi Arabia; 441203289@student.ksu.edu.sa (A.Z.A.); msanazi@ksu.edu.sa (M.A.)

2 Metabolomics Section, Department of Clinical Genomics, Center for Genomics Medicine, King Faisal Specialist Hospital and Research Centre (KFSHRC), Riyadh 11211, Saudi Arabia; rgalmalki@kfshrc.edu.sa (R.H.A.); mmogren@kfshrc.edu.sa (M.A.M.); minnie@kfshrc.edu.sa (M.J.); aalodaib@kfshrc.edu.sa (A.A.); aalfares@kfshrc.edu.sa (A.A.)

3 Department of Medical Laboratories, College of Applied Medical Sciences, Shaqra University, Shaqra 11961, Saudi Arabia; r.sebaa@su.edu.sa

4 Department of Biochemistry and Molecular Medicine, College of Medicine, Alfaisal University, Riyadh 11533, Saudi Arabia

* Correspondence: aabdelrahman46@kfshrc.edu.sa

Abstract: Currently, tandem mass spectrometry-based newborn screening (NBS), which examines targeted biomarkers, is the first approach used for the early detection of maple syrup urine disease (MSUD) in newborns, followed by confirmatory genetic mutation tests. However, these diagnostic approaches have limitations, demanding the development of additional tools for the diagnosis/screening of MUSD. Recently, untargeted metabolomics has been used to explore metabolic profiling and discover the potential biomarkers/pathways of inherited metabolic diseases. Thus, we aimed to discover a distinctive metabolic profile and biomarkers/pathways for MSUD newborns using untargeted metabolomics. Herein, untargeted metabolomics was used to analyze dried blood spot (DBS) samples from 22 MSUD and 22 healthy control newborns. Our data identified 210 altered endogenous metabolites in MSUD newborns and new potential MSUD biomarkers, particularly L-alloisoleucine, methionine, and lysoPI. In addition, the most impacted pathways in MSUD newborns were the ascorbate and aldarate pathways and pentose and glucuronate interconversions, suggesting that oxidative and detoxification events may occur in early life. Our approach leads to the identification of new potential biomarkers/pathways that could be used for the early diagnosis/screening of MSUD newborns but require further validation studies. Our untargeted metabolomics findings have undoubtedly added new insights to our understanding of the pathogenicity of MSUD, which helps us select the appropriate early treatments for better health outcomes.

Keywords: inborn errors of metabolism (IEMs); maple syrup urine disease (MSUD); genetic testing; newborn screening alloisoleucine; methionine sulfoxide; LysoPI; untargeted metabolomics; liquid chromatography high-resolution mass spectrometry

check for updates

Citation: Alotaibi, A.Z.; AlMalki, R.H.; Al Mogren, M.; Sebaa, R.; Alanazi, M.; Jacob, M.; Alodaib, A.; Alfares, A.; Abdel Rahman, A.M. Exploratory Untargeted Metabolomics of Dried Blood Spot Samples from Newborns with Maple Syrup Urine Disease. *Int. J. Mol. Sci.* **2024**, *25*, 5720. https://doi.org/10.3390/ijms25115720

Academic Editor: Andrea Bernini

Received: 15 April 2024
Revised: 20 May 2024
Accepted: 21 May 2024
Published: 24 May 2024

1. Introduction

Clinically, inborn errors of metabolism (IEMs) are a group of rare heterogeneous disorders genetically transmitted through autosomal or X-linked recessive inheritance patterns. These genetic disorders are associated with accumulations of intermediary metabolites, with subsequent shortages of metabolic products, leading to pathological consequences associated with various phenotypes [1,2]. Based on IEMbase, almost 1872 IEMs have been identified, and more disorders are expected to be discovered (http://www.iembase.org/, accessed on 2 December 2022) [3]. In particular, maple syrup urine disease (MSUD), known as an aminoacidopathy, is one of the categorized IEMs caused by a deficiency of the mitochondrial enzyme called the branched-chain ketoacid dehydrogenase (BCKD) complex, resulting from mutations in the genes that encode complex BCKD subunits including E1α, E1β, E2, and E3 [4]. According to the Human Gene Mutation Database, 259 genetic mutations have been reported to cause MSUD (http://www.hgmd.cf.ac.uk/ac/index.php, accessed on 1 October 2023). Imtiaz et al. reported, in 2017, 20 novel MSUD-related genetic mutations in *BCKDHA*, *BCKDHB*, and *DBT* in Saudi patients [5]. Generally, MSUD patients are classified into classic MSUD, accounting for 80% of the cases developed in newborns after two days of birth, and milder variants of MSUD, i.e., intermediate and intermittent cases, detected at older ages [6]. Phenotypically, MSUD patients vary in their symptoms based on their disease type, as classical MSUD has the most severe symptoms, including hypotonia, severe vomiting, seizures with encephalopathy, coma, and brain edema, while the milder variants of MSUD have lesser degrees of these clinical symptoms [7,8]. Pathologically, MSUD patients are characterized by derangements at the phenotypic, genetic, and metabolic levels. Briefly, MSUD-related genetic mutations lead to significant metabolic alterations, specifically, elevations of branched-chain amino acids (BCAAs)—including leucine, isoleucine, alloisoleucine, and valine—and branched-chain α-ketoacids (BCKAs) in the body tissues and fluids of MSUD patients. These metabolic alterations may not be seen in all types of MSUD [9]. However, among these metabolic alterations, alloisoleucine, which is derived from isoleucine in vivo by transamination, is elevated in all MSUD patient types and thus is used as a pathogenic marker for MSUD [10].

The global incidence of MSUD is estimated to be 1:185,000, affecting all ethnic populations. In Saudi Arabia, the frequency of MSUD is not well known; however, based on newborn screening (NBS), it is estimated to be 1:21,490 [5]. As expected, this prevalence is underestimated due to the possible existence of undiagnosed MSUD patients and false-positive cases [11,12], which lead to high demands for treatments and interventions that negatively affect the health system and cause financial burdens, encouraging the improvement of the accuracy of MSUD's identification in early life. The early and confirmed identification of MSUD during the neonatal period is achieved based on MSUD-related clinical phenotypes and manifestations: elevations of BCAA and BCKA detected through NBS and the detection of BCKD gene mutations via genetic testing. To further illustrate this, NBS analyses of MSUD rely on the tandem mass spectrometry (MS/MS) analysis of dried blood spots (DBSs) taken from newborns' heels to measure targeted metabolic biomarkers and their concentrations, particularly "total Leu" (Xle), which encompasses leucine (Leu), isoleucine (Ile), valine, and alloisoleucine (Allo-Ile) [10,11]. If the NBS results are positive, then follow-up genetic tests are performed using high-throughput genomic sequencing combined with Sanger sequencing for their confirmation and validation [12]. Notably, NBS and genetic testing have advanced the identification of MSUD cases [13,14], but these approaches have certain limitations that must be resolved. Additionally, genetic testing can be expensive, is occasionally accompanied by delayed results, and detects variants of uncertain significance (VUSs) [15,16]. The last-mentioned variant (VUS) requires functional analyses to examine its impact and relation to the MSUD condition. In addition to the genetic testing-related limitations, NBS also faces some obstacles. NBS can detect newborns who do not truly have MSUD (false-positive cases) or miss cases for unknown reasons. To improve the accuracy and sensitivity of both approaches and solve their limitations, additional strategies are required to diagnose and screen for true MSUD cases in early life, and these could be applied in parallel with the currently used approaches. Currently, the clinical and scientific communities have recommended utilizing untargeted metabolomics as a tool for identifying IEMs, including MSUD, for the following reasons: Firstly, since metabolism and health are strongly linked, untargeted metabolomics, which measures a wide range of small metabolites representing the metabolism status, and detects any metabolic alterations can provide valuable insights into healthy and pathological conditions.

Furthermore, untargeted metabolomics relatively measures the amounts of metabolites, the final byproducts of genes involved in metabolic pathways. According to the Human Metabolome Database (HMDB), approximately thousands of metabolites have been found [17]. These metabolites could be used as potential biomarkers for IEMs diseases. Most importantly, the great opportunity of untargeted metabolomics is that it provides comprehensive coverage of metabolites in the body system, which is in opposition to the limitations of the current biochemical or metabolic tests, which measure a limited number of metabolites, leading to them missing other altered metabolites with a significant ability to be used as disease-specific markers. Promisingly, using untargeted metabolomics can solve these issues seen with genetic testing and NBS and bridge the knowledge gap in IEMs [18].

To date, a few studies have used untargeted metabolomics in MSUD patients. In 2018, Coene et al. performed an untargeted metabolomic profiling of two MSUD patients included in their research, finding that leucine, isoleucine, ketoleucine, 2-hydroxymethyl butyric acid, and 2-hydroxy caproic acid were dysregulated significantly compared to their healthy population [19]. Another untargeted metabolomics research study performed on LC-QTOF for 21 MSUD infants revealed elevations of 3-hydroxybutyric acid, 2-oxoisovaleric acid, 2-hydroxyisovaleric acid, 2-oxo-3-methylvaleric acid, total leucine (XLeu), valine (Val), and others [20]. Furthermore, Haijes et al. (2019) evaluated seven MSUD patients' plasma, and they found that their isoleucine and alloisoleucine with keto-acids were elevated, while their 3-hydroxyisobutyrate and isovaleryl-carnitine were reduced [21]. These published findings illustrate the promising potential of using metabolomics as a diagnosis/screening tool for MSUD; however, these studies have some limitations, such as a low number of MSUD participants of various ages and the use of different analytical instruments to perform their metabolomics analyses. Thus, further metabolomics studies are required to validate their findings; to explore additional metabolic biomarkers of MSUD, especially in the neonatal period; and to standardize the workflow of metabolomics analyses by using a larger cohort of MSUD newborns, which could lead more accuracy in the early diagnosis of MSUD. Thus, we aimed to conduct an untargeted metabolomics analysis of DBS samples from MSUD newborns to discover the metabolic biomarkers and pathways that could improve the inaccurate diagnoses of MSUD caused by heterogeneous MSUD phenotypes, false NBS results, and VUS cases.

2. Results

2.1. Participant Demographics and DBS Sample Selection

In order to use appropriate samples in our study, DBS samples from newborns that were admitted to the NBS lab at King Faisal Specialized Hospital and Research Center (KFSHRC) were first checked and identified as either MSUD samples (according to the NBS-based-MSUD markers mentioned below) or normal samples showing normal NBS results. A total of 44 DBS samples were used in this study: 22 samples were collected from biochemically and genetically confirmed MSUD newborns, and the other 22 samples were age- and gender-matched healthy newborns (Table 1). The participants in the two groups were almost age- and gender-matched. The average age of the MSUD newborns was (7.63 $\pm$ 3.07 days) and for healthy newborns was (7.72 $\pm$ 3.04 days). Newborns older than 14 days or with other IEMs, based on their NBS data, were excluded from this study. MSUD newborns showed significantly increased Xleucine (581 $\pm$ 431.97 μM) and valine (431.97 $\pm$ 149.02 μM) compared to healthy newborns. For the NBS data, DBS samples from these newborns were used for untargeted metabolomics analyses.

Table 1. Demographic and clinical features of MSUD newborns and healthy controls.

Demographic and Clinical Features	MSUD (n = 22)		CTRL (n = 22)		p-Value
	Mean	SEM	Mean	SEM	
Age (Day)	7.63	3.07	7.72	3.18	0.9421
Female (%)	41	NA	41	NA	NA
Male (%)	59	NA	59	NA	NA
Xleucine (Cutoff: <245 μM)	581	431.97	<245 μM	NA	0.005 **
Valine (Cutoff: <290 μM)	425.15	149.02	<290 μM	NA	0.0042 **

For statistical analyses, an independent student *t*-test was conducted. Mean ± SEM expresses data. Student's *t*-test: Two-tailed was applied (** $p < 0.01$). MSUD Maple syrup urine disorder; CTRL Healthy control; SEM standard error of the mean; NA not applicable.

2.2. Untargeted Metabolomics Profiling of MSUD Newborns

A total of 28,769 m/z compound ions were detected (Table S1) in both positive (n = 19,336) and negative (n = 9433) ionization modes. The data were deposited in Metabolomics Workbench (ST002750). In total, 28.5% of missing values were excluded after applying a filter with a frequency > 80% to ensure quality in the data analyses, resulting in 20,568 features being retained for statistical analysis. To confirm that all depicted data have a Gaussian distribution, the median was identified, we log-transformed normalized the data, and the data were Pareto scaled to eliminate systemic variances.

A multivariate analysis using partial least squares discriminant analysis (PLS-DA), a non-supervised analysis, showed sample clustering and a clear separation between MSUD newborns (green) and healthy controls (red; Figure 1A). Additionally, orthogonal partial least squares discriminant analysis (OPLS-DA), a supervised analysis, was performed and showed a clear separation in the score plot, reflecting the difference between the two groups, with a computed $R^2Y = 0.963$ and $Q^2 = 0.353$, as shown in Figure 1B.

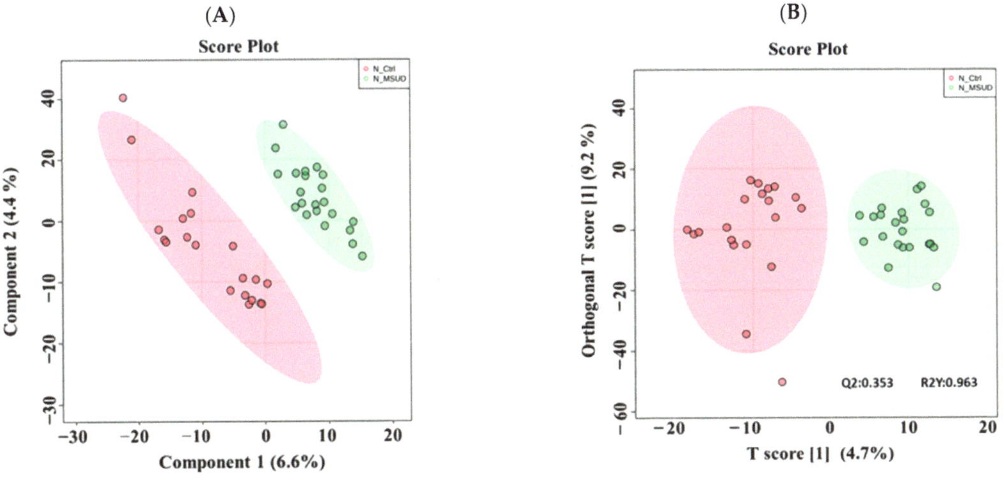

Figure 1. Sample clustering and group separation based on a group of 20,568 features. (**A**) PLS-DA shows a clear separation between the two groups: MSUD newborn and healthy control. (**B**) OPLS-DA shows a clear separation between the two groups: MSUD newborns and healthy controls. The robustness of the created models was evaluated by the fitness of the model (R^2Y: 0.963) and predictive ability (Q^2: 0.353) values.

A univariate analysis was performed after normalizing the signal and ensuring normal distribution. A volcano plot (moderated *t*-test, cut-off *p*-value $\leq$ 0.05 and fold change 1.5) was used to identify the significantly altered features between the two groups, MSUD newborns and healthy controls, and revealed 1040 significantly dysregulated features, where 303 and 737 were up- and downregulated, respectively (Figure 2; Table S2). A total of 480 features were annotated using HMDB (Table S3), and 210 of these were identified as human endogenous metabolites after excluding the exogenous molecules (i.e., environmental exposure, drugs, foods, etc.); these are listed in Table S4.

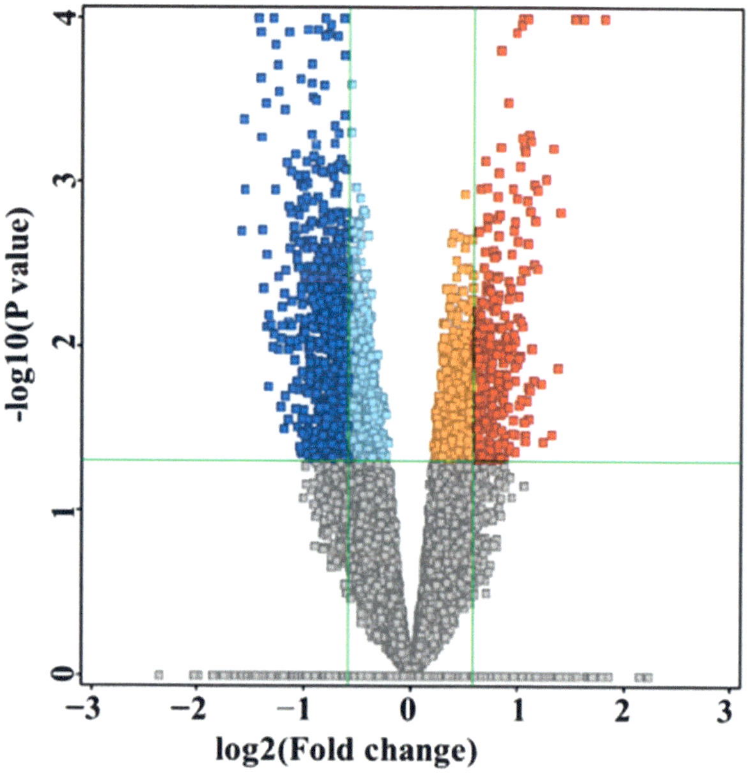

Figure 2. Volcano plot (moderated *t*-test, cut-off: $p < 0.05$, FC 1.5) between two groups: MSUD newborns and healthy controls. The heatmap revealed 1040 significantly dysregulated metabolites, where 303 (red) and 737 (blue) were up- and downregulated, respectively.

A heatmap was constructed that identified 210 as endogenous metabolites, of which 51 and 159 significantly were up- and downregulated in MSUD newborns (Figure S1; Figure 3A and 3B), respectively.

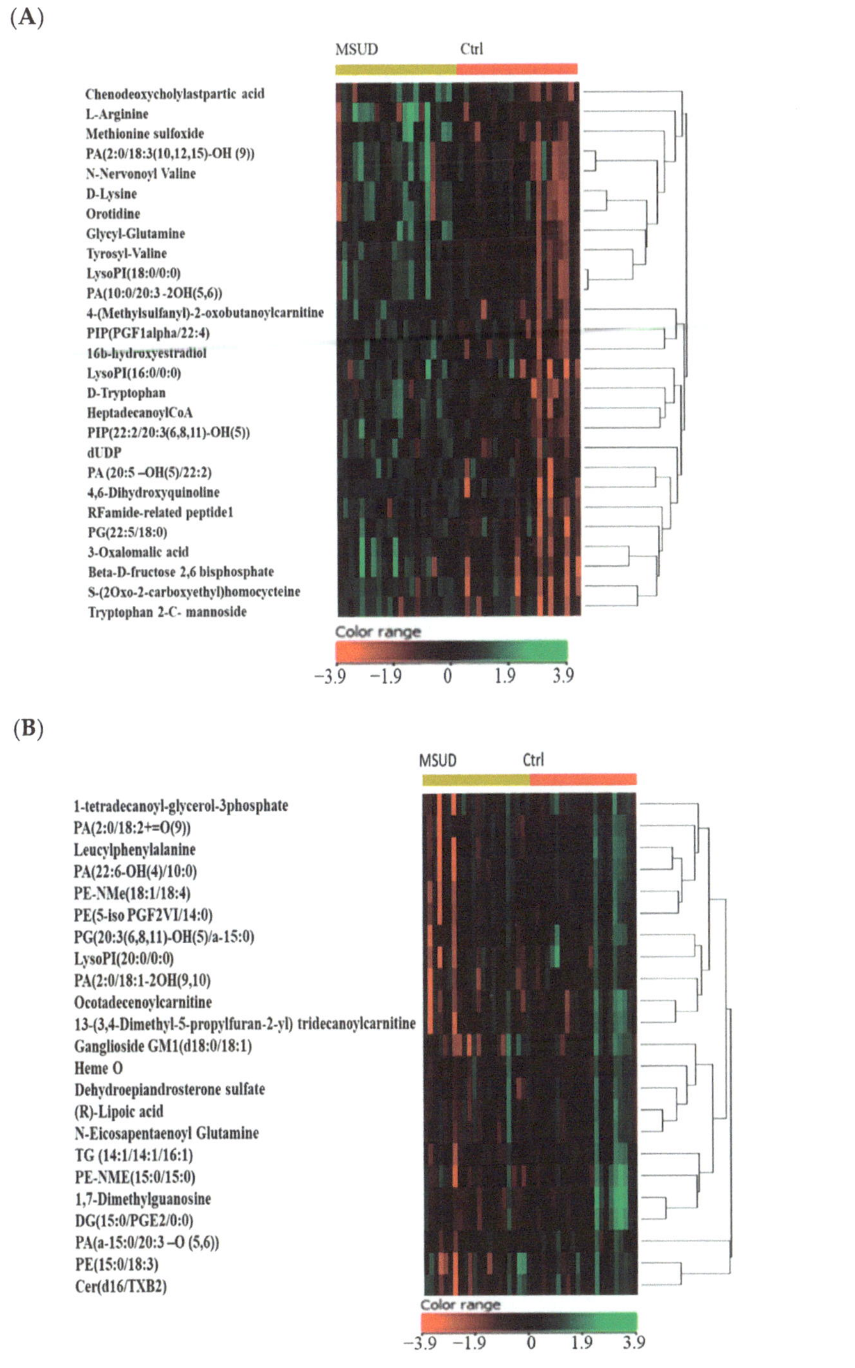

Figure 3. Hierarchical clustering (HAC) and heatmap analyses demonstrating (**A**) Upregulated metabolites. (**B**) Downregulated metabolites in MSUD newborns compared with healthy controls. The color scaled bar, green referred to upregulated metabolites, and red referred to downregulated metabolites.

2.3. Biomarkers Analysis of MSUD

MSUD biomarkers were evaluated using a receiver operating characteristics (ROC) curve analysis. As a classification and feature, the ranking approach PLS-DA was used to create a multivariate exploratory ROC analysis. Six features of the ROC curve via PLS-DA and cross-validation (CV) had area under the curve (AUC) values ranging from 0.832 to 0.922 with confidence intervals of 0.617–1 and 0.796–1 (Figure 4A). The frequency plot illustrates the 15 highest-scoring identified metabolites in the OPLS-DA model according to their level in MSUD and healthy newborns. As shown, L-alloisoleucine, methionine sulfoxide, glutathioselenol, heme O, N-eicosapentaenoyl glutamine, tryptophan 2-C-mannosidase, butenyl carnitine, and N-(1-Deoxy-1-fructosyl) isoleucine were upregulated endogenous metabolites. At the same time, PI (18:1/PGJ2) and lysoPI (16:0/0:0) were downregulated metabolites in MSUD newborns compared to the healthy control group (Figure 4B). For example, the AUC value of the ROC curves for methionine sulfoxide (Figure 4C) and L-alloisoleucine (Figure 4D) were upregulated (AUC: 0.81) and (AUC: 0.926), respectively. LysoPI was downregulated in MSUD newborns compared to their corresponding healthy controls (AUC: 0.86).

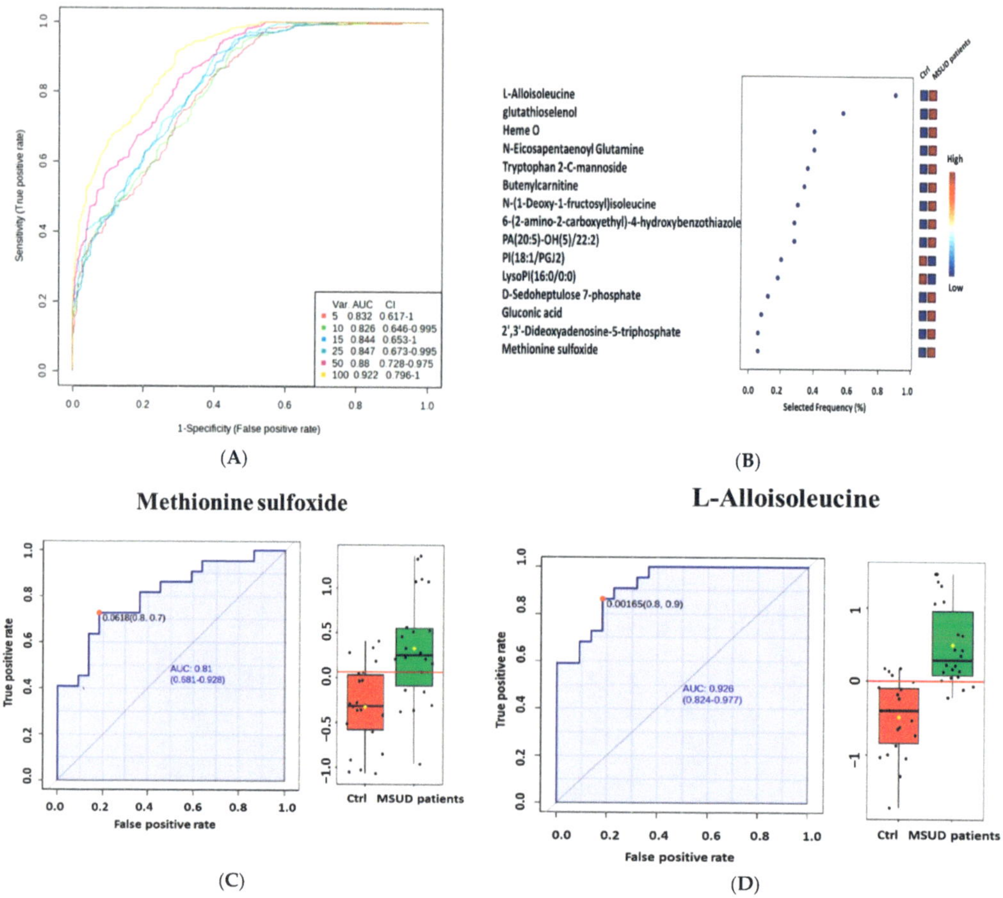

Figure 4. *Cont.*

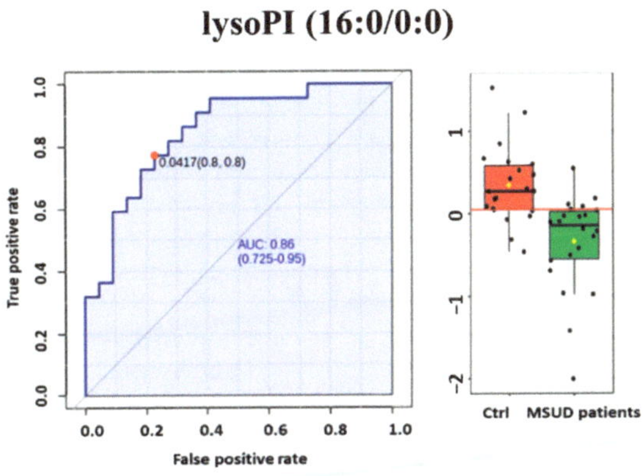

(E)

Figure 4. Metabolomics profiling and biomarker evaluation between MSUD newborn and healthy control groups. (**A**): A receiver operating characteristics (ROC) curve was created by the OPLS-DA model, with area under the curve (AUC) values calculated from the combination of 5, 10, 15, 25, 50, and 100 metabolites (**B**): The frequency plot shows the top 15 identified metabolites. (**C**,**D**): Examples of metabolites methionine sulfoxide and L-alloisoleucine were upregulated in MSUD newborn patients with (AUC:0.81) and (AUC: 0.926), respectively. (**E**): lysoPI downregulated in MSUD newborns compared to healthy control (AUC: 0.86).

2.4. Metabolomic Pathway Analysis

All these altered metabolites were subjected to a pathway analysis to identify the most affected pathways between the two study groups, which were ascorbate and aldarate metabolism and pentose and glucuronate interconversions. Other pathways were affected, such as sulfur metabolism, pyrimidine metabolism, glycerophospholipid metabolism, and pentose phosphate pathways, as illustrated in Figure 5 and detailed in Table S5.

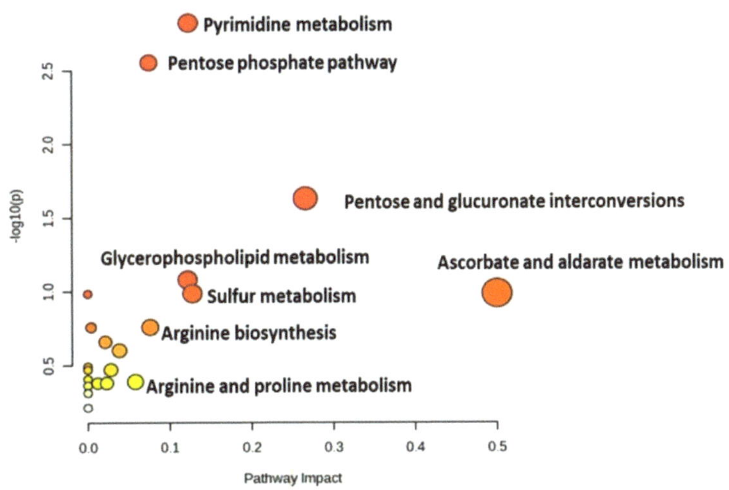

Figure 5. Pathway analysis of the significant metabolites dysregulated in MSUD newborns. In total, 210 metabolites were finally identified as human endogenous metabolites; 51 were upregulated, and 159 were downregulated. The color variation (yellow to red) shows the different significance levels of metabolites in the data.

3. Discussion

3.1. Untargeted Metabolomics as an Additional Diagnostic/Screening Tool for MSUD

In the clinical field, the early identification of MSUD is much preferred to avoid health complications and manage disease progression by providing MSUD newborns with the proper treatment in their neonatal period. Thus, MSUD diagnosis needs to be more accurate in early life. Currently, MSUD is diagnosed based on certain clinical criteria, including phenotypes/symptoms, NBS results, and genetic testing [2]. However, the last-mentioned clinical criteria have limitations, causing inaccuracy in the diagnosis of MSUD. For example, MSUD patients exhibit variable and heterogeneous phenotypes/symptoms, which cause difficulties in their early diagnosis.

Furthermore, NBS and genetic testing approaches have limited diagnostic/screening abilities in some MSUD cases, and NBS has previously led to false-positive or -negative cases of MSUD. For example, negative cases of MSUD have been seen, with samples collected from newborns at 24 h of life [22]. On the other hand, false-positive cases of MSUD have been detected in newborns diagnosed with hydroxyprolinemia, in which hydroxyproline is not distinguished from leucine and isoleucine by low-resolution mass spectrometry [23]. Additionally, during pregnancy, the maternal ingestion of sotolone-containing foods such as fenugreek can lead to false-positive cases of MSUD in newborns [24]. In addition to the limitations of NBS, genetic testing may detect VUS; the latter is detected in these analyses but does not provide detailed information about the mutation, whether disease-related or not, and further deep functional studies are required to identify the impact and relationship of VUS to the disease [25]. Some reports illustrate that some newly identified MSUD patients are found to have VUS in their MSUD-related genetic mutations and require functional analyses to confirm the pathogenicity of these variants [26,27].

Therefore, there have been increasing demands and attempts in the scientific community to solve these limitations by developing a new diagnostic tool for the better diagnosis of MSUD. Notably, untargeted metabolomics has received a lot of attention as a newly emerged diagnostic tool for IMDs because it can explore their wide metabolic profile, reflecting the metabolic status of diseases, which gives us a better understanding of IMDs. Additionally, untargeted metabolomics can uncover new metabolic biomarkers and pathways that could be added to the clinical diagnostic criteria for IMDs, improving their overall diagnosis process and increasing accuracy. A few studies have been carried out on the metabolomics of different MSUD samples, including DBS and urine, which were analyzed using MS or NMR methods [28–30]. Still, no study has performed untargeted metabolomics analyses of samples from MSUD newborns to identify its metabolic alterations and biomarkers/pathways in early life. Therefore, we performed untargeted metabolomics analyses of DBS samples from MSUD newborns, and our metabolomics results showed that MSUD newborns have an altered metabolic profile compared to healthy newborns (controls). Additionally, our findings revealed potential metabolic biomarkers and pathways that could be used for MSUD diagnosis/screening in early life, which are discussed further below.

3.2. Untargeted Metabolomics Revealed Altered Global Metabolic Profiling of MSUD Newborns

To truly study the metabolic alterations caused by this disease, we aimed to comprehensively explore the metabolic profiling of MSUD newborns who were not exposed to external factors, such as treatments or diet interventions, because the latter could modulate the true metabolic status of MSUD samples. Therefore, DBS samples were taken from MSUD newborns of an average age of around 7 days and analyzed using untargeted metabolomics. Our untargeted metabolomics analyses of MSUD DBSs revealed that MSUD had an altered metabolic profile compared to healthy newborns, with a broader range of altered metabolites. Examples of these altered metabolites include amino acids, lipid species, purine derivatives, glucuronide conjugates, and oxidative-related compounds.

3.3. Alterations in the Amino Acids in MSUD Newborns

Our data clearly showed that DBS samples from MSUD contained various altered amino acids and their modified forms, such as tryptophan, methionine, leucine, and alloisoleucine. For instance, tryptophan 2-C-mannoside was upregulated in MSUD newborns compared to healthy newborns. Tryptophan 2-C-mannoside, known as mannosyl tryptophan (CMW), is a glycosylated form of tryptophan in which the c-mannosylation site is found in the tryptophan motif located in proteins when it is targeted, it causes important protein folding, sorting, and/or secretion [31,32]. Interestingly, CMW levels were increased in the blood of patients with renal dysfunction, and they were also found to be elevated in patients with T2D [33], suggesting that CMW could act as a pathological biomarker in diseases. Given our CMW findings and the previously published studies on MCW, it is possible that CMW is elevated in various pathological conditions, including MSUD. However, further investigations are needed to deeply explore the role of MCW in the complication or progression of MSUD.

Additionally, our data revealed that methionine sulfoxide was upregulated in MSUD newborns compared to healthy controls. Methionine sulfoxide is derived from methionine and is formed post-translationally through the oxidation of methionine sulfur [34]. Methionine sulfoxide is an important metabolite that affects redox homeostasis in diseases through sulfur metabolism [35,36]. It is possible that MSUD patients could be subject to certain oxidative stress events, as is partially indicated by their elevated levels of methionine sulfoxide, which in turn could lead to the neurodegeneration and functional impairment of cells known to occur in MSUD [37]. It is suggested that methionine sulfoxide could be targeted and returned to methionine to potentially reduce some of the oxidative and degradative damages found in MSUD through reversing this post-translation modification using targeted-site modification strategies, especially as methionine is known to have antioxidant properties, leading it to ameliorate oxidative events [38]. However, further experimental work needs to be conducted to examine this suggestion in the future.

Moreover, another metabolite that was upregulated in MSUD-affected newborns was L-arginine. Biologically, arginine has various body functions, including a protective role in disposing of toxic components such as ammonia and its related derivatives, as arginine can bind to ammonia to convert it into non-toxic compounds that are eventually excreted [39]. Potentially, the elevation of arginine in MSUD newborns may play a protective role, minimizing the toxic elevations of ammonia that have been previously observed in MSUD patients and have been associated with metabolic encephalopathy [40]. Thus, we anticipated that MSUD newborns might develop defensive mechanisms during the early neonatal stage of the disease, one of which is an increased level of arginine production. It is suggested that correlation studies be performed between the levels of arginine and ammonia in MSUD newborns' DBS samples to examine the potential link between these two metabolites and help us understand the mechanism developed against the pathogenies of MSUD. Based on these amino acid-related findings, it seems that MSUD in the neonatal period leads to alterations in the structure and level of amino acids in patients. There is a need to study these amino acid alterations in patients of different ages who have or have not undergone treatment to understand these amino acid changes better.

3.4. Alterations in the Lipid Species in MSUD Newborns

Another observed metabolic alteration in our data is that various lipid species were disrupted, including the fatty acids, phospholipids, glycosphingolipids, ceramides, acylcarnitines, and glycerol derivatives. In our metabolomics data, LysoPI (16:0/0:0), Cer (d16:1(TXB2)), PE (15:0/18:3), PE-NMe (18:1/18:4), PI (18:1/PGJ2) and PA (2:0/18:2) were downregulated in MSUD newborns compared to controls. LysoPI is an endogenous ligand for G protein-coupled receptor 55(GPR55) and has various biological functions; however, its role in the central nervous system (CNS) is one that particularly affects microglial inflammatory responses. Microglia are immune cells with important functions in balancing the immune homeostasis of the CNS [41]. Additionally, Cer (d16:1/TXB2) is an oxidized ceramide, a member of the sphingolipids (SLs) or glycosylceramides. SLs are found in cell membranes, particularly in peripheral nerve cells and the cells found in the central nervous system. Impairments associated with sphingolipid metabolism are related to neurological syndromes [42]. The biosynthesis and catabolism of sphingolipids involve many intermediate metabolites and different enzymes. PE-NMe (15:0/18:3) is a monomethyl phosphatidylethanolamine, a glycerophospholipid part of phosphatidylcholine biosynthesis [43]. Phospholipids are key components of the cell lipid bilayer and are involved in metabolism and signaling. Oxidized phosphatidic acids, PA (2:0/18:2), belong to glycerolipids and can work as signaling molecules either by themselves or by indirectly interacting with other molecules. Those oxidized lipids are produced non-enzymatically through uncontrolled oxidation of free radicals, which are considered harmful to human health [44].

Furthermore, ganglioside GM1 (D18:0, 18:1) was downregulated in MSUD newborns compared to controls. Gangliosides are primarily considered the compositional components of the CNS glycome [45]. Gangliosides are glycosphingolipids made of glycan headgroups, which can engage with proteins or other glycans present on the same membrane, as well as with molecules on other cells and in the extracellular space, which results in modulation of cell signaling and communication [46,47]. GM1 exerts neuroprotective functions, specifically repairing the neuronal tissue after mechanical, biochemical, or toxic injuries [48]. Furthermore, our study showed various acylcarnitines (AC) alterations in MSUD newborns. Therefore, disrupted circulating AC may highlight the dysregulation of mitochondrial oxidation of lipids and upregulation of proinflammatory signals [49]. It could be that these alterations in lipid species worsen the phenotypes of MSUD in early life, including brain and nervous system damage.

3.5. Metabolites Involved in Oxidative Events in MSUD Newborns

Previously, it has been reported that MSUD patients have shown oxidative stress, which probably contributes to their neurological problems [50]. Regarding this state of oxidative stress state, our data showed that certain significant metabolites related to oxidative events are altered in MSUD. For example, in this study, glutathioselenol was upregulated in MSUD newborns. In the human liver, glutathioselenol reacts with glutathione to form hydrogen selenide ions, which are necessary to produce the seleno-proteins needed for biological systems, mostly as antioxidants [51,52]. Moreover, our findings showed that α-lipoic acid (ALA) was upregulated in MSUD newborns. ALA is a coenzyme of many multienzyme complexes located in the mitochondria. Mechanistically, ALA acts as an antioxidant by neutralizing free radicals and preventing oxidative damage to cells and tissues. It can work with other antioxidants such as vitamin C, vitamin E, and glutathione, enhancing the antioxidant defense system [53]. In addition, our data showed that N-acetylserotonin glucuronide (NASG) was downregulated in MSUD newborns. N-acetylserotonin glucuronide is derived from N-acetylserotonin (NAS), which controls intracellular redox states through the upregulation of enzymes involved in glutathione biosynthesis, enhancing the abundance of proteins involved in anti-oxidative defense [54]. Our metabolic finding suggested that the MSUD newborns potentially tried to develop defensive tools against the oxidative events using various mechanisms, such as glutathione- or ALA-mediated mechanisms, to reduce the health complications associated with MSUD. MSUD is a complicated case as it has many metabolic alterations, various phenotypes, and presentations, which enforce the identification of further diagnosis/screening biomarkers in addition to the standard markers used in NBS and genetic testing.

3.6. New Potential Metabolic Biomarkers/Pathways of MSUD

One of the well-known and definitive MSUD biomarkers is called "alloisoleucine", which is used globally for MSUD diagnosis/screening during NBS [10]. It was detected in our untargeted metabolomics analysis of DBSs from MSUD newborns. Based on the biomarker analyses and our metabolic data, alloisoleucine was significantly elevated in

MSUD newborns compared to healthy newborns, which makes untargeted metabolomics a trustworthy technique that could be used to validate the current standard method (NBS) and its MSUD markers. Our untargeted metabolomics analysis uncovered other new biomarkers, including methionine sulfoxide, LysoPI, and altered metabolic pathways such as ascorbate and aldarate metabolism, and pentose and glucuronate interconversions in MSUD newborns, potentially used as new additional biomarkers/pathways for MSUD. Particularly, Methionine sulfoxide was upregulated in MSUD newborns compared to healthy controls, and this could be attributed to the oxidative stress events in MSUD, making methionine sulfoxide a reliable biomarker for MSUD, needed for further validation studies. Also, LysoPI was selected as a decreased biomarker in MSUD, which merits further exploration.

As mentioned above, one of the most affected metabolic pathways in MSUD newborns is ascorbate and aldarate metabolism, compounds that are involved in the oxidative damage defense mechanism mediated by glutathione and other antioxidant molecules [55]. In addition, pentose and glucuronate interconversions were found in our data to be a highly impacted pathway in MSUD newborns. Glucuronate interconversions induce the glucuronidation process for the elimination of toxic molecules [56]. Predictably, in our study, MSUD newborns had impacted pentose and glucuronate interconversions, which may help remove the toxic substances that are a result of oxidative damage. However, further studies are required to examine the role of the glucuronidation process in MSUD.

These new metabolic biomarkers/pathways are very promising; they provide new insights into the field and should be considered additions to the diagnosis panel of MSUD in the future. However, a follow-up study is required to evaluate the discovered biomarkers' reproducibility, stability, and performance and to validate the disrupted metabolic pathways in a bigger and independent cohort, considering an appropriate study design and FDR. Moreover, a prototype targeted analytical approach will be established for the newly discovered metabolic biomarkers using reference standard materials. Additionally, for accurate diagnosis, various biological samples, not only DBSs, must be used to examine sample-type specificity. Furthermore, a cohort of MSUD patients of different ages could be studied to determine whether these new metabolic biomarkers can only be used in the neonatal period or at other ages as well. All these suggested studies would generate more information about MSUD and develop more accurate approaches for its diagnosis/screening.

4. Materials and Methods

4.1. Ethical Approval

The Institutional Review Boards at King Faisal Specialist Hospital and Research Center (KFSHRC) in Riyadh, Saudi Arabia, reviewed and approved this study procedure (RAC No. 2160 027). The leftover samples submitted for routine clinical testing were waivered from any consents.

4.2. Participants' Selection Criteria and Sample Collection

Forty-four DBS cards were collected from the Metabolomics Lab in the Center for Genomic Medicine at KFSHRC. The newborns positively screened with MSUD (n = 22) were confirmed biochemically and genetically and included in this study with healthy controls (n = 22). All healthy controls were age- and gender-matched with patients. MSUD newborns were not diagnosed with any other disease, such as IEM disorders, and less than 14 days old were included in this study. Originally, these samples were collected from newborns' heel pricks dripped from 903 Protein Saver cards (Whatman, Piscataway, NJ, USA). Then, after being dried, the DBS cards were stored at 4 °C for later biochemical and metabolomics analyses. The initial newborn screening was performed with tandem mass spectrophotometry using MassChrom kits (cat# 55000, ChromSystems, Munich, Germany), a CE-Marked diagnostic Kit available commercially. This screening test was performed routinely in our lab for clinical purposes, where the kit contains mobile phase, quality control samples, internal standard solutions labeled for the key amino acids, and acylcarnitine. In this study, this kit measured the leucine, isoleucine, valine, and MSUD markers as part of our routine laboratory practice.

4.3. Chemicals

The LC-MS graded chemicals methanol (MeOH), acetonitrile (ACN), deionized water (dH_2O), and formic acid were purchased from Fisher Scientific Company (Ottawa, ON, Canada). The reference materials used as internal standards were purchased from Sigma (Ottawa, ON, Canada).

4.4. Sample Preparation

Metabolites were extracted from the DBS of MSUD and newborn healthy controls. One punch of DBS with a size of 3.2 mm was distributed in 96 V-shaped plate wells. Then, the punch was immersed in 250 µL of (dH_2O: MeOH: ACN) (20:40:40%) as an extraction solvent. The samples were vortexed in a ThermoMixer (Eppendrof, Hamburg, Germany) at 600 rpm, 25 °C, for 2 h. Subsequently, the samples were spun down at 16,000 rpm, 4 °C, for 10 min. The supernatants were transferred into new 96-well V-shaped plates, the punches were discarded, and the samples were evaporated in a Speed-Vac (Thermo Fischer, Christ, Germany) [57].

4.5. LC-HRMS Metabolomics Analysis

All dried extracted samples were reconstituted in 50% mobile phase A (0.1% formic acid in deionized water) and mobile phase B (0.1% formic acid in (1:1) (v/v) MeOH and ACN) for an LC-MS metabolomics analysis. Initially, 5 µL of the reconstituted sample was introduced to the inlet technique, where the metabolites were separated in a reversed-phase liquid chromatography with Waters ACQUITY UPLC XSelect C18 (100 × 2.1 mm × 2.5 µm) column (Waters Ltd., Elstree, UK). The mobile phase flow rate was set to 300 µL/min, and the column was maintained at 55 °C while the sample was stored at 4 °C in the autosampler. Mobile phases A and B were pumped in a gradient mode as follows: 95–5% A (0–16 min), 5% A (16–19 min), 5–95% A (19–20 min), and 5–95% A (20–22 min). The eluted molecules from the column were ionized in the electrospray ionization source (ESI) at positive and negative modes. The gas phase ions were subjected to Xevo G2-S QTOF mass spectrometer (Waters Ltd., Elstree, UK) separation based on their m/z. The MS source temperature was fixed at 150 °C, the desolvation temperature was set at 500 °C, and the capillary voltages were kept at 3.20 kV or 3 kV for ESI+ and ESI− modes, respectively. The cone gas flow was 50 L/h, the desolvation gas flow was 800 L/h, and the cone voltage was 40 V. The collision energies for the low and high functions were set to off and 10–50 V, respectively, in the MSE data-independent acquisition (DIA) mode. As recommended by the vendor, the mass spectrometer was calibrated with sodium formate (100–1200 Da) in both ionization modes. The lock spray mass compound, MS leucine-enkephaline (an external reference to the ion m/z 556.2771 in positive mode and 554.2615 in negative mode), was constantly injected, which is responsible for switching between the sample and the reference for every 45 and 60 s in both modes, scan time was 0.5 s, the flow rate was 10 µL/min, and collision energy was 4 V and 30 V for the cone, respectively. The DIA data were gathered in continuum mode with Masslynx™ V4.1 Software (Waters Inc., Milford, MA, USA). Quality control samples (QCs) were performed gently by collecting a single 3.2 mm punch from each study sample and pooling them for extraction. After that, they were introduced to the instrument randomly to validate the system's stability [58]. After that, they were analyzed following the routine protocol. The acceptance criteria were to have all the QC samples separated from the other study groups, clustered together, and use their Relative standard deviations (RSD%) < 40%.

4.6. Metabolomics Data Processing and Statistical Analyses

The raw MS data were processed using a standard pipeline, retention time alignment, mass-to-charge ratio (m/z) correction followed by the compound peak picking. Compound signals were selected based on the quality of the peaks that were selected, where the noise peaks were excluded outside the RT range of 0.5–20 min, and fragmentation ions with intensity below 0.2% were included using Progenesis QI v.3.0 software (Waters Technologies, Milford, MA, USA; Scheme 1). Multivariate statistics was applied using MetaboAnalyst (v.5.0; McGill University, Montreal, QC, Canada; http://www.metaboanalyst.ca, accessed on 20 June 2023) [59]. All the imported data groups (compounds' names and their raw abundance information) were normalized by the median and were Pareto scaled and log-transformed, and these were used to create PLS-DA and OPLS-DA models. The generated OPLS-DA model was measured through (R^2Y) and (Q^2) values, representing the model's fitness and predictive ability, respectively [60]. A univariate analysis was applied with Mass Profiler Professional (MPP) v.15.0 software (Agilent, Santa Clara, CA, USA). A volcano plot was used to uncover significantly changed mass features based on a moderated t-test, cut-off: $p < 0.05$, fold change 1.5. Compared to the controls, heatmap analysis for altered features in MSUD newborns was performed using the Pearson distance measure according to the Pearson similarity test [61]. Pathway and biomarker analyses, linked with MSUD-linked biomarkers, and receiver operating characteristic (ROC) curves were created using the PLS-DA approach in the MetaboAnalyst v 5.0 for globe analysis to specify potential biomarkers.

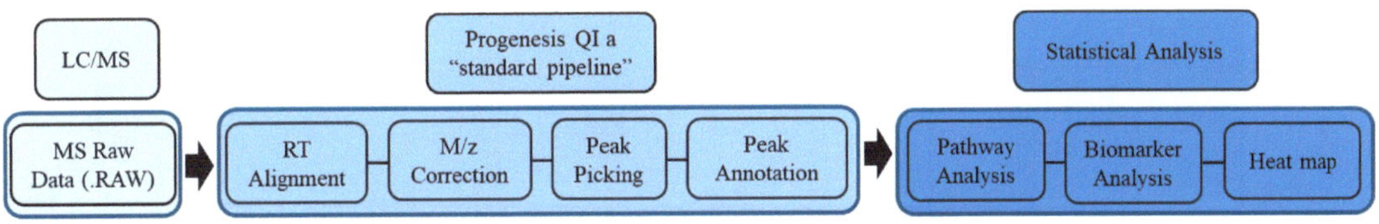

Scheme 1. The workflow of data analyses performed in this study.

4.7. Metabolite Identification (Peak Annotation)

All the statistically significant features between the study groups were selected using Progenesis QI v.3.0 software (Waters Technologies, Milford, MA, USA) for peak annotation. The precursor and product ions were annotated based on accurate mass, fragmentation pattern, and isotopic distributions in the Human Metabolome Database (HMDB) with a 5 ppm mass error [62]. Exogenous metabolites, such as food additives, pharmaceuticals, and exposomes, were removed from the finalized list.

5. Conclusions

This study has presented valuable information about the metabolic alterations in amino acids, lipids, and other molecules and the potentially induced oxidative events in the neonatal period of MSUD. Additionally, this study revealed its capability to measure the standard MSUD biomarker "alloisoleucine" and uncovered new metabolic biomarkers/pathways (methionine sulfoxide; LysoPI; ascorbate and aldarate metabolism, and pentose and glucuronate interconversions) that will be able to be used in its diagnosis/screening in the near further, after further validation studies. These future studies will strengthen our findings and help improve the accuracy of the current diagnostic tools, NBS and genetic tests, for better health outcomes and disease treatments.

Supplementary Materials: The supporting information can be downloaded at: https://www.mdpi.com/article/10.3390/ijms25115720/s1.

Author Contributions: Conceptualization, A.M.A.R., A.Z.A. and R.H.A.; methodology, R.H.A. and A.M.A.R.; software, A.Z.A., M.A.M. and R.H.A.; validation, R.S., M.J. and A.Z.A.; formal analysis, A.Z.A. and R.H.A. investigation, A.A. (Ahmad Alfares), A.A. (Ahamd Alodaib) and M.A.; data curation, A.M.A.R., A.Z.A. and R.H.A.; writing—original draft preparation, A.Z.A., R.S. and R.H.A.; writing—review and editing, A.M.A.R., R.S., A.Z.A., A.A. (Ahmad Alfares), A.A. (Ahamd Alodaib) and R.H.A.; visualization, M.J.; supervision, A.M.A.R. and M.A.; project administration, A.M.A.R.; funding acquisition, M.A. All authors have read and agreed to the published version of the manuscript.

Funding: The Deanship of Scientific Research, King Saud, funded this research through the Vice Deanship of Scientific Research Chair.

Institutional Review Board Statement: The study was conducted following the Declaration of Helsinki and approved by the Institutional Review Board reviewed and approved (RAC No. 2160 027) on 1 January 2016.

Informed Consent Statement: Informed consent was obtained from all subjects involved in the study.

Data Availability Statement: The raw data of this study were deposited at Metabolomics Workbench on 18 July 2023 and can be accessed under accession number ST002750.

Acknowledgments: The authors extend their appreciation to the Deanship of Scientific Research, King Saud University, for funding through the Vice Deanship of Scientific Research Chairs and the Genome Research Chair.

Conflicts of Interest: The authors declare no conflicts of interest.

References

1. Bharadwaj, A.; Wahi, N.; Saxena, A. Occurrence of Inborn Errors of Metabolism in Newborns, Diagnosis and Prophylaxis. *Endocr. Metab. Immune Disord. Drug Targets* **2021**, *21*, 592–616. [CrossRef] [PubMed]
2. Guerrero, R.B.; Salazar, D.; Tanpaiboon, P. Laboratory diagnostic approaches in metabolic disorders. *Ann. Transl. Med.* **2018**, *6*, 470. [CrossRef] [PubMed]
3. IEMbase. Inborn Errors of Metabolism Knowledgebase. Available online: http://www.iembase.org/ (accessed on 2 December 2022).
4. Pontoizeau, C.; Simon-Sola, M.; Gaborit, C.; Nguyen, V.; Rotaru, I.; Tual, N.; Colella, P.; Girard, M.; Biferi, M.G.; Arnoux, J.B.; et al. Neonatal gene therapy achieves sustained disease rescue of maple syrup urine disease in mice. *Nat. Commun.* **2022**, *13*, 3278. [CrossRef] [PubMed]
5. Imtiaz, F.; Al-Mostafa, A.; Allam, R.; Ramzan, K.; Al-Tassan, N.; Tahir, A.I.; Al-Numair, N.S.; Al-Hamed, M.H.; Al-Hassnan, Z.; Al-Owain, M.; et al. Twenty novel mutations in BCKDHA, BCKDHB and DBT genes in a cohort of 52 Saudi Arabian patients with maple syrup urine disease. *Mol. Genet. Metab. Rep.* **2017**, *11*, 17–23. [CrossRef] [PubMed]
6. Puliyanda, D.P.; Harmon, W.E.; Peterschmitt, M.J.; Irons, M.; Somers, M.J.G. Utility of hemodialysis in maple syrup urine disease. *Pediatr. Nephrol.* **2002**, *17*, 239–242. [CrossRef]
7. Amaral, A.U.; Leipnitz, G.; Fernandes, C.G.; Seminotti, B.; Schuck, P.F.; Wajner, M. α-Ketoisocaproic acid and leucine provoke mitochondrial bioenergetic dysfunction in rat brain. *Brain Res.* **2010**, *1324*, 75–84. [CrossRef] [PubMed]
8. Amaral, A.U.; Wajner, M. Pathophysiology of maple syrup urine disease: Focus on the neurotoxic role of the accumulated branched-chain amino acids and branched-chain α-keto acids. *Neurochem. Int.* **2022**, *157*, 105360. [CrossRef] [PubMed]
9. Strauss, K.A.; Carson, V.J.; Soltys, K.; Young, M.E.; Bowser, L.E.; Puffenberger, E.G.; Brigatti, K.W.; Williams, K.B.; Robinson, D.L.; Hendrickson, C.; et al. Branched-chain α-ketoacid dehydrogenase deficiency (maple syrup urine disease): Treatment, biomarkers, and outcomes. *Mol. Genet. Metab.* **2020**, *129*, 193–206. [CrossRef]
10. Schadewaldt, P.; Bodner-Leidecker, A.; Hammen, H.W.; Wendel, U. Significance of L-alloisoleucine in plasma for diagnosis of maple syrup urine disease. *Clin. Chem.* **1999**, *45*, 1734–1740. [CrossRef]
11. Chen, T.; Lu, D.; Xu, F.; Ji, W.; Zhan, X.; Gao, X.; Qiu, W.; Zhang, H.; Liang, L.; Gu, X.; et al. Newborn screening of maple syrup urine disease and the effect of early diagnosis. *Clin. Chim. Acta* **2023**, *548*, 117483. [CrossRef]
12. Ghosh, A.; Schlecht, H.; Heptinstall, L.E.; Bassett, J.K.; Cartwright, E.; Bhaskar, S.S.; Urquhart, J.; Broomfield, A.; Morris, A.A.; Jameson, E.; et al. Diagnosing childhood-onset inborn errors of metabolism by next-generation sequencing. *Arch. Dis. Child.* **2017**, *102*, 1019–1029. [CrossRef]
13. Fingerhut, R. Recall rate and positive predictive value of MSUD screening is not influenced by hydroxyproline. *Eur. J. Pediatr.* **2009**, *168*, 599–604. [CrossRef] [PubMed]
14. Fang, X.; Zhu, X.; Feng, Y.; Bai, Y.; Zhao, X.; Liu, N.; Kong, X. Genetic analysis by targeted next-generation sequencing and novel variation identification of maple syrup urine disease in Chinese Han population. *Sci. Rep.* **2021**, *11*, 18939. [CrossRef] [PubMed]
15. Bodian, D.L.; Klein, E.; Iyer, R.K.; Wong, W.S.; Kothiyal, P.; Stauffer, D.; Huddleston, K.C.; Gaither, A.D.; Remsburg, I.; Khromykh, A.J.G.i.M. Utility of whole-genome sequencing for detection of newborn screening disorders in a population cohort of 1696 neonates. *Anesth. Analg.* **2016**, *18*, 221–230. [CrossRef] [PubMed]

16. Tarini, B.A.; Goldenberg, A.J. Ethical Issues with Newborn Screening in the Genomics Era. *Annu. Rev. Genom. Hum. Genet.* **2012**, *13*, 381–393. [CrossRef] [PubMed]
17. Wishart, D.S.; Feunang, Y.D.; Marcu, A.; Guo, A.C.; Liang, K.; Vázquez-Fresno, R.; Sajed, T.; Johnson, D.; Li, C.; Karu, N.; et al. HMDB 4.0: The human metabolome database for 2018. *Nucleic Acids Res.* **2018**, *46*, D608–D617. [CrossRef] [PubMed]
18. Thistlethwaite, L.R.; Li, X.; Burrage, L.C.; Riehle, K.; Hacia, J.G.; Braverman, N.; Wangler, M.F.; Miller, M.J.; Elsea, S.H.; Milosavljevic, A. Clinical diagnosis of metabolic disorders using untargeted metabolomic profiling and disease-specific networks learned from profiling data. *Sci. Rep.* **2022**, *12*, 6556. [CrossRef]
19. Coene, K.L.M.; Kluijtmans, L.A.J.; van der Heeft, E.; Engelke, U.F.H.; de Boer, S.; Hoegen, B.; Kwast, H.J.T.; van de Vorst, M.; Huigen, M.; Keularts, I.; et al. Next-generation metabolic screening: Targeted and untargeted metabolomics for the diagnosis of inborn errors of metabolism in individual patients. *J. Inherit. Metab. Dis.* **2018**, *41*, 337–353. [CrossRef] [PubMed]
20. Janeckova, H.; Kalivodova, A.; Najdekr, L.; Friedecký, D.; Hron, K.; Bruheim, P.; Adam, T. Untargeted metabolomic analysis of urine samples in the diagnosis of some inherited metabolic disorders. *Biomed. Pap.* **2015**, *159*, 582–585. [CrossRef]
21. Haijes, H.A.; Willemsen, M.; Van der Ham, M.; Gerrits, J.; Pras-Raves, M.L.; Prinsen, H.C.M.T.; Van Hasselt, P.M.; De Sain-van der Velden, M.G.M.; Verhoeven-Duif, N.M.; Jans, J.J.M. Direct Infusion Based Metabolomics Identifies Metabolic Disease in Patients' Dried Blood Spots and Plasma. *Metabolites* **2019**, *9*, 12. [CrossRef]
22. Adam, M.P.; Feldman, J.; Mirzaa, G.M.; Pagon, R.A.; Wallace, S.E.; Bean, L.J.H.; Gripp, K.W.; Amemiya, A. *GeneReviews(®)*; University of Washington: Seattle, WA, USA, 1993.
23. Staufner, C.; Haack, T.B.; Feyh, P.; Gramer, G.; Raga, D.E.; Terrile, C.; Sauer, S.; Okun, J.G.; Fang-Hoffmann, J.; Mayatepek, E.; et al. Genetic cause and prevalence of hydroxyprolinemia. *J. Inherit. Metab. Dis.* **2016**, *39*, 625–632. [CrossRef] [PubMed]
24. Korman, S.H.; Cohen, E.; Preminger, A. Pseudo-maple syrup urine disease due to maternal prenatal ingestion of fenugreek. *J. Paediatr. Child. Health* **2001**, *37*, 403–404. [CrossRef] [PubMed]
25. Burke, W.; Parens, E.; Chung, W.K.; Berger, S.M.; Appelbaum, P.S. The Challenge of Genetic Variants of Uncertain Clinical Significance: A Narrative Review. *Ann. Intern. Med.* **2022**, *175*, 994–1000. [CrossRef]
26. Sajeev, M.; Chin, S.; Ho, G.; Bennetts, B.; Sankaran, B.P.; Gutierrez, B.; Devanapalli, B.; Tolun, A.A.; Wiley, V.; Fletcher, J.; et al. Challenges in Diagnosing Intermediate Maple Syrup Urine Disease by Newborn Screening and Functional Validation of Genomic Results Imperative for Reproductive Family Planning. *Int. J. Neonatal Screen.* **2021**, *7*, 25. [CrossRef]
27. Abadingo, M.E.; Abacan, M.A.R.; Basas, J.R.U.; Padilla, C.D. Pregnancy in an adolescent with maple syrup urine disease: Case report. *Mol. Genet. Metab. Rep.* **2021**, *27*, 100745. [CrossRef]
28. Constantinou, M.A.; Papakonstantinou, E.; Benaki, D.; Spraul, M.; Shulpis, K.; Koupparis, M.A.; Mikros, E. Application of nuclear magnetic resonance spectroscopy combined with principal component analysis in detecting inborn errors of metabolism using blood spots: A metabonomic approach. *Anal. Chim. Acta* **2004**, *511*, 303–312. [CrossRef]
29. Douglas, T.D.; Newby, L.K.; Eckstrand, J.; Wixted, D.; Singh, R.H. Lipid changes in the metabolome of a single case study with maple syrup urine disease (MSUD) after five days of improved diet adherence of controlled branched-chain amino acids (BCAA). *Mol. Genet. Metab. Rep.* **2020**, *25*, 100651. [CrossRef]
30. Miller, M.J.; Kennedy, A.D.; Eckhart, A.D.; Burrage, L.C.; Wulff, J.E.; Miller, L.A.; Milburn, M.V.; Ryals, J.A.; Beaudet, A.L.; Sun, Q.; et al. Untargeted metabolomic analysis for the clinical screening of inborn errors of metabolism. *J. Inherit. Metab. Dis.* **2015**, *38*, 1029–1039. [CrossRef] [PubMed]
31. Minakata, S.; Manabe, S.; Inai, Y.; Ikezaki, M.; Nishitsuji, K.; Ito, Y.; Ihara, Y. Protein C-Mannosylation and C-Mannosyl Tryptophan in Chemical Biology and Medicine. *Molecules* **2021**, *26*, 5258. [CrossRef]
32. Ihara, Y.; Inai, Y.; Ikezaki, M.; Matsui, I.-S.L.; Manabe, S.; Ito, Y. C-Mannosylation: Modification on Tryptophan in Cellular Proteins. In *Glycoscience: Biology and Medicine*; Taniguchi, N., Endo, T., Hart, G.W., Seeberger, P.H., Wong, C.-H., Eds.; Springer Japan: Tokyo, Japan, 2015; pp. 1091–1099.
33. Morita, S.; Inai, Y.; Minakata, S.; Kishimoto, S.; Manabe, S.; Iwahashi, N.; Ino, K.; Ito, Y.; Akamizu, T.; Ihara, Y. Quantification of serum C-mannosyl tryptophan by novel assay to evaluate renal function and vascular complications in patients with type 2 diabetes. *Sci. Rep.* **2021**, *11*, 1946. [CrossRef]
34. Schnekenburger, M.; Diederich, M. Chapter 18-Nutritional Epigenetic Regulators in the Field of Cancer: New Avenues for Chemopreventive Approaches. In *Epigenetic Cancer Therapy*; Gray, S.G., Ed.; Academic Press: Boston, MA, USA, 2015; pp. 393–425.
35. Qin, D.; Lei, Y.; Xie, W.; Zheng, Q.; Peng, Z.; Liu, Y.; Dai, B.; Ma, T.; Wei, P.; Gao, C.; et al. Methionine sulfoxide suppresses adipogenic differentiation by regulating the mitogen-activated protein kinase signaling pathway. *Cell Biol. Int.* **2023**, *47*, 648–659. [CrossRef] [PubMed]
36. Bennet, S.; Kaufmann, M.; Takami, K.; Sjaarda, C.; Douchant, K.; Moslinger, E.; Wong, H.; Reed, D.E.; Ellis, A.K.; Vanner, S.; et al. Small-molecule metabolome identifies potential therapeutic targets against COVID-19. *Sci. Rep.* **2022**, *12*, 10029. [CrossRef] [PubMed]
37. Barschak, A.G.; Sitta, A.; Deon, M.; Busanello, E.N.; Coelho, D.M.; Cipriani, F.; Dutra-Filho, C.S.; Giugliani, R.; Wajner, M.; Vargas, C.R. Amino acids levels and lipid peroxidation in maple syrup urine disease patients. *Clin. Biochem.* **2009**, *42*, 462–466. [CrossRef] [PubMed]
38. Levine, R.L.; Mosoni, L.; Berlett, B.S.; Stadtman, E.R. Methionine residues as endogenous antioxidants in proteins. *Proc. Natl. Acad. Sci. USA* **1996**, *93*, 15036–15040. [CrossRef]

39. Jobgen, W.S.; Wu, G. L-Arginine increases AMPK phosphorylation and the oxidation of energy substrates in hepatocytes, skeletal muscle cells, and adipocytes. *Amino Acids* **2022**, *54*, 1553–1568. [CrossRef] [PubMed]
40. Dereymaeker, A.; Jansen, K.; Vervisch, J.; Naulaers, G.; Regal, L. P98–3040: Early diagnosis of maple syrup urine disease (MSUD) by means of neonatal EEG and MRI in a newborn with encephalopathy. *Eur. J. Paediatr. Neurol.* **2015**, *19*, S121–S122. [CrossRef]
41. Minamihata, T.; Takano, K.; Moriyama, M.; Nakamura, Y. Lysophosphatidylinositol, an Endogenous Ligand for G Protein-Coupled Receptor 55, Has Anti-inflammatory Effects in Cultured Microglia. *Inflammation* **2020**, *43*, 1971–1987. [CrossRef]
42. Alaamery, M.; Albesher, N.; Aljawini, N.; Alsuwailm, M.; Massadeh, S.; Wheeler, M.A.; Chao, C.C.; Quintana, F.J. Role of sphingolipid metabolism in neurodegeneration. *J. Neurochem.* **2021**, *158*, 25–35. [CrossRef]
43. Calzada, E.; Onguka, O.; Claypool, S.M. Phosphatidylethanolamine Metabolism in Health and Disease. *Int. Rev. Cell Mol. Biol.* **2016**, *321*, 29–88. [CrossRef]
44. Hajeyah, A.A.; Griffiths, W.J.; Wang, Y.; Finch, A.J.; O'Donnell, V.B. The Biosynthesis of Enzymatically Oxidized Lipids. *Front. Endocrinol.* **2020**, *11*, 591819. [CrossRef]
45. Varki, A.; Cummings, R.D.; Esko, J.D.; Stanley, P.; Hart, G.W.; Aebi, M.; Mohnen, D.; Kinoshita, T.; Packer, N.H.; Prestegard, J.H. *Essentials of Glycobiology [Internet]*; Cold Spring Harbor Laboratory Press: Cold Spring Harbor, NY, USA, 2022.
46. Fukuda, M.; Rutishauser, U.; Schnaar, R.L. *Neuroglycobiology*; Oxford University Press: New York, NY, USA, 2005.
47. de Chaves, E.P.; Sipione, S. Sphingolipids and gangliosides of the nervous system in membrane function and dysfunction. *FEBS Lett.* **2010**, *584*, 1748–1759. [CrossRef] [PubMed]
48. Svennerholm, L. Gangliosides—A new therapeutic agent against stroke and Alzheimer's disease. *Life Sci.* **1994**, *55*, 2125–2134. [CrossRef]
49. Dambrova, M.; Makrecka-Kuka, M.; Kuka, J.; Vilskersts, R.; Nordberg, D.; Attwood, M.M.; Smesny, S.; Sen, Z.D.; Guo, A.C.; Oler, E.; et al. Acylcarnitines: Nomenclature, Biomarkers, Therapeutic Potential, Drug Targets, and Clinical Trials. *Pharmacol. Rev.* **2022**, *74*, 506–551. [CrossRef]
50. Sitta, A.; Ribas, G.S.; Mescka, C.P.; Barschak, A.G.; Wajner, M.; Vargas, C.R. Neurological Damage in MSUD: The Role of Oxidative Stress. *Cell. Mol. Neurobiol.* **2014**, *34*, 157–165. [CrossRef] [PubMed]
51. Fairweather-Tait, S.J.; Bao, Y.; Broadley, M.R.; Collings, R.; Ford, D.; Hesketh, J.E.; Hurst, R. Selenium in human health and disease. *Antioxid. Redox Signal.* **2011**, *14*, 1337–1383. [CrossRef] [PubMed]
52. Goff, J.P. Invited review: Mineral absorption mechanisms, mineral interactions that affect acid–base and antioxidant status, and diet considerations to improve mineral status. *J. Dairy Sci.* **2018**, *101*, 2763–2813. [CrossRef]
53. Sharifi-Rad, M.; Anil Kumar, N.V.; Zucca, P.; Varoni, E.M.; Dini, L.; Panzarini, E.; Rajkovic, J.; Tsouh Fokou, P.V.; Azzini, E.; Peluso, I.; et al. Lifestyle, Oxidative Stress, and Antioxidants: Back and Forth in the Pathophysiology of Chronic Diseases. *Front. Physiol.* **2020**, *11*, 694. [CrossRef]
54. Chattoraj, A.; Liu, T.; Zhang, L.S.; Huang, Z.; Borjigin, J. Melatonin formation in mammals: In vivo perspectives. *Rev. Endocr. Metab. Disord.* **2009**, *10*, 237–243. [CrossRef]
55. Smirnoff, N. Ascorbic acid metabolism and functions: A comparison of plants and mammals. *Free Radic. Biol. Med.* **2018**, *122*, 116–129. [CrossRef]
56. Perreault, M.; Białek, A.; Trottier, J.; Verreault, M.; Caron, P.; Milkiewicz, P.; Barbier, O. Role of glucuronidation for hepatic detoxification and urinary elimination of toxic bile acids during biliary obstruction. *PLoS ONE* **2013**, *8*, e80994. [CrossRef]
57. Sebaa, R.; AlMalki, R.H.; Alseraty, W.; Abdel Rahman, A.M. A Distinctive Metabolomics Profile and Potential Biomarkers for Very Long Acylcarnitine Dehydrogenase Deficiency (VLCADD) Diagnosis in Newborns. *Metabolites* **2023**, *13*, 725. [CrossRef] [PubMed]
58. AlMalki, R.H.; Sebaa, R.; Al-Ansari, M.M.; Al-Alwan, M.; Alwehaibi, M.A.; Rahman, A.M.A. *E. coli* Secretome Metabolically Modulates MDA-MB-231 Breast Cancer Cells' Energy Metabolism. *Int. J. Mol. Sci.* **2023**, *24*, 4219. [CrossRef] [PubMed]
59. Pang, Z.; Chong, J.; Zhou, G.; de Lima Morais, D.A.; Chang, L.; Barrette, M.; Gauthier, C.; Jacques, P.-É.; Li, S.; Xia, J. MetaboAnalyst 5.0: Narrowing the gap between raw spectra and functional insights. *Nucleic Acids Res.* **2021**, *49*, W388–W396. [CrossRef] [PubMed]
60. Worley, B.; Powers, R. Multivariate Analysis in Metabolomics. *Curr. Metabolomics* **2013**, *1*, 92–107. [CrossRef] [PubMed]
61. Gu, X.; Al Dubayee, M.; Alshahrani, A.; Masood, A.; Benabdelkamel, H.; Zahra, M.; Li, L.; Abdel Rahman, A.M.; Aljada, A. Distinctive Metabolomics Patterns Associated with Insulin Resistance and Type 2 Diabetes Mellitus. *Front. Mol. Biosci.* **2020**, *7*, 609806. [CrossRef]
62. Wishart, D.S.; Guo, A.; Oler, E.; Wang, F.; Anjum, A.; Peters, H.; Dizon, R.; Sayeeda, Z.; Tian, S.; Lee, B.L.; et al. HMDB 5.0: The Human Metabolome Database for 2022. *Nucleic Acids Res.* **2021**, *50*, D622–D631. [CrossRef]

International Journal of *Molecular Sciences*

MDPI

Article

Gastric Cancer and Intestinal Metaplasia: Differential Metabolic Landscapes and New Pathways to Diagnosis

Seong Ji Choi [1,2,†], Hyuk Soon Choi [2,†], Hyunil Kim [3], Jae Min Lee [2], Seung Han Kim [2], Jai Hoon Yoon [1], Bora Keum [2], Hyo Jung Kim [2], Hoon Jai Chun [2,*] and Youngja H. Park [3,*]

[1] Department of Internal Medicine, Hanyang University College of Medicine, Seoul 04763, Republic of Korea; drcoolandy@gmail.com (S.J.C.); jaihoonyoon@hanyang.ac.kr (J.H.Y.)

[2] Division of Gastroenterology and Hepatology, Department of Internal Medicine, Korea University College of Medicine, Seoul 02841, Republic of Korea; mdkorea@gmail.com (H.S.C.); jmlee1202@gmail.com (J.M.L.); kimseunghan09@gmail.com (S.H.K.); borakeum@hanmail.net (B.K.); hjkimmd@korea.ac.kr (H.J.K.)

[3] EN BIO, Cheongju-si 28494, Republic of Korea; kimhyunil7071@gmail.com

* Correspondence: hjchan@korea.ac.kr (H.J.C.); youngja.hwang@en-bio.com (Y.H.P.); Tel.: +82-43-217-1357 (Y.H.P.)

† These authors equally contributed to this work as co-first.

Abstract: Gastric cancer (GC) is the fifth most common cause of cancer-related death worldwide. Early detection is crucial for improving survival rates and treatment outcomes. However, accurate GC-specific biomarkers remain unknown. This study aimed to identify the metabolic differences between intestinal metaplasia (IM) and GC to determine the pathways involved in GC. A metabolic analysis of IM and tissue samples from 37 patients with GC was conducted using ultra-performance liquid chromatography with tandem mass spectrometry. Overall, 665 and 278 significant features were identified in the aqueous and 278 organic phases, respectively, using false discovery rate analysis, which controls the expected proportion of false positives among the significant results. sPLS-DA revealed a clear separation between IM and GC samples. Steroid hormone biosynthesis, tryptophan metabolism, purine metabolism, and arginine and proline metabolism were the most significantly altered pathways. The intensity of 11 metabolites, including N1, N2-diacetylspermine, creatine riboside, and N-formylkynurenine, showed significant elevation in more advanced GC. Based on pathway enrichment analysis and cancer stage-specific alterations, we identified six potential candidates as diagnostic biomarkers: aldosterone, N-formylkynurenine, guanosine triphosphate, arginine, S-adenosylmethioninamine, and creatine riboside. These metabolic differences between IM and GC provide valuable insights into gastric carcinogenesis. Further validation is needed to develop noninvasive diagnostic tools and targeted therapies to improve the outcomes of patients with GC.

Keywords: gastric cancer; intestinal metaplasia; metabolomics; metabolite profiling; biomarker

Citation: Choi, S.J.; Choi, H.S.; Kim, H.; Lee, J.M.; Kim, S.H.; Yoon, J.H.; Keum, B.; Kim, H.J.; Chun, H.J.; Park, Y.H. Gastric Cancer and Intestinal Metaplasia: Differential Metabolic Landscapes and New Pathways to Diagnosis. *Int. J. Mol. Sci.* **2024**, *25*, 9509. https://doi.org/10.3390/ijms25179509

Academic Editor: Nam Deuk Kim

Received: 30 June 2024
Revised: 16 August 2024
Accepted: 19 August 2024
Published: 1 September 2024

1. Introduction

Gastric cancer (GC) is the fifth most common cancer type and the fifth leading cause of cancer-related deaths globally [1]. In the Republic of Korea, GC imposes a significant socioeconomic burden, particularly among individuals who do not participate in the national cancer screening program [2]. GC is more common in men; in developed countries, men are 2.2 times more likely to be diagnosed with GC than women. Owing to the asymptomatic nature of the disease, patients are often diagnosed at advanced stages when the disease becomes inoperable, either surgically or endoscopically. The 5-year survival rate varies widely depending on the cancer stage at diagnosis, ranging from less than 20% in late stages to >90% in early stages [3]. Endoscopy is currently the most reliable and widely used method for diagnosing GC. However, it has several disadvantages, including invasiveness, significant miss rates, risk of infection, and complications [4,5]. Miss rates are particularly high in patients with small-sized or non-elevated GC morphology [6].

Therefore, the early detection and prediction of GC are crucial for improving outcomes and preventing disease.

To address these challenges, various studies have focused on identifying effective diagnostic methods and developing novel biomarkers. Carbohydrate antigen family members (72-4, 19-9, and 125) and carcinoembryonic antigen (CEA) have been used as serum-based biomarkers for GC detection [7,8]. However, these markers are not exclusive to GC and are altered in other types of cancers [7,9]. The development of GC-specific biomarkers requires high disease specificity. Proteomic [10] and genomic approaches [11] have also been explored to identify GC; however, metabolites provide the most accurate presentation of the molecular phenotype. Metabolite-associated biomarkers have been identified in various cancer types such as breast, brain, ovarian, and prostate cancers and are currently used for diagnosis and prognosis prediction [12–15]. Prognostic biomarkers, including the neutrophil-to-lymphocyte ratio, have also been proposed, particularly in patients with advanced GC treated with Programmed Cell Death Protein 1 (PD-1) and Programmed Death-Ligand 1 (PD-L1) agents [16]. In this study, we aimed to identify diagnostic biomarkers specific to GC using metabolomics.

It is widely accepted that GC develops through a multistep cascade: gastritis, atrophy, intestinal metaplasia (IM), dysplasia, and cancer [17]. As IM is considered the most significant stage in GC development [18], the phenotypic transition from IM to GC is a key step in gastric carcinogenesis. Previous studies have applied metabolomics to assess patients with GC using gastric tissue, blood, urine, or gastric juice samples [19]. Among the studies using tissues, most compared GC tissue metabolites with those from normal tissue [20–22], non-cancerous [23], and chronic superficial gastritis tissues [24].

In this study, we aimed to identify the metabolic landscapes of IM and GC and discover diagnostic biomarkers for the progression from IM to GC by comparing their metabolic differences.

2. Results

2.1. Patient Characteristics

We enrolled 37 patients and collected two GC tissue samples and two corresponding IM tissue samples from each patient, investigating the aqueous and organic metabolites obtained from these tissues (Figure 1a). The cancer location, stage, pathological differentiation, and Lauren classification were used to obtain detailed information on the participants. Based on the American Joint Committee on Cancer Tumor-Node-Metastasis (TNM) system, 15 patients with stage I, 7 with stage II, 4 with stage III, and 11 with stage IV were included in this study. The history of *Helicobacter pylori* infection and family history of GC were also considered. Table 1 shows the characteristics of patients with GC.

Table 1. Clinical characteristics of the participants.

Variables	Total (n = 37)
Age, median	62.5 ± 13.9
Sex, n (%)	
Male	23 (62.2)
Female	14 (37.8)
Location, n (%)	
Upper	5 (13.5)
Middle	11 (29.7)
Lower	21 (56.8)
Stage, n (%)	
I	15 (40.5)
II	7 (18.9)
III	4 (10.8)
IV	11 (29.7)

Table 1. *Cont.*

Variables	Total (n = 37)
Pathologic differentiation	
Poorly	22 (59.5)
Moderately	13 (35.1)
Not applicable	2 (5.4)
Lauren classification	
Intestinal	17 (45.9)
Diffuse	12 (32.4)
Mixed	8 (21.6)
Current *Helicobacter pylori* infection	26 (70.3)
Family history of gastric cancer	5 (13.5)

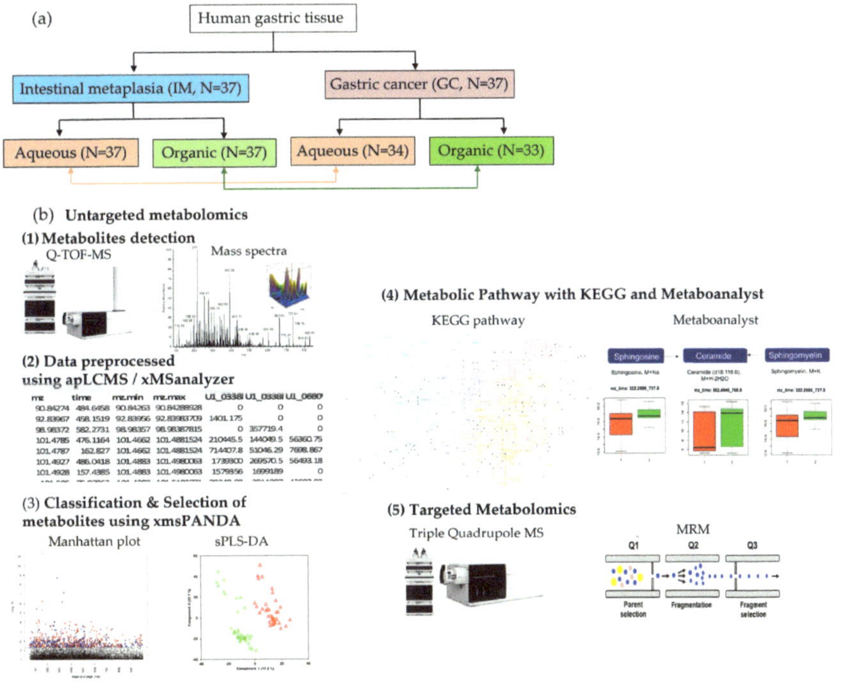

Figure 1. Schematic overview of the biomarker identification criteria applied in this study. (**a**) The metabolic overview started with the extraction of metabolites from 37 gastric cancer (GC) and intestinal metaplasia (IM) tissues. (**b**) The scheme of untargeted metabolomics analysis. (1) Metabolites detection was done using Q-TOF-MS, generating mass spectra. (2) Data preprocessed using apLCMS version 6.3.8 and xMSanalyzer version 2.0.6.1. (3) Metabolic profiling with Manhattan plot with false discovery rate (FDR) analysis and sparse partial least square discriminant analysis (sPLS-DA). (4) Pathway analysis with Kyoto Encyclopedia of Genes and Genomes (KEGG). (5) Biomarker quantification using multiple reaction monitoring (MRM). apLCMS: adaptive processing of liquid chromatography-mass spectrometry. FDR, false discovery rate; GC, gastric cancer; IM, intestinal metaplasia; KEGG, Kyoto Encyclopedia of Genes and Genomes; MS, mass spectrometery; Q-TOF-MS, quadrupole time-of-flight mass spectrometry; sPLS-DA: sparse partial least square discriminant analysis.

2.2. Analytical Performance Evaluation

A schematic overview of the biomarker identification criteria is shown in Figure 1b. Data were extracted using adaptive processing of liquid chromatography-mass spectrometry and subjected to the xMSanalyzer. To assess the performance of the analytical system, the processed data from the xMSanalyzer (24,591 features in the aqueous phase and 18,980 features in the organic phase) were input into xmsPANDA version 1.3.2. A total of 665 significant features from the aqueous phase and 278 features from the organic phase were selected based on a false discovery rate FDR q-value of $\leq$0.05 using the Limma R package. With significant features identified in the FDR analysis (q $\leq$ 0.05), principal component analysis (PCA) showed a clear separation between case samples (yellow and red) and pool samples (green), in both aqueous and organic phases (Figure 2).

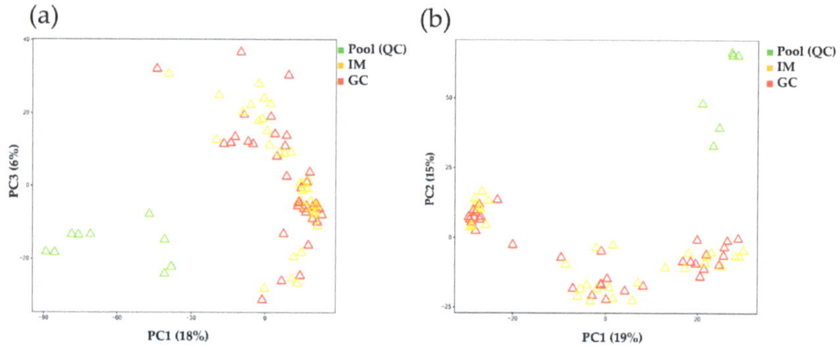

Figure 2. Analytical performance evaluation by comparing pool and case samples. (**a**) PCA using 665 significant features (FDR q $\leq$ 0.05) from aqueous extraction data. (**b**) PCA using 278 significant features (FDR q $\leq$ 0.05) from organic extraction data. FDR, false discovery rate; GC, gastric cancer; IM, intestinal metaplasia; PCA, principal component analysis; PC1, principal component 1; PC2, principal component 2; PC3, principal component 3; QC, quality control.

2.3. *Uni/Multivariate Observation of Metabolic Alterations between IM and GC*

2.3.1. Comparison of Aqueous Data

For a detailed investigation of metabolic differences caused by GC, the aqueous extracted data from the IM and GC groups were analyzed using Manhattan plot (FDR q $\leq$ 0.05), principal component analysis (PCA), and sparse partial least squares discriminant analysis (sPLS-DA). A total of 665 features (FDR q $\leq$ 0.05)) were selected as discriminatory features from the aqueous extract data using a *t*-test, and presented as colored dots in the Manhattan plot (Figure 3a). $-\log_{10}(p)$ values are shown on the y-axis, whereas the mass/charge ratio (m/z) are displayed on the x-axis (Figure 3a).

The dashed line (FDR q $\leq$ 0.05) separates the significant metabolites from the input features. Above the line, the blue dots represent the features that are highly expressed in GC, while the red dots represent the features that are poorly expressed in GC. Using 665 features (FDR q $\leq$ 0.05), hierarchical cluster analysis (HCA) did not clearly separate the two groups (Figure 3b). Therefore, to obtain clear separation, sPLS-DA was applied, which resulted in better separation compared with PCA (Figure 3c).

2.3.2. Comparison of Organic Data

Following the analyses of the aqueous data, a Manhattan plot, PCA, and sPLS-DA were employed to assess the metabolic differences in the organic data. In total, 278 features (FDR q $\leq$ 0.05), shown as colored dots in the Manhattan plot, were selected as discriminatory features from the organic extract data (Figure 3d). An explanation of the Manhattan plot was provided in the previous section. Using 278 features (FDR q $\leq$ 0.05), the two groups were not clearly separated by HCA (Figure 3e). Therefore, to obtain clear separation, sPLS-DA was applied, which resulted in better separation compared with PCA (Figure 3f).

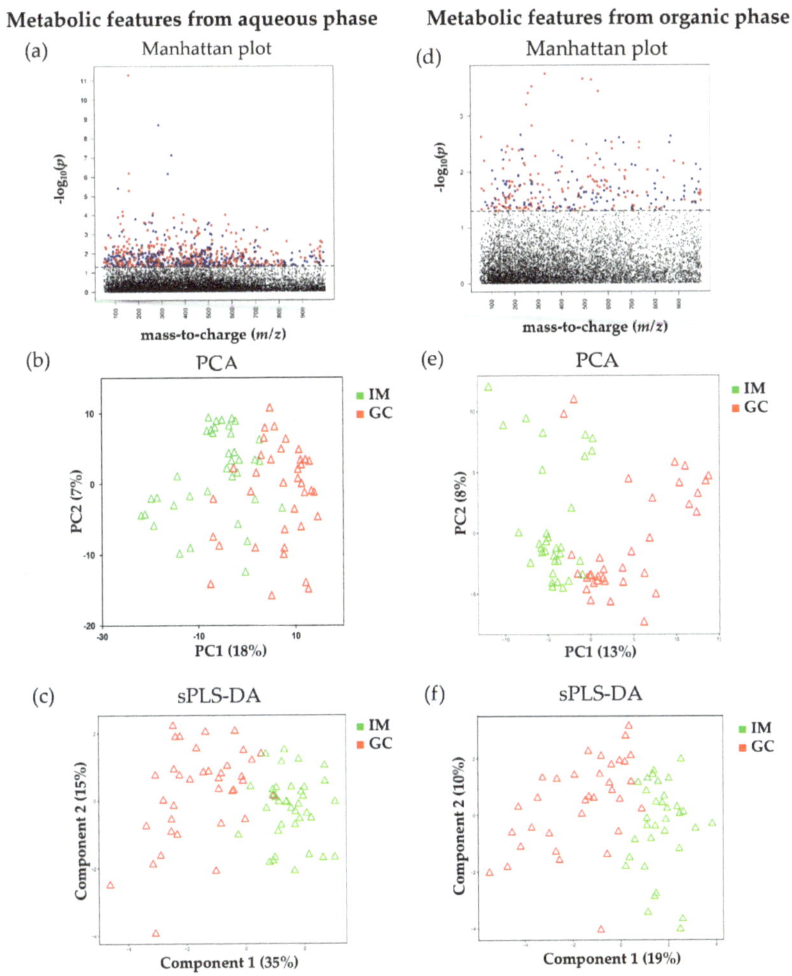

Figure 3. Manhattan plot, PCA, and sPLS-DA between IM and GC. The Manhattan plot presents the significant features (FDR q $\leq$ 0.05) as colored dots, while their distribution is expressed in m/z. (**a**) Manhattan plot showing 665 significant features (FDR q $\leq$ 0.05) derived from the aqueous data. (**d**) Manhattan plot with 278 significant features (FDR q $\leq$ 0.05) derived from the organic data. PCA shows the separation of samples: (**b**) aqueous data and (**e**) organic data. sPLS-DA shows the separation of samples: (**c**) aqueous data and (**f**) organic data. PCA, principal component analysis; PC1, principal component 1; PC2, principal component 2, sPLS-DA, sparse partial least squares discriminant analysis.

2.4. Pathway Enrichment Analysis between IM and GC

The 665 and 278 significant features (FDR q $\leq$ 0.05) identified from the comparison of aqueous data and organic data, respectively, were matched to the Kyoto Encyclopedia of Genes and Genomes (KEGG) IDs using xMSannotator version 1.3.2. The matched KEGG IDs that coped with the intensity table were input into the Metaboanalyst 5.0 (www.metaboanalyst.ca (accessed on 12 April 2024)) for pathway analysis. As a result, highly affected pathways, including pathway name, $-\log_{10}(p)$, pathway hits, and FDR, were determined (Figure 4), including steroid hormone biosynthesis [$-\log_{10}(p)$ = 8.13, impact = 0.24], tryptophan metabolism [$-\log_{10}(p)$ = 2.27, impact = 0.38], purine metabolism

[$-\log_{10}(p)$ = 1.56, impact = 0.21], and arginine and proline metabolism [$-\log_{10}(p)$ = 0.75, impact = 0.24].

(a)

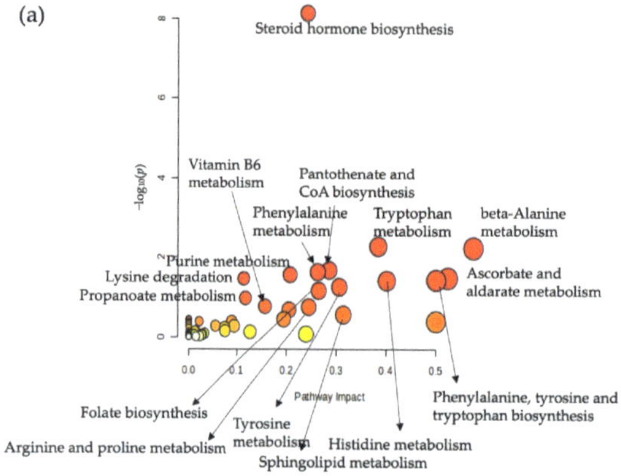

(b)

Color key ($-\log_{10}(p)$) Min Max — Size key (Impact) Min Max

Pathway Name	Match Status	p	$-\log_{10}(p)$	Impact
Steroid hormone biosynthesis	25/87	0.00	8.13	0.24
Tryptophan metabolism	9/41	0.01	2.27	0.38
beta-Alanine metabolism	6/21	0.01	2.23	0.57
Pantothenate and CoA biosynthesis	5/20	0.02	1.67	0.29
Phenylalanine metabolism	3/8	0.02	1.63	0.26
Purine metabolism	11/70	0.03	1.56	0.21
Ascorbate and aldarate metabolism	3/9	0.03	1.48	0.52
Lysine degradation	6/30	0.03	1.47	0.11
Phenylalanine, tyrosine and tryptophan biosynthesis	2/4	0.04	1.43	0.50
Histidine metabolism	4/16	0.04	1.41	0.40
Tyrosine metabolism	7/42	0.06	1.25	0.31
Folate biosynthesis	5/27	0.07	1.16	0.26
Propanoate metabolism	4/22	0.11	0.98	0.12
Vitamin B6 metabolism	2/9	0.17	0.77	0.16
Arginine and proline metabolism	5/36	0.18	0.75	0.24
Sphingolipid metabolism	4/32	0.28	0.28	0.31

Figure 4. Overview of the pathway analysis of the significant metabolites extracted from the combined aqueous and organic phases. (**a**) The bubble plot shows the pathways by impact (x-axis) and $-\log_{10}(p)$ (y-axis). The color and size of each bubble represent the $-\log_{10}(p)$ and impact, represented as color and size keys, respectively. (**b**) The top 16 pathways based on the $-\log_{10}(p)$ are listed alongside their match status, indicating the hit metabolites to whole metabolites involved in each pathway, *p*-value, and impact.

2.5. Relative Intensities of Metabolites Significantly Altered by GC

We used molecular tolerance within the range of 10 ppm to annotate the ion features, and fragmented the ion features to validate them with standard chemicals. Several significantly altered metabolites were identified in the significant pathways. In steroid hormone biosynthesis, metabolite alterations start with two metabolites: 21-deoxycortisol (*m/z*: 347.2185, $[M+H]^+$) and androst-4-ene-3,17-dione (Figure 5a and 5b, respectively). The levels of 21-deoxycortisol and metabolites derived from 21-deoxycortisol, including corticosterone (*m/z*: 347.2185, $[M+H]^+$), aldosterone (*m/z*: 361.2033, $[M+K]^+$), 11β, 21 dihydroxy-5β-pregnane-3,20-dione (*m/z*: 387.1921, $[M+K]^+$), and 21-dihydroxy-5β-pregnane-3,11,20-trione (*m/z*: 347.2185, $[M+H]^+$) were significantly lower in GC than in IM. The levels of metabolites derived from androst-4-ene-3,17-dione, including 11β-hydroxyandrost-4-ene-3,17-dione (*m/z*: 267.1751, $[M+H-2H_2O]^+$), adrenosterol (*m/z*: 283.1713, $[M+H-H_2O]^+$), etiocholan-3α-ol-17-one (*m/z*: 329.1866, $[M+K]^+$), 19-hydroxyandrost-4-ene-3,17-dione (*m/z*: 267.1751, $[M+H-2H_2O]^+$), 19-oxoandrost-4-ene-3,17-dione (*m/z*: 283.1713, $[M+H-2H_2O]^+$), and androsterone (*m/z*: 329.1866, $[M+K]^+$), were significantly altered by GC. Among these, the levels of adrenosterone and 19-oxoandrost-4-ene-3,17-dione were significantly higher in the GC group than in the IM group, while those of 11β, 21-dihydroxy-5β-pregnane-3,20-dione, etiocholan-3α-ol-17-one, 19-hydroxyandrost-4-ene-3,17-dione, and androsterone were significantly lower in the GC group than in the IM group.

In tryptophan metabolism (Figure 6), the levels of L-tryptophan (*m/z*: 205.0964, $[M+H]^+$) and its metabolic derivatives, including indole (*m/z*: 188.0702, $[M+H]^+$), indole acetaldehyde (*m/z*: 160.0750, $[M+H]^+$), N-formyl kynurenine (*m/z*: 254.1141, $[M+NH_4]^+$), indoleacetate (*m/z*: 170.0600, $[M+H]^+$), and L-kynurenine (*m/z*: 191.0820, $[M+H]^+$) were significantly altered. N-formylkynurenine was significantly higher in the GC group than in the IM group, while those of indole, L-tryptophan, indole acetaldehyde, indoleacetate, and L-kynurenine were significantly lower in the GC group than in the IM group.

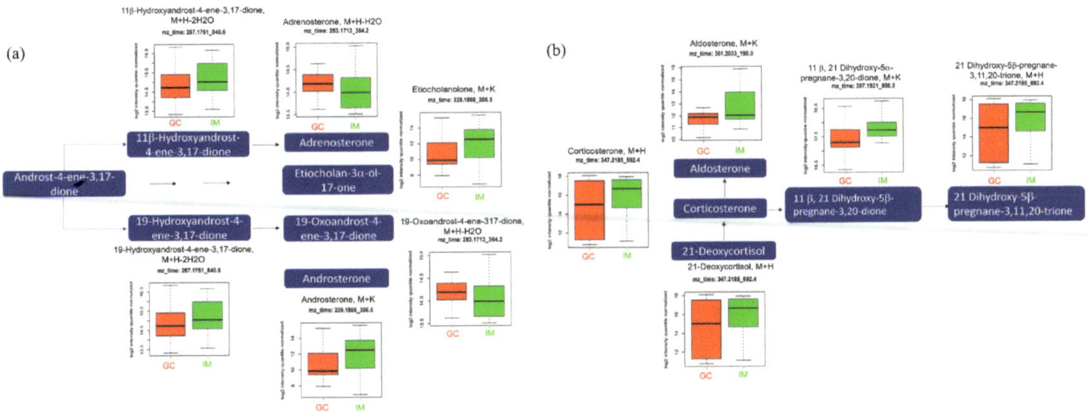

Figure 5. Analysis of significantly altered metabolites between GC and IM in the steroid hormone biosynthesis pathway from the KEGG pathway. (**a**) Pathway of androgen steroid and (**b**) pathway of mineralocorticoid. All metabolites in the figure a and b were significantly altered (FDR, q ≤ 0.05). Boxplots illustrate the upper quartile, median (dashed bar), and lower quartile, with whiskers indicating the maximum and minimum values. FDR, false discovery rate; GC, gastric cancer; IM, intestinal metaplasia; KEGG, Kyoto Encyclopedia of Genes and Genomes.

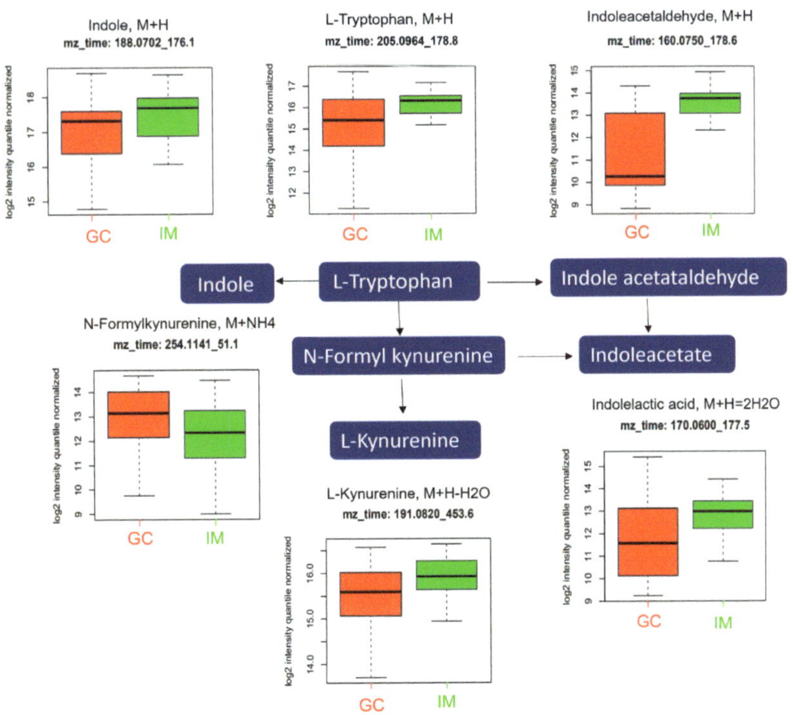

Figure 6. Analysis of significantly altered metabolites between GC and IM in the tryptophan metabolism pathway from the KEGG pathway. All metabolites in the figure were significantly altered (FDR, q ≤ 0.05). Boxplots illustrate the upper quartile, median (dashed bar), and lower quartile, with whiskers indicating the maximum and minimum values. FDR, false discovery rate; GC, gastric cancer; IM, intestinal metaplasia; KEGG, Kyoto Encyclopedia of Genes and Genomes.

In purine metabolism (Figure 7), the expression levels of guanosine 3′,5′-bis(diphosphate) (*m*/*z*: 437.3722, $[M+Na]^+$) and its metabolic derivatives (including guanosine triphosphate (GTP) (*m*/*z*: 523.9958, $[M+H]^+$), inosine (*m*/*z*: 307.0433, $[M+K]^+$), hypoxanthine (*m*/*z*: 119.0350, $[M+H-H_{tyl}0]^+$), guanosine (*m*/*z*: 322.0538, $[M+K]^+$), and guanine (*m*/*z*: 152.0580, $[M+H]^+$)) and 5′-phosphoribosyl-N-formylglycinamide (*m*/*z*: 279.0390, $[M+H-2H_20]^+$) and its metabolic derivatives (including 2-(formamido)-N1-(5′-phosphoribosyl)acetamidine (*m*/*z*: 296.0659, $[M+H-H_20]^+$) and aminoimidazole ribotide (*m*/*z*: 296.0659, $[M+H]^+$)) were significantly altered. The expression level of GTP was significantly higher in the GC group than in the IM group, while those of guanosine 3′,5′-bis(diphosphate), inosine, hypoxanthine, guanosine, guanine, 5′-Phosphoribosyl-N-formylglycinamide, 2-(formamido)-N1-(5′-phosphoribosyl)acetamidine, and aminoimidazole ribotide were significantly lower in the GC group than in the IM group.

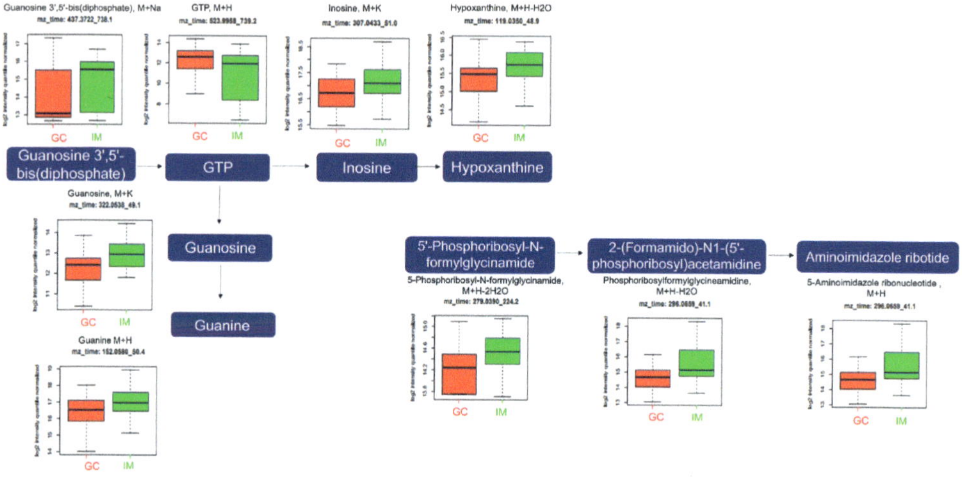

Figure 7. Analysis of significantly altered metabolites between GC and IM in the purine metabolism pathway from the KEGG pathway. All metabolites in the figure were significantly altered (FDR, $q \leq 0.05$). Boxplots illustrate the upper quartile, median (dashed bar), and lower quartile, with whiskers indicating the maximum and minimum values. FDR, false discovery rate; GC, gastric cancer; GTP, guanosine triphosphate; IM, intestinal metaplasia; KEGG, Kyoto Encyclopedia of Genes and Genomes.

In arginine and proline metabolism (Figure 8), the expression levels of S-adenosylmethioninamine (*m*/*z*: 150.0865, $[M+Na]^+$), N-acetylputrescine (*m*/*z*: 113.1071, $[M+H-H_2O]^+$), N_4-acetylaminobutanal (*m*/*z*: 152.0694, $[M+Na]^+$), and N-carbamoyl sarcosine (*m*/*z*: 150.0865, $[M+NH_4]^+$) were significantly higher in the GC group than in the IM group, while those of arginine (*m*/*z*: 175.1194, $[M+H]^+$), creatine (*m*/*z*: 132.0766, $[M+H]^+$), sarcosine (*m*/*z*: 90.0543, $[M+H]^+$), N-methylhydantoin (*m*/*z*:, $[M+NH_4]^+$), and creatinine (*m*/*z*: 136.0467, $[M+Na]^+$) were significantly lower in GC than in the GC group than in the IM group.

2.6. Alteration of Metabolite Intensities According to GC Stage

Figure 9 shows the significantly altered metabolite intensities according to stage. The significance of all metabolites was calculated by the analysis of variance (ANOVA) with a Tukey post hoc test ($p \leq 0.05$), using GraphPad Prism. The expression levels of 11 metabolites, including N1, N2-diacetylspermine (*m*/*z*: 287.2430, $[M+H]^+$), creatine riboside (*m*/*z*: 281.1446, $[M+NH_4]^+$), (3z,6z)-3,6-nonadienal (*m*/*z*: 161.0922, $[M+H]^+$), N-formyl kynurenine (*m*/*z*: 237.0870, $[M+H]^+$), S-adenosylmethioninamine (*m*/*z*: 339.1629, $[M+H-H_2O]^+$), methionyl-leucine (*m*/*z*: 263.1399, $[M+H]^+$), norepinephrine (*m*/*z*: 187.1073, $[M+NH_4]^+$),

S-(formylmethyl)glutathione (*m*/*z*: 357.1258, $[M+NH_4]^+$), methionylphenylalanine (*m*/*z*: 297.1287, $[M+H]^+$), oxindole-3-acetate (*m*/*z*: 192.0654, $[M+H]^+$), and N1-acetylspermine (*m*/*z*: 230.0954, $[M+H]^+$), significantly increased according to stage.

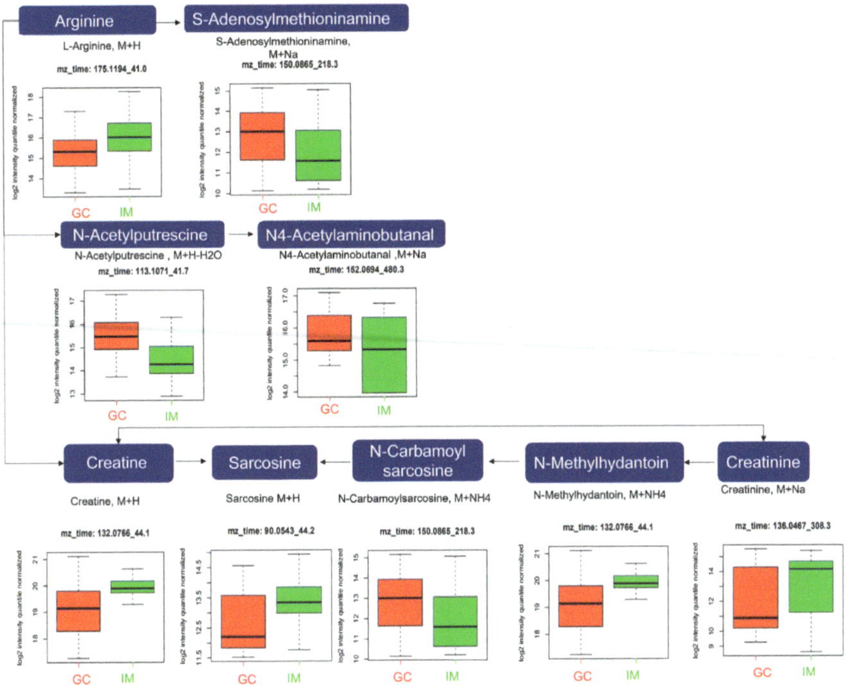

Figure 8. Analysis of significantly altered metabolites between GC and IM in the arginine and proline metabolism pathways from the KEGG pathway. All metabolites in the figure were significantly altered (FDR, $q \leq 0.05$). Boxplots illustrate the upper quartile, median (dashed bar), and lower quartile, with whiskers indicating the maximum and minimum values. FDR, false discovery rate; GC, gastric cancer; IM, intestinal metaplasia; KEGG, Kyoto Encyclopedia of Genes and Genomes.

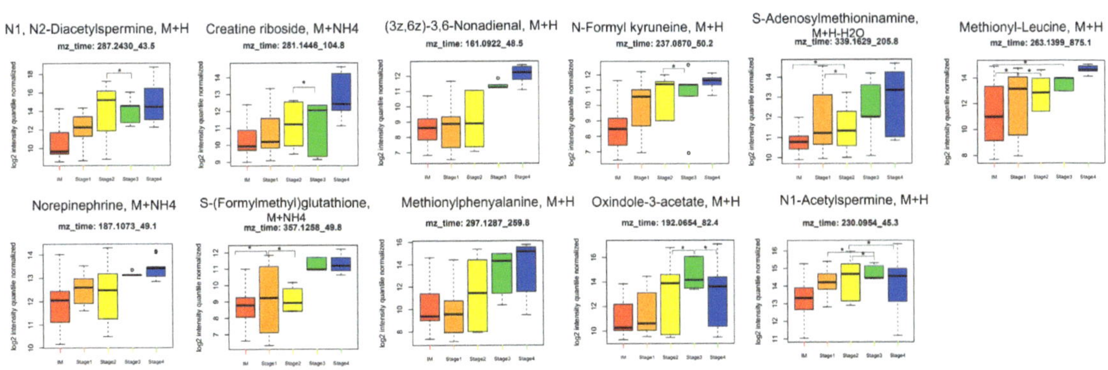

Figure 9. Relative metabolite intensities in tissues across various stages of GC at diagnosis. Boxplots illustrate the upper quartile, median (dashed bar), and lower quartile, with whiskers indicating the maximum and minimum values. Relative intensities of significant compounds correlated with four stages of GC. * $p \leq 0.05$. GC, gastric cancer.

3. Discussion

Most patients with GC are diagnosed at an advanced stage, resulting in poor prognosis and limited treatment options [1,25]. Current biomarkers for GC diagnosis and prognosis have low sensitivity and specificity [26]. Hence, most of the current diagnoses are based on invasive endoscopy. Thus, less invasive diagnostic tools and more specific biomarkers must be developed for the early detection of GC [27]. This study aimed to identify the metabolic alterations caused by GC. PLS-DA showed significant differences in the metabolic phenotypes during the progression of GC. Our analysis identified key metabolites, including aldosterone, N-formylkynurenine, GTP, arginine, and creatine riboside, which exhibited distinct alterations from high-impact pathways or stage-dependent analysis. These metabolites have potential as diagnostic biomarkers for GC, serving as candidates for the identification and prediction of the disease. To the best of our knowledge, this is the first metabolomic study to differentiate between GC and IM using tissue samples.

In this study, we extracted 943 significant metabolites from both aqueous and organic phases. Through pathway enrichment analysis and ANOVA tests with cancer stages, we aimed to identify potential candidates for discriminating GC from IM samples.

In the pathway analysis, the steroid hormone biosynthesis pathway had the highest number of hits. Accordingly, it has been hypothesized that the higher GC incidence rates in men than in women are due to the higher levels of sex steroid hormones [28]. Our results showed significant alterations in the levels of 11beta-ehydroxyandrost-4-ene-3,17-dione, adrenosterone, etiocholan-3α-ol-17-one 19-hydroxyandrost-4-ene-3,17-dione, 19-oxoandrost-4-ene-3,17-dione, and cortodoxone. The outcomes of this study regarding sex hormones showed mixed results; however, the levels of several metabolites from the corticosteroid pathway were consistently lower in GC patients. The level of aldosterone, which is included in the renin–angiotensin–aldosterone system (RAAS) as an adrenal component, was also lower in patients with GC in our study. The RAAS influences cell signaling, migration, death, and metastasis in cancer by balancing multiple receptor pathways [29]. Lee et al. reported that the use of RAAS inhibitors may be beneficial in gastrointestinal cancer prevention [30]. Busada et al. have suggested that glucocorticoids prevent gastric metaplasia by suppressing spontaneous inflammation [31]. They explored the role of glucocorticoids in the development of gastric inflammation and metaplasia. Our results suggest that glucocorticoids are essential for maintaining gastric homeostasis, and glucocorticoid deficiency may lead to GC development.

In tryptophan metabolism, the level of N-formylkynurenine was significantly higher in the GC group than in the IM group, while those of indole, L-tryptophan, indole acetaldehyde, indoleacetate, and L-kynurenine were significantly lower in the GC group than in the IM group. Tryptophan metabolism is altered in several cancer types [32], including GC [33]. L-tryptophan, an essential amino acid, has been studied in line with the kynurenine pathway, and its depletion is highly associated with cellular function and survival [34]. Tryptophan metabolism induces cancer progression via immunosuppressive responses; a recent study by Luo et al. concluded that tryptophan metabolism-associated genes could predict GC prognosis [34]. The level of indoleamine 2, 3-dioxygenase 1, which is a rate-limiting enzyme that converts tryptophan to kynurenine, is upregulated in multiple types of cancer, suggesting a possible role in carcinogenesis and as a potential biomarker [35]. In addition, the lower levels of indole, indole-acetaldehyde, and indoleacetate were observed in GC; indole metabolites could be formed in the stomach from tryptophan catabolism by the gut microbiota [36]. Moreover, several novel indole derivatives inhibit NEDDylation and MAPK pathways in GC cells [37].

Several metabolic alterations have been observed with purine derivatives. Cao et al. recently performed a pathway analysis using plasma metabolites and proposed that purine and arachidonic acid metabolism may play important roles in GC progression [38]. In our study, the GTP levels were significantly higher in patients with GC. Guanylate-binding protein 5, a member of the GTPase family, upregulates and promotes the proliferation and migration of GC cells, while the active GTP-bound form of Ras homolog family member

A promotes tumorigenesis [39]. In addition, guanine nucleotide-binding protein subunit beta-4 plays a crucial role in the initiation and progression of cancers, including GC [40]. Purine derivatives, including inosine, hypoxanthine, and 5-aminoimidazole ribotide, are precursors of uric acid that are reduced in hyperuricemia. Uric acid plays an important role in carcinogenesis owing to its pro- and antioxidant properties, and high levels of uric acid are associated with cancer [41]. Our results support this finding; however, the connection between them remains unclear.

Several metabolites involved in arginine and proline metabolism, including S-adenosylmethioninamine, N-acetylputrescine, N_4-acetylaminobutanal, N-carbamoyl sarcosine, arginine, creatine, sarcosine, N-methylhydantoin, and creatinine, were significantly altered in the GC group. Several studies have shown a correlation between abnormal amino acid metabolism and GC; arginine systemically decreased in GC, suggesting that it is a potential biomarker [42,43]. Creatine plays a key role in the recycling of adenosine triphosphate (ATP) [44], which is considered an energy currency for all cell behaviors, particularly in cancer cells, owing to its highly proliferative properties. In colorectal cancer cells, creatine is phosphorylated by creatine kinase, and phosphocreatine is utilized to provide energy for cell survival. Furthermore, creatine plays a role in macrophage polarization by inhibiting the M1-like phenotype and promoting the M2-like phenotype in macrophages [45]. The M2-like phenotype promotes metastasis and epithelial mesenchymal transition in GC cells [28]. Therefore, GC cells accelerate the physiological process leading to creatine depletion.

Several notable metabolites were identified through ANOVA testing with cancer stages, including N1, N2-diacetylspermine, creatine riboside, and N-formylkynurenine. Increasing levels of polyamines such as diacetylspermidine have been suggested as useful tumor markers for various cancer types, including GC [46]. Although not significant in the arginine and proline metabolic pathways in our study, creatine riboside is a cancer cell-derived metabolite that is associated with various cancer types, such as cervical, lung, and liver cancer, and has shown potential as a poor prognosis biomarker. This observation is consistent with the findings of the present study, in which the intensity of creatine riboside increased with increasing stage [47]. N-formylkynurenine was a significant metabolite in tryptophan metabolism between the GC and IM and showed higher intensity with increasing stage.

Figure 9 also shows that S-adenosylmethioninamine, methionyl-leucine, and S-(formylmethyl)glutathione stand out as promising candidates for the early detection of GC. S-adenosylmethioninamine is one of the significantly altered metabolites between GC and IM in the arginine and proline metabolism pathways from the KEGG pathway. S-adenosylmethioninamine, produced by the decarboxylation of S-adenosylmethionine, serves as a substrate necessary for the biosynthesis of polyamines such as spermine and spermidine. Polyamines are essential for the growth, differentiation, and development of eukaryotic cells. Therefore, an increase in S-adenosylmethioninamine levels promotes polyamine production, thereby promoting cancer progression [48]. S-(formylmethyl)glutathione is formed by the conjugation of bioactive compounds, for example, bromoacetaldehyde to glutathione, facilitated by glutathione S-(formylmethyl)transferase. Bromoacetaldehyde, a highly reactive compound, is derived from the carcinogen 1,2-dibromoethane [49]. Under conditions of glutathione deficiency, there is a possibility that the aldehyde intermediate, 2-bromoacetaldehyde, may interact with macromolecules, potentially contributing to cancer progression. When comparing IM and early GC, these differences were statistically significant (ANOVA $p \leq 0.05$). However, as GC progresses, these differences tend to become more pronounced. Therefore, metabolites that show significant differences in the late stage may also become significant biomarkers for the early stage when validated in a larger cohort. Thus, we aim to explore the biomarker potential of these metabolites through further validation studies, leveraging existing knowledge to guide our selection and exploration process.

There are several studies on metabolic profiling in gastric cancer (GC), but direct comparisons with our study are challenging due to differences in study design and control samples. Hirayama et al. [20] obtained tumor and surrounding grossly normal-appearing tissues from 12 stomach cancer patients after surgical treatment, while our study involved 37 patients who underwent esophagogastroduodenoscopy for evaluation and biopsy of both intestinal metaplasia (IM) and GC lesions. Despite the differences in control tissue samples, the metabolic profiling of amino acids, particularly tryptophan and arginine metabolism, was consistent. Other studies have used varied methodologies and control samples, complicating direct comparisons. For instance, one study focused on nuclear magnetic resonance spectroscopy to determine macromolecules mainly in urine with some gastric tissues [21], while another utilized gas chromatography/mass spectrometry metabonomics to fingerprint tumor tissues and matched normal mucosae, identifying significant lipid metabolites [22]. Additionally, research on gastric cardia cancer employed proteomics and metabolomics to explore different metabolites in cancerous and non-cancerous tissues [23]. Lastly, a study examining metabolites in tissues and plasma from GC patients, postoperative GC patients, and control patients with chronic superficial gastritis (CSG) focused on the balance profile of metabolites between the tumor microenvironment and the systemic environment, further complicating direct comparisons with this study [24]. These variations highlight the complexity of comparing metabolic profiling results across different studies and emphasize the need for careful consideration of study design and control selection.

Our study has several limitations. Despite efforts to control for factors that could influence the metabolomics results, tissue specificity may have affected the concentrations of molecules. The heterogeneity of the study group and composition of the cancer tissue could not be avoided. The small sample size and limited ethnic diversity, primarily Koreans, also posed significant limitations. Expanding our study to include diverse ethnicities, healthy controls and a larger sample size could enhance its potential to identify new biomarkers for GC. There were several significant metabolites that could not be discussed due to the lack of literature describing them. To overcome these limitations, we would combine the study results with the gene variations using data from The Cancer Genome Atlas to provide more accurate and meaningful results in a further study.

4. Materials and Methods

4.1. Study Design and Patient Enrollment

The study protocol was reviewed and approved by the Institutional Review Board of Korea University Anam Hospital, Seoul, Republic of Korea (16 July 2018; 2018AN0265). Written informed consent was obtained from all patients prior to their participation in this study. This study was registered in the open registry of the Clinical Research Information Service (http://cris.nih.go.kr; KCT0005597 (accessed on 15 August 2024)). Forty patients with pathologically confirmed GC were recruited from Korea University Anam Hospital between August 2018 and December 2019. Written informed consent was obtained from all participants in accordance with the Declaration of Helsinki. The inclusion criteria were as follows: (1) pathologically diagnosed with GC and (2) aged between 20 and 80 years. The exclusion criteria were as follows: (1) previously treated for GC, (2) previously treated for *Helicobacter pylori* infection, and (3) with inability to discontinue gastrointestinal medication. Data on demographics, medical history, family history, medications, and laboratory tests were obtained. Accordingly, two patients with inadequate biopsy specimens and one patient whose IM were not confirmed during the pathologic test were excluded from the enrolled cohort. All three patients refused to undergo a restudy. Hence, a total of 37 patients were analyzed.

4.2. Endoscopic Sampling

All 37 patients underwent esophagogastroduodenoscopy for the evaluation and biopsy of both IM and GC lesions. Endoscopy was performed between 8 a.m. and 10 a.m. after more than 8 h of fasting. Gastric mucosal biopsies were performed using standard gastroscopic forceps to obtain samples measuring 1–2 mm. In each patient, 4 biopsy samples were collected, 2 from each of the pathology-confirmed GC lesions and another 2 from the IM lesions, at least 5 cm away from the cancer site. Tissues obtained from odd-numbered orders were sent to the pathology department for pathological confirmation, and tissues obtained from even-numbered orders were immediately frozen in liquid nitrogen and stored at −80 °C for further metabolite analysis. The samples were re-evaluated pathologically to ensure that the biopsies were performed in lesion areas where histological changes indicative of cancer and IM were present.

4.3. Quality Control Samples

A standard quality control (QC) strategy was employed to ensure reproducibility and stability. Initially, a pooled (QC) sample of all reconstituted tissue extracts was loaded and run more than ten times to condition the column before analyzing the study samples, as described previously [50]. Separate runs were conducted for QC samples of aqueous and organic extracts.

4.4. Sample Treatment for Metabolite Extraction

Tissue samples (5–10 mg) were added to pre-chilled 50% methanol (MeOH) to obtain aqueous metabolites. The volume of the solution was adjusted according to the weight of the sample starting with a maximum weight of 30 μL/mg. Tissue was sonicated and centrifuged at 13,000× g for 20 min, at 4 °C. Supernatant aliquots were transferred to Eppendorf tubes. Samples were spun in a vacuum concentrator for 3 h at 45 °C until dry, and stored at −40 °C until analysis. The aqueous pellet was redissolved in 60 μL of solvent mixture of acetonitrile (ACN) and water in 95:5 ratio and transferred into a new Eppendorf tube after centrifugation for 20 min at 13,000× g at 4 °C.

Organic extracts were harvested from the residual pellets of the aqueous extract. Following aqueous extraction, a solution of pre-chilled dichloromethane/MeOH (3:1) was added to the residual pellet. The volume of the solution was proportional to the sample weight (as described in the previous paragraph; aqueous extraction). The samples were then centrifuged at 13,000× g for 20 min, and the organic phase supernatant was aliquoted into glass vials. The samples were allowed to evaporate overnight at room temperature in an extractor hood and stored at −40 °C until analysis. The dry residue was redissolved in 60 μL of water/ACN/isopropanol (1:1:2). The supernatant was centrifuged (5000× g, 4 °C, 10 min), as previously described [51].

4.5. Analysis of Metabolites by Liquid Chromatography with Tandem Mass Spectrometry

An ultra-performance liquid chromatography system (UPLC; Agilent 1260 Infinity Quaternary, Santa Clara, CA, USA) coupled with an Agilent liquid chromatography with tandem mass spectrometry (LC-MS/MS) was performed using a Q-TOF 6550 iFunnel Q-TOF mass spectrometer (Agilent) for metabolomic profiling. The samples were analyzed using C18 Synchronis aQ (1.9 μm, 100 × 2.1 mm; Thermo Fisher Scientific, Inc., Waltham, MA, USA). The autosampler and column temperatures were maintained at 10 °C and 45 °C, respectively. Solvent A comprised 0.1% formic acid in water (HPLC grade, Tedia, OH, USA), while solvent B comprised 0.1% formic acid in acetonitrile (HPLC grade, Tedia, OH, USA). The injection volume and flow rate were 5 μL/min and 0.4 mL/min, respectively. The HPLC gradient was programmed as follows: 95% water for 1 min, a linear decrease to 55% water over 8 min, a descending gradient to 10% water over 3 min, hold for 1.5 min, and a return to 95% water over 0.1 min. The electrospray ionization detector was operated with a curtain gas of 35 psi at 250 °C, supplied at 14 mL/min, and a sheath gas temperature of 250 °C, supplied at a flow rate of 11 mL/. Ion detection was set from 50 to 1000 with a resolution of 20,000 over 15 min. All samples were run in triplicate, and data for each ionization technique were acquired in the positive ion mode.

4.6. Alteration of Metabolite Intensities Caused by GC Corrected with Stage

Multivariate analysis was conducted using apLCMS within an ion range of 50 to 1000 based on the mass spectral data [52], the R package xMSanalyzer [53], and xmsPANDA [54,55] for aqueous and organic extracts. In the Manhattan plot, the y-axis represents the negative log of the *p*-values, while the x-axis represents m/z. Significant metabolites were selected using an FDR q-value of $\leq$0.05 with the Manhattan plot. Prior to model fitting, the features were normalized to the median, $\log_2$ transformed, and Pareto scaled values. Unsupervised PCA and supervised sPLS-DA were performed to visualize the metabolic differences between the IM and GC. In the sPLS-DA, the first component of the variable importance in projection (VIP) value > 1.5 from sPLS-DA was considered influential for the separation of samples. The pathway analysis was considered significant if five or more m/z values were changed. The annotation of m/z values was matched to metabolites from mass spectrometry databases, including xMSannotator, Human Metabolome Database, and KEGG [56–60]. MetaboAnalyst 5.0 software (www.metaboanalyst.ca) was used to determine the influence of these metabolites on the carcinogenic transformation from IM to GC. The number of pathway hits was calculated, and four or more hits were considered significant.

5. Conclusions

In this study, we successfully identified significant metabolic differences between IM and GC using a comprehensive metabolomic approach. The steroid hormone biosynthesis, tryptophan metabolism, purine metabolism, and arginine and proline metabolism were altered significantly. Aldosterone, N-formylkynurenine, GTP, arginine, S-adenosylmethioninamine, and creatine riboside showed distinct alterations, highlighting their potential as diagnostic biomarkers for GC. Additionally, the correlation of certain metabolite intensities with the stage of GC progression suggests their utility not only in early diagnosis but also in monitoring disease advancement.

These findings provide valuable insights into the metabolic alterations associated with gastric carcinogenesis and offer promising biomarkers for the early detection and prognosis of GC. While our current study utilized invasive endoscopic sampling to identify these key metabolites, subsequent studies should aim to validate these biomarkers using noninvasive methods such as blood or urine tests. Furthermore, validation in larger and more diverse cohorts is essential to explore their potential as noninvasive diagnostic tools. The identification and application of these biomarkers could improve patient outcomes through early detection and targeted therapeutic strategies.

Author Contributions: H.J.C. and Y.H.P.; Formal analysis, H.K.; Investigation, J.H.Y. and B.K.; Methodology, S.J.C., H.S.C., J.M.L., S.H.K. and H.J.K.; Supervision, H.J.C. and Y.H.P.; Writing—original draft, S.J.C. and H.K.; Writing—review and editing, Y.H.P. All authors have read and agreed to the published version of the manuscript.

Funding: This work was supported by the Basic Science Research Program through the National Research Foundation of Korea (NRF) funded by the Ministry of Education (2018R1D1A1B07048202), by the National Research Foundation of Korea (NRF) grant funded by the Korean government (MSIT) (NRF-2020R1A2C2103067, NRF-2021M3E5D1A01015177, and RS-2024-00341814), and by the research fund of Hanyang University (HY-202200000001107).

Institutional Review Board Statement: This study was conducted in accordance with the guidelines of the Declaration of Helsinki, reviewed and approved by the Korea University Anam Hospital (2018AN0265), and performed in accordance with the guidelines of the Ethical Committee of the Korea University Anam Hospital. The study was registered in the open registry of the Clinical Research Information Service (http://cris.nih.go.kr (accessed on 15 August 2024); KCT0005597).

Informed Consent Statement: Informed consent was obtained from all patients involved in this study.

Data Availability Statement: The data presented in this study are available upon request from the corresponding author.

Conflicts of Interest: The authors declare no conflicts of interest.

References

1. Bray, F.; Laversanne, M.; Sung, H.; Ferlay, J.; Siegel, R.L.; Soerjomataram, I.; Jemal, A. Global cancer statistics 2022: GLOBOCAN estimates of incidence and mortality worldwide for 36 cancers in 185 countries. *CA Cancer J. Clin.* **2024**, *74*, 229–263. [CrossRef] [PubMed]
2. Suh, Y.S.; Lee, J.; Woo, H.; Shin, D.; Kong, S.H.; Lee, H.J.; Shin, A.; Yang, H.K. National cancer screening program for gastric cancer in Korea: Nationwide treatment benefit and cost. *Cancer* **2020**, *126*, 1929–1939. [CrossRef]
3. Park, Y.H.; Kim, N. Review of atrophic gastritis and intestinal metaplasia as a premalignant lesion of gastric cancer. *J. Cancer Prev.* **2015**, *20*, 25–40. [CrossRef]
4. ASGE Standards of Practice Committee; Ben-Menachem, T.; Decker, G.A.; Early, D.S.; Evans, J.; Fanelli, R.D.; Fisher, D.A.; Fisher, L.; Fukami, N.; Hwang, J.H.; et al. Adverse events of upper GI endoscopy. *Gastrointest. Endosc.* **2012**, *76*, 707–718. [CrossRef]
5. Menon, S.; Trudgill, N. How commonly is upper gastrointestinal cancer missed at endoscopy? A meta-analysis. *Endosc. Int. Open* **2014**, *2*, E46–E50. [CrossRef]
6. Kim, S.J.; Choi, C.W. Common Locations of Gastric Cancer: Review of Research from the Endoscopic Submucosal Dissection Era. *J. Korean Med. Sci.* **2019**, *34*, e231. [CrossRef]
7. Emoto, S.; Ishigami, H.; Yamashita, H.; Yamaguchi, H.; Kaisaki, S.; Kitayama, J. Clinical significance of CA125 and CA72-4 in gastric cancer with peritoneal dissemination. *Gastric Cancer* **2012**, *15*, 154–161. [CrossRef]
8. Li, Y.; Yang, Y.; Lu, M.; Shen, L. Predictive value of serum CEA, CA19-9 and CA72.4 in early diagnosis of recurrence after radical resection of gastric cancer. *Hepatogastroenterology* **2011**, *58*, 2166–2170. [CrossRef]
9. Gao, Y.; Wang, J.; Zhou, Y.; Sheng, S.; Qian, S.Y.; Huo, X. Evaluation of Serum CEA, CA19-9, CA72-4, CA125 and Ferritin as Diagnostic Markers and Factors of Clinical Parameters for Colorectal Cancer. *Sci. Rep.* **2018**, *8*, 2732. [CrossRef] [PubMed]
10. Wu, D.; Zhang, P.; Ma, J.; Xu, J.; Yang, L.; Xu, W.; Que, H.; Chen, M.; Xu, H. Serum biomarker panels for the diagnosis of gastric cancer. *Cancer Med.* **2019**, *8*, 1576–1583. [CrossRef]
11. Cho, J.; Ahn, S.; Son, D.S.; Kim, N.K.; Lee, K.W.; Kim, S.; Lee, J.; Park, S.H.; Park, J.O.; Kang, W.K.; et al. Bridging genomics and phenomics of gastric carcinoma. *Int. J. Cancer* **2019**, *145*, 2407–2417. [CrossRef] [PubMed]
12. Khan, A.; Choi, S.A.; Na, J.; Pamungkas, A.D.; Jung, K.J.; Jee, S.H.; Park, Y.H. Noninvasive Serum Metabolomic Profiling Reveals Elevated Kynurenine Pathway's Metabolites in Humans with Prostate Cancer. *J. Proteome Res.* **2019**, *18*, 1532–1541. [CrossRef] [PubMed]
13. Kos, Z.; Dabbs, D.J. Biomarker assessment and molecular testing for prognostication in breast cancer. *Histopathology* **2016**, *68*, 70–85. [CrossRef]
14. Pandey, R.; Caflisch, L.; Lodi, A.; Brenner, A.J.; Tiziani, S. Metabolomic signature of brain cancer. *Mol. Carcinog.* **2017**, *56*, 2355–2371. [CrossRef]
15. Zeleznik, O.A.; Eliassen, A.H.; Kraft, P.; Poole, E.M.; Rosner, B.A.; Jeanfavre, S.; Deik, A.A.; Bullock, K.; Hitchcock, D.S.; Avila-Pacheco, J.; et al. A Prospective Analysis of Circulating Plasma Metabolites Associated with Ovarian Cancer Risk. *Cancer Res.* **2020**, *80*, 1357–1367. [CrossRef] [PubMed]
16. Moschetta, M.; Uccello, M.; Kasenda, B.; Mak, G.; McClelland, A.; Boussios, S.; Forster, M.; Arkenau, H.T. Dynamics of Neutrophils-to-Lymphocyte Ratio Predict Outcomes of PD-1/PD-L1 Blockade. *BioMed Res. Int.* **2017**, *2017*, 1506824. [CrossRef]
17. Correa, P.; Piazuelo, M.B. The gastric precancerous cascade. *J. Dig. Dis.* **2012**, *13*, 2–9. [CrossRef]
18. Yoon, H.; Kim, N. Diagnosis and management of high risk group for gastric cancer. *Gut Liver* **2015**, *9*, 5–17. [CrossRef]
19. Xiao, S.; Zhou, L. Gastric cancer: Metabolic and metabolomics perspectives (Review). *Int. J. Oncol.* **2017**, *51*, 5–17. [CrossRef]
20. Hirayama, A.; Kami, K.; Sugimoto, M.; Sugawara, M.; Toki, N.; Onozuka, H.; Kinoshita, T.; Saito, N.; Ochiai, A.; Tomita, M.; et al. Quantitative metabolome profiling of colon and stomach cancer microenvironment by capillary electrophoresis time-of-flight mass spectrometry. *Cancer Res.* **2009**, *69*, 4918–4925. [CrossRef]
21. Jung, J.; Jung, Y.; Bang, E.J.; Cho, S.I.; Jang, Y.J.; Kwak, J.M.; Ryu, D.H.; Park, S.; Hwang, G.S. Noninvasive diagnosis and evaluation of curative surgery for gastric cancer by using NMR-based metabolomic profiling. *Ann. Surg. Oncol.* **2014**, *21* (Suppl. S4), S736–S742. [CrossRef] [PubMed]
22. Song, H.; Wang, L.; Liu, H.L.; Wu, X.B.; Wang, H.S.; Liu, Z.H.; Li, Y.; Diao, D.C.; Chen, H.L.; Peng, J.S. Tissue metabolomic fingerprinting reveals metabolic disorders associated with human gastric cancer morbidity. *Oncol. Rep.* **2011**, *26*, 431–438. [CrossRef] [PubMed]
23. Cai, Z.; Zhao, J.S.; Li, J.J.; Peng, D.N.; Wang, X.Y.; Chen, T.L.; Qiu, Y.P.; Chen, P.P.; Li, W.J.; Xu, L.Y.; et al. A combined proteomics and metabolomics profiling of gastric cardia cancer reveals characteristic dysregulations in glucose metabolism. *Mol. Cell. Proteom.* **2010**, *9*, 2617–2628. [CrossRef]
24. Aa, J.; Yu, L.; Sun, M.; Liu, L.; Li, M.; Cao, B.; Shi, J.; Xu, J.; Cheng, L.; Zhou, J.; et al. Metabolic features of the tumor microenvironment of gastric cancer and the link to the systemic macroenvironment. *Metabolomics* **2012**, *8*, 164–173. [CrossRef]
25. Venerito, M.; Ford, A.C.; Rokkas, T.; Malfertheiner, P. Review: Prevention and management of gastric cancer. *Helicobacter* **2020**, *25* (Suppl. S1), e12740. [CrossRef] [PubMed]
26. Mi, L.; Ji, X.; Ji, J. Prognostic biomarker in advanced gastric cancer. *Transl. Gastrointest. Cancer* **2016**, *5*, 16–29.
27. Necula, L.; Matei, L.; Dragu, D.; Neagu, A.I.; Mambet, C.; Nedeianu, S.; Bleotu, C.; Diaconu, C.C.; Chivu-Economescu, M. Recent advances in gastric cancer early diagnosis. *World J. Gastroenterol.* **2019**, *25*, 2029–2044. [CrossRef]

28. Li, W.; Zhang, X.; Wu, F.; Zhou, Y.; Bao, Z.; Li, H.; Zheng, P.; Zhao, S. Gastric cancer-derived mesenchymal stromal cells trigger M2 macrophage polarization that promotes metastasis and EMT in gastric cancer. *Cell Death Dis.* **2019**, *10*, 918. [CrossRef]
29. Hassani, B.; Attar, Z.; Firouzabadi, N. The renin-angiotensin-aldosterone system (RAAS) signaling pathways and cancer: Foes versus allies. *Cancer Cell Int.* **2023**, *23*, 254. [CrossRef]
30. Lee, S.H.; Park, J.; Park, R.W.; Shin, S.J.; Kim, J.; Sung, J.D.; Kim, D.J.; Yang, K. Renin-Angiotensin-Aldosterone System Inhibitors and Risk of Cancer: A Population-Based Cohort Study Using a Common Data Model. *Diagnostics* **2022**, *12*, 263. [CrossRef]
31. Busada, J.T.; Ramamoorthy, S.; Cain, D.W.; Xu, X.; Cook, D.N.; Cidlowski, J.A. Endogenous glucocorticoids prevent gastric metaplasia by suppressing spontaneous inflammation. *J. Clin. Investig.* **2019**, *129*, 1345–1358. [CrossRef]
32. Brochez, L.; Chevolet, I.; Kruse, V. The rationale of indoleamine 2,3-dioxygenase inhibition for cancer therapy. *Eur. J. Cancer* **2017**, *76*, 167–182. [CrossRef]
33. Choi, J.M.; Park, W.S.; Song, K.Y.; Lee, H.J.; Jung, B.H. Development of simultaneous analysis of tryptophan metabolites in serum and gastric juice—An investigation towards establishing a biomarker test for gastric cancer diagnosis. *Biomed. Chromatogr.* **2016**, *30*, 1963–1974. [CrossRef] [PubMed]
34. Luo, P.; Chen, G.; Shi, Z.; Yang, J.; Wang, X.; Pan, J.; Zhu, L. Comprehensive multi-omics analysis of tryptophan metabolism-related gene expression signature to predict prognosis in gastric cancer. *Front. Pharmacol.* **2023**, *14*, 1267186. [CrossRef]
35. Zhai, L.; Ladomersky, E.; Lenzen, A.; Nguyen, B.; Patel, R.; Lauing, K.L.; Wu, M.; Wainwright, D.A. IDO1 in cancer: A Gemini of immune checkpoints. *Cell. Mol. Immunol.* **2018**, *15*, 447–457. [CrossRef] [PubMed]
36. Lamas, B.; Richard, M.L.; Leducq, V.; Pham, H.P.; Michel, M.L.; Da Costa, G.; Bridonneau, C.; Jegou, S.; Hoffmann, T.W.; Natividad, J.M.; et al. CARD9 impacts colitis by altering gut microbiota metabolism of tryptophan into aryl hydrocarbon receptor ligands. *Nat. Med.* **2016**, *22*, 598–605. [CrossRef]
37. Fu, D.J.; Cui, X.X.; Zhu, T.; Zhang, Y.B.; Hu, Y.Y.; Zhang, L.R.; Wang, S.H.; Zhang, S.Y. Discovery of novel indole derivatives that inhibit NEDDylation and MAPK pathways against gastric cancer MGC803 cells. *Bioorganic Chem.* **2021**, *107*, 104634. [CrossRef]
38. Cao, K.; Lyu, Y.; Chen, J.; He, C.; Lyu, X.; Zhang, Y.; Chen, L.; Jiang, Y.; Xiang, J.; Liu, B.; et al. Prognostic Implication of Plasma Metabolites in Gastric Cancer. *Int. J. Mol. Sci.* **2023**, *24*, 12774. [CrossRef]
39. Cao, F.Y.; Wang, C.H.; Li, X.; Ma, M.Z.; Tao, G.C.; Yang, C.; Li, K.; He, X.B.; Tong, S.L.; Zhao, Q.C.; et al. Guanylate binding protein 5 accelerates gastric cancer progression via the JAK1-STAT1/GBP5/CXCL8 positive feedback loop. *Am. J. Cancer Res.* **2023**, *13*, 1310–1328. [PubMed]
40. Gao, J.; Pan, H.; Zhu, Z.; Yu, T.; Huang, B.; Zhou, Y. Guanine nucleotide-binding protein subunit beta-4 promotes gastric cancer progression via activating Erk1/2. *Acta Biochim. Biophys. Sin.* **2020**, *52*, 975–987. [CrossRef]
41. Allegrini, S.; Garcia-Gil, M.; Pesi, R.; Camici, M.; Tozzi, M.G. The Good, the Bad and the New about Uric Acid in Cancer. *Cancers* **2022**, *14*, 4959. [CrossRef]
42. Miyagi, Y.; Higashiyama, M.; Gochi, A.; Akaike, M.; Ishikawa, T.; Miura, T.; Saruki, N.; Bando, E.; Kimura, H.; Imamura, F.; et al. Plasma free amino acid profiling of five types of cancer patients and its application for early detection. *PLoS ONE* **2011**, *6*, e24143. [CrossRef]
43. Bednarz-Misa, I.; Fleszar, M.G.; Fortuna, P.; Lewandowski, L.; Mierzchala-Pasierb, M.; Diakowska, D.; Krzystek-Korpacka, M. Altered L-Arginine Metabolic Pathways in Gastric Cancer: Potential Therapeutic Targets and Biomarkers. *Biomolecules* **2021**, *11*, 1086. [CrossRef]
44. Wallimann, T.; Tokarska-Schlattner, M.; Schlattner, U. The creatine kinase system and pleiotropic effects of creatine. *Amino Acids* **2011**, *40*, 1271–1296. [CrossRef]
45. Kazak, L.; Cohen, P. Creatine metabolism: Energy homeostasis, immunity and cancer biology. *Nat. Rev. Endocrinol.* **2020**, *16*, 421–436. [CrossRef] [PubMed]
46. Tatsuki, N.; Fumiaki, S.; Takashi, S.; Yoko, O.; Satoshi, Y.; Yuichiro, O.; Hideaki, O.; Kimihiko, F.; Hideaki, S. Urinary N1, N12-Diacetylspermine Level in the Patients with Various Cancer; A Pilot Study in Seven Types of Cancer. *Toho J. Med.* **2023**, *9*, 29–35.
47. Parker, A.L.; Toulabi, L.; Oike, T.; Kanke, Y.; Patel, D.; Tada, T.; Taylor, S.; Beck, J.A.; Bowman, E.; Reyzer, M.L.; et al. Creatine riboside is a cancer cell-derived metabolite associated with arginine auxotrophy. *J. Clin. Investig.* **2022**, *132*, e157410. [CrossRef] [PubMed]
48. Sari, I.N.; Setiawan, T.; Kim, K.S.; Wijaya, Y.T.; Cho, K.W.; Kwon, H.Y. Metabolism and function of polyamines in cancer progression. *Cancer Lett.* **2021**, *519*, 91–104. [CrossRef]
49. Allemang, A.; Lester, C.; Roth, T.; Pfuhler, S.; Peuschel, H.; Kosemund, K.; Mahony, C.; Bergeland, T.; O'Keeffe, L. Assessing the genotoxicity and carcinogenicity of 2-chloroethanol through structure activity relationships and in vitro testing approaches. *Food Chem. Toxicol.* **2022**, *168*, 113290. [CrossRef] [PubMed]
50. Want, E.J.; Masson, P.; Michopoulos, F.; Wilson, I.D.; Theodoridis, G.; Plumb, R.S.; Shockcor, J.; Loftus, N.; Holmes, E.; Nicholson, J.K. Global metabolic profiling of animal and human tissues via UPLC-MS. *Nat. Protoc.* **2013**, *8*, 17–32. [CrossRef]
51. Vorkas, P.A.; Isaac, G.; Anwar, M.A.; Davies, A.H.; Want, E.J.; Nicholson, J.K.; Holmes, E. Untargeted UPLC-MS profiling pipeline to expand tissue metabolome coverage: Application to cardiovascular disease. *Anal. Chem.* **2015**, *87*, 4184–4193. [CrossRef] [PubMed]
52. Yu, T.; Park, Y.; Johnson, J.M.; Jones, D.P. apLCMS—Adaptive processing of high-resolution LC/MS data. *Bioinformatics* **2009**, *25*, 1930–1936. [CrossRef] [PubMed]

53. Uppal, K.; Soltow, Q.A.; Strobel, F.H.; Pittard, W.S.; Gernert, K.M.; Yu, T.; Jones, D.P. xMSanalyzer: Automated pipeline for improved feature detection and downstream analysis of large-scale, non-targeted metabolomics data. *BMC Bioinform.* **2013**, *14*, 15. [CrossRef]
54. Uppal, K.; Soltow, Q.A.; Promislow, D.E.; Wachtman, L.M.; Quyyumi, A.A.; Jones, D.P. MetabNet: An R Package for Metabolic Association Analysis of High-Resolution Metabolomics Data. *Front. Bioeng. Biotechnol.* **2015**, *3*, 87. [CrossRef] [PubMed]
55. Uppal, K.; Walker, D.I.; Liu, K.; Li, S.; Go, Y.M.; Jones, D.P. Computational Metabolomics: A Framework for the Million Metabolome. *Chem. Res. Toxicol.* **2016**, *29*, 1956–1975. [CrossRef]
56. Smith, C.A.; O'Maille, G.; Want, E.J.; Qin, C.; Trauger, S.A.; Brandon, T.R.; Custodio, D.E.; Abagyan, R.; Siuzdak, G. METLIN: A metabolite mass spectral database. *Ther. Drug Monit.* **2005**, *27*, 747–751. [CrossRef] [PubMed]
57. Ogata, H.; Goto, S.; Fujibuchi, W.; Kanehisa, M. Computation with the KEGG pathway database. *Biosystems* **1998**, *47*, 119–128. [CrossRef]
58. Guijas, C.; Montenegro-Burke, J.R.; Domingo-Almenara, X.; Palermo, A.; Warth, B.; Hermann, G.; Koellensperger, G.; Huan, T.; Uritboonthai, W.; Aisporna, A.E.; et al. METLIN: A Technology Platform for Identifying Knowns and Unknowns. *Anal. Chem.* **2018**, *90*, 3156–3164. [CrossRef]
59. Wishart, D.S.; Feunang, Y.D.; Marcu, A.; Guo, A.C.; Liang, K.; Vazquez-Fresno, R.; Sajed, T.; Johnson, D.; Li, C.; Karu, N.; et al. HMDB 4.0: The human metabolome database for 2018. *Nucleic Acids Res.* **2018**, *46*, D608–D617. [CrossRef]
60. Kanehisa, M.; Furumichi, M.; Tanabe, M.; Sato, Y.; Morishima, K. KEGG: New perspectives on genomes, pathways, diseases and drugs. *Nucleic Acids Res.* **2016**, *45*, D353–D361. [CrossRef]

International Journal of
Molecular Sciences

MDPI

Review

The Role of MicroRNAs in Progressive Supranuclear Palsy—A Systematic Review

Aleksandra Ćwiklińska [1,†], Grzegorz Procyk [2,3,†], Dariusz Koziorowski [1] and Stanisław Szlufik [1,*]

[1] Department of Neurology, Faculty of Health Sciences, Medical University of Warsaw, 03-242 Warsaw, Poland; s079989@student.wum.edu.pl (A.Ć.); dariusz.koziorowski@wum.edu.pl (D.K.)
[2] 1st Chair and Department of Cardiology, Medical University of Warsaw, Banacha 1A, 02-097 Warsaw, Poland; grzegorz.procyk@wum.edu.pl
[3] Doctoral School, Medical University of Warsaw, 02-091 Warsaw, Poland
* Correspondence: stanislaw.szlufik@wum.edu.pl
† These authors contributed equally to this work and are both first co-authors (A.Ć. and G.P.).

Abstract: Progressive supranuclear palsy (PSP) is a rare, neurodegenerative movement disorder. Together with multiple system atrophy (MSA), Dementia with Lewy bodies (DLB), and corticobasal degeneration (CBD), PSP forms a group of atypical parkinsonisms. The latest diagnostic criteria, published in 2017 by the Movement Disorders Society, classify PSP diagnosis into defined, probable, and possible categories based on clinical examination. However, no single test is specific and sensitive for this disease. Microribonucleic acids (miRNAs) are promising molecules, particularly in the case of diseases that lack appropriate diagnostic and treatment tools, which supports exploring their role in PSP. We aimed to systematically review the current knowledge about the role of miRNAs in PSP. This study was registered in the Open Science Framework Registry, and the protocol is available online. Primary original studies, both clinical and preclinical, written in English and assessing miRNAs in PSP were included. Systematic reviews, meta-analyses, reviews, case reports, letters to editors, commentaries, conference abstracts, guidelines/statements, expert opinions, preprints, and book chapters were excluded. The following five databases were searched: Embase, Medline Ultimate, PubMed, Scopus, and Web of Science. Each database was last searched on 18 June 2024. Eventually, nine original studies relevant to the discussed area were included. The risk of bias was not assessed. The selected research suggests that miRNAs may be considered promising biomarkers in PSP. However, the exact involvement of miRNAs in the pathogenesis of PSP is still to be determined. Several microRNAs were found to be dysregulated in patients with PSP. This applies to both brain tissue and fluids like cerebrospinal fluid CSF or blood. Several miRNAs were found that could potentially be helpful in differentiating among PSP patients, PD patients, and healthy individuals. Although some correlations and alterations have already been found, this field requires much more research. MicroRNAs are exciting and promising small molecules, and their investigation into many diseases, including PSP, may lead to significant discoveries.

Keywords: progressive supranuclear palsy; microRNAs; biomarkers; atypical parkinsonian disorders

Citation: Ćwiklińska, A.; Procyk, G.; Koziorowski, D.; Szlufik, S. The Role of MicroRNAs in Progressive Supranuclear Palsy—A Systematic Review. *Int. J. Mol. Sci.* **2024**, *25*, 8243. https://doi.org/10.3390/ijms25158243

Academic Editor: Andrea Bernini

Received: 25 June 2024
Revised: 24 July 2024
Accepted: 26 July 2024
Published: 28 July 2024

1. Introduction

1.1. Progressive Nuclear Palsy Diagnostic Criteria

Progressive supranuclear palsy (PSP) is a rare, neurodegenerative movement disorder. Together with multiple system atrophy (MSA), Dementia with Lewy bodies (DLB), and corticobasal degeneration (CBD), PSP forms a group of atypical parkinsonisms [1]. Neurodegeneration is caused by intracellular amyloidogenic protein aggregation. Parkinson's Disease (PD), MSA, and DLB are referred to as synucleinopathies resulting from abnormal deposition of α-synuclein in brain cells, while PSP and CBD are referred to as tauopathies resulting from abnormal deposition of tau protein [2]. The first description of PSP was published in 1964. Even after 60 years, the diagnosis of this condition is still primarily based on clinical examination, with definitive confirmation achievable only through neuropathological examination of brain tissue, during which the aggregation of tau protein can be detected [3]. PSP is often underdiagnosed and frequently misdiagnosed as PD, especially in the early stages of the disease [4]. Despite some overlapping symptoms, PSP is characterized by distinct syndromes essential to distinguishing it from PD [5]. The latest diagnostic criteria, published in 2017 by the Movement Disorders Society, classify PSP diagnosis into defined, probable, and possible categories based on clinical examination [6]. PSP has many subtypes that differ in their clinical presentation, especially in the early stages [7]. The most common is PSP-Richardson syndrome, with vertical ocular motor dysfunction and the early beginning of postural instability [8]. Others include PSP-OM with an initial presentation in ocular motor dysfunction, PSP-P with a very similar clinical presentation to PD, PSP-PI with an early onset of postural instability, PSP-F with initial symptoms such as frontotemporal dementia, PSP-CBS with corticobasal syndrome, PSP-PGF with progressive gait freezing, and PSP-SL with an early manifestation of speech/language disorders [3,6,9–12]. The criteria primarily concern reaching high specificity and sensitivity in the diagnosis of PSP in general, and then also describe criteria for specific subtypes. The core diagnostic criteria include the most characteristic clinical symptoms of the disease including the following: ocular motor dysfunction, postural instability, akinesia, and cognitive dysfunction. The supportive features include levodopa resistance, a characteristic of atypical parkinsonism, hypokinesia, spastic dysarthria, dysphagia, and photophobia. Additionally, imaging tests can reveal midbrain atrophy [7]. In PSP, the grey matter of the brain is particularly affected. Studies describe grey matter reductions in subcortical and cortical areas in the frontal motor cortices, in the medial and lateral frontal cortices, and in the insula, striatum, thalamus, anterior cerebellum, and midbrain [13]. Degeneration of the midbrain can be found with MRI, which can help make the diagnosis [14].

1.2. Epidemiology

Diagnostic difficulties and low disease awareness make it difficult to create high-quality epidemiological data. The prevalence of this disease is approximately 7 per 100,000 people and increases with age [15]. Additionally, the average age of diagnosis is 72 years old with no gender predominance [16]. Because of the non-specific onset of the disease, diagnosis is usually delayed by 3–4 years from the onset of the first symptoms [6]. Most often, PSP is misdiagnosed as PD, Alzheimer's Disease (AD), cognitive disorders, or depression [15]. As a fatal, incurable condition, the median survival of PSP patients is estimated at 5–8 years and depends on the phenotype of the disease [17,18]. However, survival data may vary because of differences in the definition of survival, either from symptom onset to death or from date of diagnosis to death [19]. The only confirmed risk factor for the disease is age, while some environmental factors, such as exposure to metals or chemicals, stress, and hypertension, are described as additional potential risk factors [20].

1.3. Problems in Clinical Practice

Lack of knowledge about PSP among general practitioners leads to delays in diagnosis, which poses difficulties in precisely describing the course of the disease and potential premorbid or early-stage factors and biomarkers [21]. However, even if the disease is diagnosed early, no disease-modifying therapies are yet available [22]. Furthermore, symptomatic treatment in most cases, especially in advanced stages, is also insufficient. Among patients with PSP, the effect of using levodopa is weak; it slightly relieves symptoms and has no impact on disease duration. Treatment focuses on alleviating symptoms separately, e.g., toxinum botulinum is used for eyelid opening [23]. Deep brain stimulation (DBS), commonly used in PD, was tested in PSP patients and showed no benefits [24]. In addition, some clinical trials tested tau-targeted therapy, but all were in the early stages [25]. PSP-specific biomarkers and targeted, effective treatment need to be further explored.

1.4. MicroRNAs as Novel Biomarkers

Microribonucleic acids (miRNAs, miRs) are small non-coding RNA molecules consisting of about 22 nucleotides [26]. They play an essential role in the regulation of gene expression. They can bind to an mRNA molecule, inhibiting mRNA translation or even leading to mRNA degradation [27]. Since their discovery in 1993 [28], they have been eagerly investigated in various conditions. MiRNA expression levels are altered in multiple diseases, including cardiovascular diseases, e.g., myocarditis [29–31] or aortic stenosis [32], Duchenne Muscular Dystrophy [33], and, more importantly, also neurodegenerative diseases like Alzheimer's Disease, Parkinson's Disease, and triplet repeat disorders [34]. MiRNAs can be found in solid tissues and fluids, e.g., blood, cerebrospinal fluid, and saliva [35]. This makes them promising potential biomarkers that could serve as diagnostic tools, particularly useful in conditions for which diagnosis is made in advanced stages of the disease, such as in PSP. MicroRNAs were also reported to help distinguish among the severities of various diseases [36,37]. Moreover, they can potentially be used to identify good- and poor-responders before treatment implementation [38]. Whether alterations in miRNA levels are causative of the investigated diseases or are only the result of a given condition must be determined for each disease individually. If such a cause-and-effect relationship is established, it opens the door for novel therapies based on synthetic oligonucleotides [39]. MicroRNAs are promising molecules, particularly in the case of diseases that lack appropriate diagnostic and treatment tools, which supports exploring their role in PSP.

We aimed to systematically review the current knowledge about the role of miRNAs in PSP. We aimed to answer the following questions: (1) Can miRNAs differentiate patients with PSP from healthy people? (2) Can miRNAs differentiate patients with PSP from patients with other forms of atypical parkinsonism? (3) Can miRNAs differentiate patients with PSP from patients with Parkinson's disease? (4) Are miRNA levels altered in an animal model of progressive supranuclear palsy?

2. Methods

This systematic review was conducted according to the PRISMA 2020 Statement [40]. This study was registered in the Open Science Framework Registry, and the is protocol available online (DOI: https://doi.org/10.17605/OSF.IO/YSBQV).

Primary original studies, both clinical and preclinical, written in English and assessing miRNAs in PSP were included. Systematic reviews, meta-analyses, reviews, case reports, letters to editors, commentaries, conference abstracts, guidelines/statements, expert opinions, preprints, and book chapters were excluded. The following five databases were searched: Embase, Medline Ultimate, PubMed, Scopus, and Web of Science, by the following query: "("miRNA" OR "microRNA") AND ("PSP" OR "progressive supranuclear palsy" OR "progresive supranuclear palsy" OR "progressive supranucleal palsy" OR "progressive supranuclear palssy" OR "Steele–Richardson–Olszewski syndrome")", which yielded a total of 262 records. Each database was last searched on 18 June 2024. For each screening stage, we used 2 screeners working together (not independently). Final decisions in arguable cases were reached by a consensus between the 2 screeners. Data from each included work were extracted by a single extractor independently. In the case of any doubts, the issue was discussed with the other extractor.

The following data were extracted from each included paper: study characteristics: sample sizes, methodology used, year of publication, changes in miRNA levels among compared groups, and results of receiver operating characteristic (ROC) analysis (area under the ROC curve [AUC] with 95% CI). The risk of bias was not assessed.

After removing 121 duplicates, the remaining 141 records were screened by title and abstract. This yielded 16 records that met the inclusion/exclusion criteria. All 16 studies were retrieved in complete form. The complete data reports were evaluated for eligibility, excluding 7 studies as follows: 6 because of paper type (only abstract/poster as a post-conference material) and 1 because of inappropriateness. Eventually, 9 original studies relevant to the discussed area were included (Figure 1). The small number of included research papers allowed us to discuss and summarize all the existing knowledge in this field thoroughly.

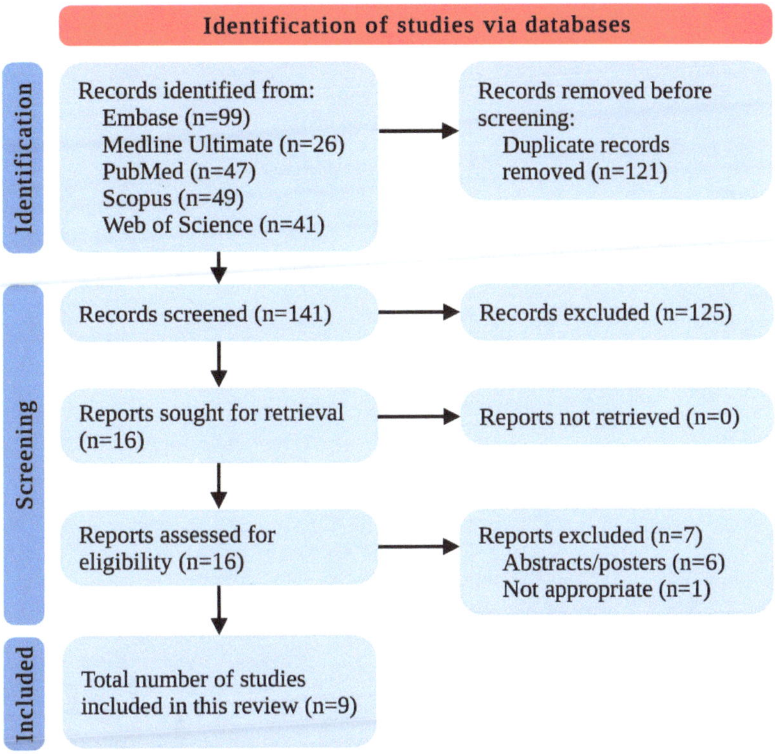

Figure 1. The flowchart for the selection process; n—number of studies.

We divided these studies into the following parts: (i) miRNA changes in a PSP animal model, (ii) miRNA changes in human brain tissues of PSP patients, and (iii) miRNA changes in the body fluids (CSF, blood/serum/plasma) of PSP patients.

3. Results and Discussion

3.1. miRNA Changes in a Progressive Supranuclear Palsy Animal Model

Lauretti et al. investigated miRNA changes in the mice model of tauopathy, which also encompasses PSP. They used wild-type C57BL/6 and hTau mice, which do not express endogenous mice tau but express all six isoforms of human tau protein. First, they compared the expression levels of several miRNAs in brain tissues (cortex, hippocampus, cerebellum) of hTau mice to their expression levels in wild-type mice at different time points. They found that miR-22-3p, miR-132-3p, miR-146a-5p, and miR-455-5p in the hippocampus and miR-132-3p and miR-146a-5p in the cerebellum were increased when assessed at the age of 12 months. At six months, miR-132-3p and miR-146a-5p in the cerebellum were also increased, but in the hippocampus, only miR-146a-5p was increased. At three months, there was no difference between hTau and wild-type mice. In addition, the authors assessed the time-related changes in hTau and wild-type mice. Exciting patterns were found in the cortex of wild-type mice; mir-22-3p, miR-132-3p, and miR-146a-5p increased over time, while no change was seen in hTau mice. In contrast, in the hippocampus of hTau mice, miR-132-3p and miR-146a-5p increased over time, while no change was found in wild-type mice. In the cerebellum of both hTau and wild-type mice, miR 22-3p decreased over time, while miR-146a decreased only in wild-type mice [41]. The studies discussed in this subsection with additional data are summarized in Table 1.

Table 1. Summary of recent studies regarding miRNA changes in a progressive supranuclear palsy animal model.

Ref.	Year	Population	Comparison	miRNA	Outcome	Methodology
[41]	2021	hTau mice (n = 3–8/group depending on the experiment)	wild-type C57BL/6J mice (n = 3–8/group depending on the experiment)	miR-22-3p, miR-132-3p, miR-146a-5p, miR-455-5p	12 months of age: ↑ miR-22-3p, miR-132-3p, miR-146a-5p, miR-455-5p in the hippocampus and ↑ miR-132-3p, miR-146a-5p in the the cerebellum of hTau mice 6 months of age: ↑ miR-146a-5p in the hippocampus and ↑ miR-132-3p, miR-146a-5p in the the cerebellum of hTau mice 3 months of age: no difference	miRNA in brain tissues (cortex, hippocampus, cerebellum) by qRT-PCR

↑—increased, miR/miRNA—microRNA, n—number of individuals, qRT-PCR—quantitative reverse transcription-polymerase chain reaction, ref.—reference, RNA—ribonucleic acid.

3.2. miRNA Changes in Human Brain Tissues of Progressive Supranuclear Palsy Patients

Smith et al. examined the following five miRNAs: miR-9, miR-124, miR-132, miR-137, and miR-153. MiRNA levels were measured in brain samples from PSP patients and healthy controls. Specifically, in the temporal, parietal, and prefrontal lobes. MiR-9, miR-124, miR-132, miR-137, and miR-153 increase the 3R-tau level, while miR-9, miR-132, and miR-137 decrease 4R-tau. The study confirmed that the 4R:3R-tau ratio was elevated in PSP patients compared with the healthy control group. MiR-132 expression was significantly decreased in the temporal lobes of PSP patients compared with healthy controls, and there were no differences for the other miRNAs. Moreover, the correlation between miR-132 levels and polypyrimidine tract binding protein 2 (PTB2) was examined. The results showed that miR-132 targets PTB2, which is increased in PSP patients and may lead to aberrant splicing of tau exon 10, increasing the risk of tauopathy [42].

Tatura et al. investigated miRNA in PSP in two steps. First, a microarray was performed to find dysregulated miRNA in the frontal lobe tissue of PSP patients compared to healthy controls. The test showed that four miRNAs were significantly dysregulated. MiR-147a and miR-518e were significantly upregulated in the PSP patients' forebrains, while miR-504 and miR525-3p were significantly decreased. Confirmation of the results was performed using quantitative reverse transcription-polymerase chain reaction (qRT-PCR), which showed significant changes only in the levels of miR-147a and miR-518e, with no significant changes in the other two. In addition, the expression of genes targeted by miR-147a—NF1, ACLY, and ALG12—and by miR-518e—CPEB1 and JAZF1—was also examined. The results showed repression of these genes in the frontal lobe tissue of PSP patients, which correlated with increased expression of miR-147a and miR-518e. The targeted genes play a role in various cellular functions, so researchers suggested that their downregulation may be associated with disease occurrence [43]. Both studies discussed in this subsection, along with additional data, are summarized in Table 2.

Table 2. Summary of recent studies regarding miRNA changes in human brain tissues of progressive supranuclear palsy patients.

Ref.	Year	Population	Comparison	miRNA	Outcome	Methodology
[42]	2011	8 PSP pts	8 HCs	miR-9, miR-124, miR-132, miR-137, miR-153	↓ miR-132 in the temporal lobes of PSP patients compared with HCs	miRNA in brain tissues by qRT-PCR
[43]	2016	20 PSP pts	20 HCs	miR-147, miR-518e, miR-504, miR-525-3p	↑ miR-147a and miR-518e in the frontal lobes of PSP patients compared with HCs	miRNA in frontal lobe tissue by qRT-PCR

↑—increased, ↓—decreased, HCs—healthy controls, miR/miRNA—microRNA, PD—Parkinson's Disease, PSP—progressive supranuclear palsy, pts—patients, qRT-PCR—quantitative reverse transcription-polymerase chain reaction, ref.—reference, RNA—ribonucleic acid.

3.3. miRNA Changes in the Body Fluids (Cerebrospinal Fluid, Blood/Serum/Plasma) of Progressive Supranuclear Patients

Starhof et al. evaluated the differences in the expression of 46 various miRNAs between PD patients with two atypical parkinsonism diseases—MSA and PSP—and a healthy control group. They measured miRNA expression levels in CSF and plasma. The results of the CSF examination showed that miR-106b-5p could differentiate between PD and PSP effectively. Similarly, miR-218-5p measured in plasma separated PSP and PD with good efficacy. However, no similarities were found between the results obtained in plasma and CSF. Furthermore, the relationship between α-synuclein and miRNA levels was examined, but no significant correlation was found [44].

Manna et al. evaluated the expression levels of several miRNAs in cohorts of PSP and PD patients and healthy controls. They found that miR-22-3p and miR-425-5p were upregulated in PSP patients compared with healthy controls, but this difference appeared non-significant after appropriate adjustments. However, the most efficient miRNA profile (miR-425-5p, miR-21-3p, miR-223-5p, miR-22-3p, miR-29a-3p, miR-483-5p) presented an AUC of 0.90 in differentiating between PSP patients and healthy controls. Furthermore, the authors found that miR-425-5p, miR-21-3p, and miR-199a-5p decreased in PSP patients compared with PD patients, and the differences remained significant even after adjustments. ROC analysis of these three miRNAs showed an AUC of 0.86 in discriminating between PSP and PD patients. Additionally, the best miRNA profile (consisting of miR-21-3p, miR-199a-5p, miR-425-5p, miR-483-5p, miR-22-3p, and miR-29a-3p) had an AUC of 0.91 in this setting. The authors analyzed the potential pathways of dysregulated miRNAs, and they found that the pathways most involved were fatty acid biosynthesis, ECM-receptor interaction, fatty acid metabolism, and the Hippo signaling pathway [45].

Nonaka et al. undertook an interesting study that included PSP patients and age- and sex-matched controls. They assessed miRNA levels in CSF, which makes it potentially useful in a clinical setting. The authors used a microarray technique capable of determining 2632 various miRNAs, of which 1104 were detectable in samples from both groups. They found that 38 different miRNAs were increased, while miR-6840-5p was decreased in PSP patients compared with controls. Unfortunately, the authors did not validate their results using the qRT-PCR method. However, they selected the two most upregulated and the most downregulated miRNAs in early-stage PSP patients (which were defined as patients who underwent neurological evaluation within two years of symptom onset) and then compared the levels of these miRNAs among early-stage PSP, advanced-stage PSP, and controls. Upregulation of miR-204-3p and miR-873-3p and downregulation of miR-6840-5p were found in early-stage PSP patients compared with controls. Similarly, miR-204-3p and miR-873-3p were increased, and miR-6840-5p was increased in advanced-stage PSP patients compared with controls. None of these miRNAs had altered expression levels between early-stage and advanced-stage PSP patients [46]

Ramaswamy et al. conducted a multi-stage study investigating plasma miRNAs' potential role in PSP diagnosis. First, they performed miRNA profiling and found 28 dysregulated miRNAs in PSP patients; 23 were upregulated, while five were downregulated compared with healthy age-matched controls. Then, they evaluated miR-19b-3p, miR-33a-5p, miR-130b-3p, miR-136-3p, and miR-210-3p using the qPCR method. As expected, all the above miRNAs were upregulated in PSP patients compared to controls, consistent with the profiling results. The authors also studied the utility of these miRNAs in PSP diagnosis with ROC analysis. They found that miR-19b-3p, miR-33a-5p, miR-130b-3p, miR-136-3p, and miR-210-3p had an AUC of 0.7059, 0.8578, 0.7778, 0.7882, and 0.7810, respectively. They also assessed the diagnostic value of the combination of all five miRNAs and calculated an AUC of 0.7817 with a specificity of 66.67% and a sensitivity of 72.41%. Surprisingly, this multi-miRNA panel presented a lower AUC than, e.g., miR-33a-5p alone, which suggests not incorporating some of these miRNAs into such a panel. The authors also investigated the predicted target genes of the evaluated miRNAs, and they found 48 different pathways involved, of which 12 were targeted by at least two different miRNAs. Listing all the genes potentially targeted by these miRNAs is out of the scope of this review [47].

Simoes et al. investigated various non-coding RNAs as potential diagnostic tools in PSP patients. In addition to miRNAs, which we discuss below, they also measured piwi-interacting RNAs (piRs) and transfer RNAs. The authors measured miRNA concentrations in both serum and CSF. Nevertheless, the miRNAs that presented enough amplification to be analyzed differed in these two fluids. The only repeated one was piR-31068, but no correlation was found in its expression level between serum and CSF. Nonetheless, miR-92a-3p and miR-626 were shown to be downregulated in the serum of PSP patients compared with healthy controls. On the other hand, let-7a-5p was found to be upregulated in the CSF of PSP patients. The authors searched for potential target genes of dysregulated miRNAs. For miR-92a-3p, they identified the extracellular matrix–receptor interaction and regulation of the actin cytoskeleton as potentially involved pathways, while for miR-626, the epidermal growth factor receptor pathway, estrogen signaling pathway, and the phosphatidylinositol signaling system were identified. The potential targets for let-7a-5p were involved in the following pathways: cell cycle, lysine degradation, hippo signaling pathway, oocyte meiosis, extracellular matrix–receptor interaction, adherens junctions, and thyroid hormone signaling pathways [48].

Pavelka et al. conducted a study that included PSP patients, PD patients, and healthy controls. They performed miRNA profiling in whole blood samples with microarrays. They evaluated the differences in expression levels among the studied groups, and the potential diagnostic utility was assessed with ROC analysis. It was found that when compared with healthy controls, PSP patients presented upregulation of nine miRNAs, namely, miR-2115-5p, miR-4270, miR-505-3p, miR-769-5p, miR-3065-3p, miR-4638-5p, miR-197-3p, let-7d-3p, and miR-1225-5p, and downregulation of seven miRNAs, specifically, miR-4762-3p, miR-7975, miR-1233-5p, miR-6085, miR-125a-3p, miR-4465, and miR-564. Interestingly, the authors found no differences between PSP and PD patients in miRNA expression when adjusted for sex and age. Analysis with ROC revealed that miRNA panels were helpful in differentiating between PSP patients and healthy controls. In differentiating PSP and PD patients, miRNAs did not show high utility. However, miRNAs were potentially more useful in distinguishing between early-stage PSP and early-stage PD patients. In the analysis of predicted target genes, the most prominently affected pathway was the "BioCarta natural killer (NK) cell pathway" [49]. All studies discussed in this subsection with additional data are summarized in Table 3.

Table 3. Summary of recent studies regarding miRNA changes in body fluids (cerebrospinal fluid, blood/serum/plasma) of progressive supranuclear patients.

Ref.	Year	Population	Comparison	miRNA	Outcome	Methodology
[44]	2019	32 PSP pts	37 PD pts 29 MSA pts 23 HCS	46 various miRNAs	ROC analysis (PD vs. PSP) CSF: miR-106b-5p AUC 0.85 (95% CI: 0.757–0.945) plasma: miR-218-5p AUC 0.71 (95% CI: 0.594–0.826)	miRNA in CSF and plasma by qRT-PCR
[45]	2021	20 PSP pts	40 PD pts 33 HCs	miR-425-5p, miR-21-3p, miR-223-5p, miR-22-3p, let-7i-5p, miR-199a-5p, miR-29a-3p, miR-483-5p	↓ miR-425-5p, miR-21-3p, and miR-199a-5p in PSP pts compared with PD pts ROC analysis (PSP vs. PD) combination of miR-425-5p, miR-21-3p, and miR 199a-5p with AUC 0.86 (95% CI: 0.74-0.97) combination of miR-21-3p, miR-199a-5p, miR-425-5p, miR-483-5p, miR-22-3p, and miR-29a-3p with AUC 0.91 (95% CI: 0.82–1.00)	exosomal miRNA in serum by qRT-PCR
[46]	2022	11 PSP pts	8 age- and sex-matched controls	2632 various miRNAs	↑ 38 various miRNAs and ↓ miR-6840-5p in PSP pts compared with controls ↑ miR-204-3p and miR-873-3p and ↓ miR-6840-5p in early-stage PSP pts compared with controls ↑ miR-204-3p and miR-873-3p and ↓ miR-6840-5p in advanced-stage PSP pts compared with controls	miRNA in CSF by microarray analysis
[47]	2022	18 PSP pts	17 age-matched HCs	miR-19b-3p, miR-33a-5p, miR-130b-3p, miR-136-3p, miR-210-3p	↑ miR-19b-3p, miR-33a-5p, miR-130b-3p, miR-136-3p, miR-210-3p in PSP pts ROC analysis: miR-19b-3p AUC 0.7059 miR-33a-5p AUC 0.8578 miR-130b-3p AUC 0.7778 miR-136-3p AUC 0.7882 miR-210-3p AUC 0.7810 combination of all five miRNAs: AUC 0.7817 (95% CI: 0.7126–0.8508)	miRNA in plasma by qPCR
[48]	2022	31 PSP pts	20 age- and sex-matched HCs	let-7a-5p, let-7b-5p, let-7f-2-3p, miR-1-3p, miR-16-5p, miR-92a-3p, miR-148a-3p, miR-626, miR-3168	↓ miR-92a-3p, miR-626 in serum of PSP pts ↑ let-7a-5p in CSF of PSP pts	miRNA in CSF and serum by qRT-PCR
[49]	2024	35 PSP pts	367 PD pts 416 HCs	2549 various miRNAs	↑ miR-2115-5p, miR-4270, miR-505-3p, miR-769-5p, miR-3065-3p, miR-4638-5p, miR-197-3p, let-7d-3p, and miR-1225-5p and ↓ miR-4762-3p, miR-7975, miR-1233-5p, miR-6085, miR-125a-3p, miR-4465, and miR-564 in PSP pts compared with HCs no difference in miRNA expression between PSP and PD pts	miRNA in whole blood by microarray analysis

↑—increased, ↓—decreased, AUC—area under the ROC curve, CI—confidence interval, CSF—cerebrospinal fluid, HCs—healthy controls, miR/miRNA—microRNA, MSA—multiple system atrophy, qPCR—quantitative polymerase-chain-reaction, PD—Parkinson's Disease, PSP—progressive supranuclear palsy, pts—patients, qRT-PCR—quantitative reverse transcription-polymerase chain reaction, ref.—reference, RNA—ribonucleic acid, ROC—receiver operating characteristic.

4. Discussion

Multiple microRNAs were found to be dysregulated in patients with PSP. This applies to both brain tissue and fluids like CSF or blood. Nevertheless, in general, miRNAs found to be altered in PSP do not overlap among the studies. This can be explained by the intrinsic limitations of research on miRNAs. In most studies, the researchers first explore potentially altered miRNAs in a small cohort of patients using microarrays. Then, they choose several miRNAs whose concentrations are altered the most significantly and perform qRT-PCR validation of these miRNAs on a bigger cohort of patients. Potentially, all miRNAs could be assessed with qRT-PCR, but this would be unreasonable in terms of the costs and time needed to perform such a study. Therefore, miRNAs that enter the validation step may substantially differ among studies since the first step is most often performed on four to five patients, and the effect size of miRNA alteration may significantly differ among studies even if the direction is consistent. The potential solution to overcome this limitation is to include miRNAs known to be altered from previous studies in the validation step; however, it is not always performed.

Notably, the measurement methods differed significantly among the analyzed studies. Only a few studies used qRT-PCR as a validation tool, while the others utilized microarrays alone, making it challenging to draw unambiguous conclusions.

However, miR-132 is worth mentioning in this context since it was upregulated in the cerebellum and hippocampus of mice models and downregulated in the temporal lobe of PSP patients. The opposite alteration could have been caused by (1) different brain parts where miRNAs were measured, (2) the fact that it was a mice model of tauopathy and not of PSP per se, and (3) the possibility that the mice model does not represent PSP in humans ideally. Interestingly, miR-132 is also downregulated in the brain of patients suffering from Alzheimer's Disease, which is also a tauopathy like PSP [50].

As presented throughout this review, many studies not only compared the expression levels between PSP patients and healthy controls or PD patients, but they also performed an ROC analysis, which found several miRNAs that are potentially helpful in differentiating among PSP patients, healthy individuals, and PD patients. Nevertheless, we did not find a study that compared PSP patients with patients suffering from other forms of atypical parkinsonism. Although the study by Starhof et al. [43] included patients with MSA, the authors did not compare miRNA levels between PSP and MSA patients. Therefore, we cannot appropriately answer the question (2) Can miRNAs differentiate patients with PSP from patients with other forms of atypical parkinsonism?

Furthermore, although there are several subtypes of PSP, only a few studies presented exact patient characterizations with the number of each subtype. However, even in those studies, there was no analysis comparing the miRNA levels among various PSP subtypes.

5. Conclusions and Future Perspectives

The discussed studies suggest that miRNAs may be considered promising biomarkers in PSP. However, the exact involvement of miRNAs in the pathogenesis of PSP is still to be determined. Nevertheless, the observed alterations in miRNA concentrations, if validated by further research, may lead to the creation of a multi-biomarker panel. Including more miRNAs in such a panel would be associated with higher sensitivity and specificity but, on the other hand, it would lead to higher costs. Therefore, cost-effectiveness analysis would be needed to create a precise and affordable panel for use in daily clinical practice.

Although some correlations and alterations have already been found, this field requires much more research. Nevertheless, there are some difficulties, e.g., the fact that this condition is rare. For future research, we recommend validating results with qRT-PCR. It would be of great value to perform a study that includes healthy individuals from whom blood samples are collected at baseline and then correlate the baseline miRNA expression levels with PSP occurrence. This would require a prominent follow-up period of about 20-40 years. However, this could be performed using biobanks like the U.K. Biobank.

Another interesting research field is comparing various PSP subtypes. Finding molecular biomarkers could even lead to reclassification, which is currently based primarily on clinical presentation.

MicroRNAs are exciting and promising small molecules, and their investigation into many diseases, including PSP, may lead to significant discoveries. Figure 2 presents a comprehensive summary of the current knowledge about the role of miRNAs in PSP.

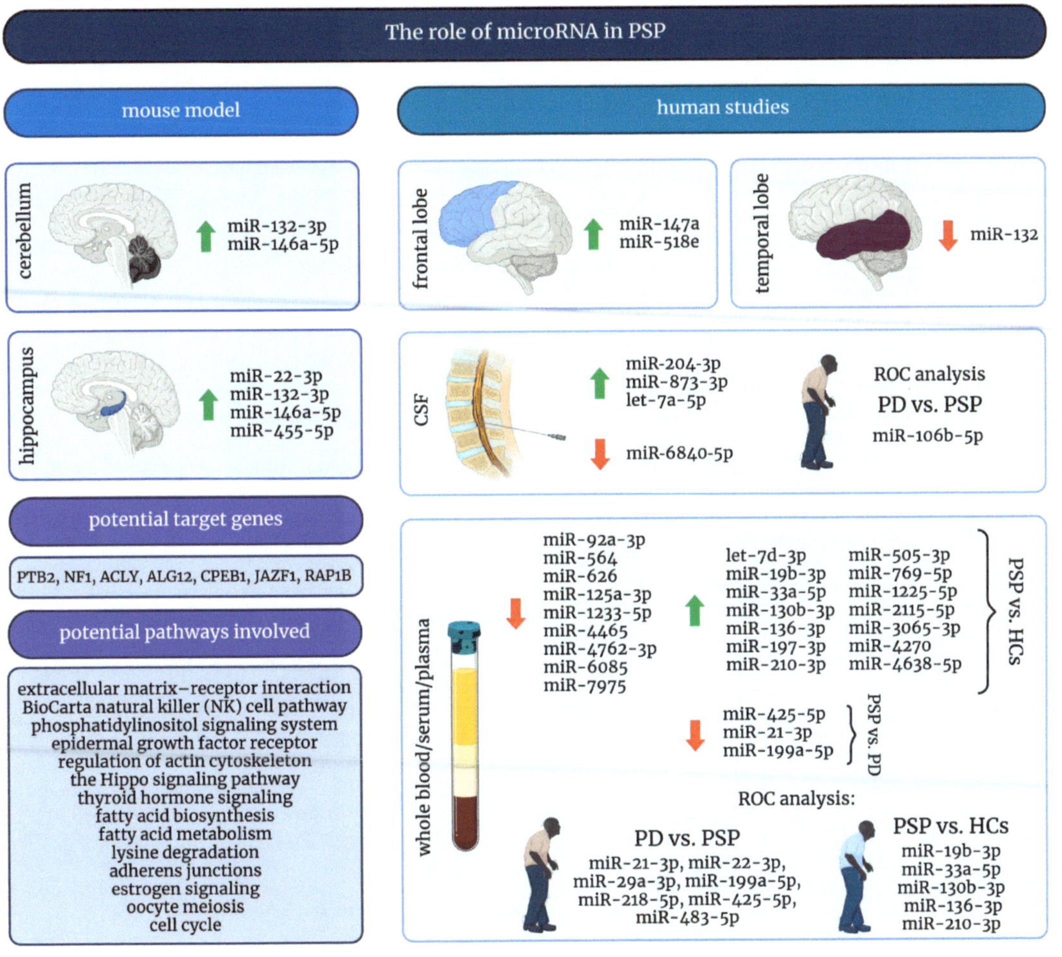

Figure 2. A graphical summary of the role of microRNAs in progressive supranuclear palsy. ↑—increased; ↓—decreased; HCs—healthy controls; miRNAs/miRs—micro-ribonucleic acids; PD—Parkinson's Disease; PSP—progressive supranuclear palsy; ROC—receiver operating characteristic.

6. Limitations

We must disclose a few limitations of this systematic review. First, this study reviewed only miRNAs and no other non-coding RNAs. Nevertheless, such a narrowing of the research topic allowed us to discuss it exhaustively within the space expected of a review paper. Secondly, this systematic review included different study types. However, this can be explained by a very limited amount of research in the field. Last, we did not assess the risk of bias, which may limit drawing firm conclusions. However, this might not be pivotal to such a small number of studies available.

Author Contributions: Conceptualization, G.P. and A.Ć.; writing—original draft preparation, A.Ć. and G.P.; writing—review and editing, G.P., D.K. and S.S.; visualization, G.P.; supervision, D.K. and S.S.; funding acquisition, A.Ć., G.P., D.K. and S.S. All authors have read and agreed to the published version of the manuscript.

Funding: This research received no external funding.

Data Availability Statement: All data will be made available from the authors upon request.

Acknowledgments: The figures were created with BioRender.com, under a licensed version of G.P.

Conflicts of Interest: The authors declare no conflicts of interest.

References

1. Bhatia, K.P.; Stamelou, M. Nonmotor Features in Atypical Parkinsonism. *Int. Rev. Neurobiol.* **2017**, *134*, 1285–1301. [CrossRef]
2. Levin, J.; Kurz, A.; Arzberger, T.; Giese, A.; Höglinger, G.U. The Differential Diagnosis and Treatment of Atypical Parkinsonism. *Dtsch Arztebl. Int.* **2016**, *113*, 61–69. [CrossRef] [PubMed]
3. Litvan, I.; Agid, Y.; Calne, D.; Campbell, G.; Dubois, B.; Duvoisin, R.C.; Goetz, C.G.; Golbe, L.I.; Grafman, J.; Growdon, J.H.; et al. Clinical research criteria for the diagnosis of progressive supranuclear palsy (Steele-Richardson-Olszewski syndrome): Report of the NINDS-SPSP international workshop. *Neurology* **1996**, *47*, 1–9. [CrossRef]
4. Kawashima, M.; Miyake, M.; Kusumi, M.; Adachi, Y.; Nakashima, K. Prevalence of progressive supranuclear palsy in Yonago, Japan. *Mov. Disord.* **2004**, *19*, 1239–1240. [CrossRef] [PubMed]
5. McFarland, N.R. Diagnostic Approach to Atypical Parkinsonian Syndromes. *Continuum* **2016**, *22*, 1117–1142. [CrossRef]
6. Höglinger, G.U.; Respondek, G.; Stamelou, M.; Kurz, C.; Josephs, K.A.; Lang, A.E.; Mollenhauer, B.; Müller, U.; Nilsson, C.; Whitwell, J.L.; et al. Clinical diagnosis of progressive supranuclear palsy: The movement disorder society criteria. *Mov. Disord.* **2017**, *32*, 853–864. [CrossRef]
7. Kato, N.; Arai, K.; Hattori, T. Study of the rostral midbrain atrophy in progressive supranuclear palsy. *J. Neurol. Sci.* **2003**, *210*, 57–60. [CrossRef] [PubMed]
8. Williams, D.R.; de Silva, R.; Paviour, D.C.; Pittman, A.; Watt, H.C.; Kilford, L.; Holton, J.L.; Revesz, T.; Lees, A.J. Characteristics of two distinct clinical phenotypes in pathologically proven progressive supranuclear palsy: Richardson's syndrome and PSP-parkinsonism. *Brain* **2005**, *128*, 1247–1258. [CrossRef]
9. Respondek, G.; Stamelou, M.; Kurz, C.; Ferguson, L.W.; Rajput, A.; Chiu, W.Z.; van Swieten, J.C.; Troakes, C.; Al Sarraj, S.; Gelpi, E.; et al. The phenotypic spectrum of progressive supranuclear palsy: A retrospective multicenter study of 100 definite cases. *Mov. Disord.* **2014**, *29*, 1758–1766. [CrossRef]
10. Birdi, S.; Rajput, A.H.; Fenton, M.; Donat, J.R.; Rozdilsky, B.; Robinson, C.; Macaulay, R.; George, D. Progressive supranuclear palsy diagnosis and confounding features: Report on 16 autopsied cases. *Mov. Disord.* **2002**, *17*, 1255–1264. [CrossRef]
11. Tsuboi, Y.; Josephs, K.A.; Boeve, B.F.; Litvan, I.; Caselli, R.J.; Caviness, J.N.; Uitti, R.J.; Bott, A.D.; Dickson, D.W. Increased tau burden in the cortices of progressive supranuclear palsy presenting with corticobasal syndrome. *Mov. Disord.* **2005**, *20*, 982–988. [CrossRef] [PubMed]
12. Boeve, B.; Dickson, D.; Duffy, J.; Bartleson, J.; Trenerry, M.; Petersen, R. Progressive nonfluent aphasia and subsequent aphasic dementia associated with atypical progressive supranuclear palsy pathology. *Eur. Neurol.* **2003**, *49*, 72–78. [CrossRef] [PubMed]
13. Pan, P.; Liu, Y.; Zhang, Y.; Zhao, H.; Ye, X.; Xu, Y. Brain gray matter abnormalities in progressive supranuclear palsy revisited. *Oncotarget* **2017**, *8*, 80941–80955. [CrossRef]
14. Koziorowski, D.; Figura, M.; Milanowski, Ł.M.; Szlufik, S.; Alster, P.; Madetko, N.; Friedman, A. Mechanisms of Neurodegeneration in Various Forms of Parkinsonism—Similarities and Differences. *Cells* **2021**, *10*, 656. [CrossRef]
15. Barer, Y.; Chodick, G.; Cohen, R.; Grabarnik-John, M.; Ye, X.; Zamudio, J.; Gurevich, T. Epidemiology of Progressive Supranuclear Palsy: Real World Data from the Second Largest Health Plan in Israel. *Brain Sci.* **2022**, *12*, 1126. [CrossRef] [PubMed]
16. Viscidi, E.; Litvan, I.; Dam, T.; Juneja, M.; Li, L.; Krzywy, H.; Eaton, S.; Hall, S.; Kupferman, J.; Höglinger, G.U. Clinical Features of Patients With Progressive Supranuclear Palsy in an US Insurance Claims Database. *Front. Neurol.* **2021**, *12*, 571800. [CrossRef] [PubMed]
17. Mahale, R.R.; Krishnan, S.; Divya, K.P.; Jisha, V.T.; Kishore, A. Subtypes of PSP and Prognosis: A Retrospective Analysis. *Ann. Indian Acad. Neurol.* **2021**, *24*, 56–62. [CrossRef]
18. Vöglein, J.; Kostova, I.; Arzberger, T.; Roeber, S.; Schmitz, P.; Simons, M.; Ruf, V.; Windl, O.; Herms, J.; Dieterich, M.; et al. First symptom guides diagnosis and prognosis in neurodegenerative diseases-a retrospective study of autopsy proven cases. *Eur. J. Neurol.* **2021**, *28*, 1801–1811. [CrossRef] [PubMed]
19. Williams, D.R.; Lees, A.J. Progressive supranuclear palsy: Clinicopathological concepts and diagnostic challenges. *Lancet Neurol.* **2009**, *8*, 270–279. [CrossRef]
20. Park, H.K.; Ilango, S.D.; Litvan, I. Environmental Risk Factors for Progressive Supranuclear Palsy. *J. Mov. Disord.* **2021**, *14*, 103–113. [CrossRef]
21. Moore, T.; Guttman, M. Challenges Faced by Patients With Progressive Supranuclear Palsy and their Families. *Mov. Disord. Clin. Pract.* **2014**, *1*, 188–193. [CrossRef] [PubMed]
22. Höllerhage, M.; Klietz, M.; Höglinger, G.U. Disease modification in Parkinsonism: Obstacles and ways forward. *J. Neural Transm.* **2022**, *129*, 1133–1153. [CrossRef]
23. Piccione, F.; Mancini, E.; Tonin, P.; Bizzarini, M. Botulinum toxin treatment of apraxia of eyelid opening in progressive supranuclear palsy: Report of two cases. *Arch. Phys. Med. Rehabil.* **1997**, *78*, 525–529. [CrossRef] [PubMed]
24. Scelzo, E.; Lozano, A.M.; Hamani, C.; Poon, Y.Y.; Aldakheel, A.; Zadikoff, C.; Lang, A.E.; Moro, E. Peduncolopontine nucleus stimulation in progressive supranuclear palsy: A randomised trial. *J. Neurol. Neurosurg. Psychiatry* **2017**, *88*, 613–616. [CrossRef] [PubMed]

25. Boxer, A.L.; Yu, J.T.; Golbe, L.I.; Litvan, I.; Lang, A.E.; Höglinger, G.U. Advances in progressive supranuclear palsy: New diagnostic criteria, biomarkers, and therapeutic approaches. *Lancet Neurol.* **2017**, *16*, 552–563. [CrossRef] [PubMed]
26. Ha, M.; Kim, V.N. Regulation of microRNA biogenesis. *Nat. Rev. Mol. Cell Biol.* **2014**, *15*, 509–524. [CrossRef] [PubMed]
27. Krol, J.; Loedige, I.; Filipowicz, W. The widespread regulation of microRNA biogenesis, function and decay. *Nat. Rev. Genet.* **2010**, *11*, 597–610. [CrossRef] [PubMed]
28. Lee, R.C.; Feinbaum, R.L.; Ambros, V.; The, C. elegans heterochronic gene lin-4 encodes small RNAs with antisense complementarity to lin-14. *Cell* **1993**, *75*, 843–854. [CrossRef]
29. Grodzka, O.; Procyk, G.; Gąsecka, A. The Role of MicroRNAs in Myocarditis-What Can We Learn from Clinical Trials? *Int. J. Mol. Sci.* **2022**, *23*, 16022. [CrossRef] [PubMed]
30. Grodzka, O.; Procyk, G.; Wrzosek, M. A Narrative Review of Preclinical In Vitro Studies Investigating microRNAs in Myocarditis. *Curr. Issues Mol. Biol.* **2024**, *46*, 1413–1423. [CrossRef]
31. Procyk, G.; Grodzka, O.; Procyk, M.; Gąsecka, A.; Głuszek, K.; Wrzosek, M. MicroRNAs in Myocarditis-Review of the Preclinical In Vivo Trials. *Biomedicines* **2023**, *11*, 2723. [CrossRef]
32. Krauze, A.; Procyk, G.; Gąsecka, A.; Garstka-Pacak, I.; Wrzosek, M. The Role of MicroRNAs in Aortic Stenosis-Lessons from Recent Clinical Research Studies. *Int. J. Mol. Sci.* **2023**, *24*, 13095. [CrossRef]
33. Kiełbowski, K.; Bakinowska, E.; Procyk, G.; Ziętara, M.; Pawlik, A. The Role of MicroRNA in the Pathogenesis of Duchenne Muscular Dystrophy. *Int. J. Mol. Sci.* **2024**, *25*, 6108. [CrossRef] [PubMed]
34. Nelson, P.T.; Wang, W.X.; Rajeev, B.W. MicroRNAs (miRNAs) in neurodegenerative diseases. *Brain Pathol.* **2008**, *18*, 130–138. [CrossRef] [PubMed]
35. Weber, J.A.; Baxter, D.H.; Zhang, S.; Huang, D.Y.; How Huang, K.; Jen Lee, M.; Galas, D.J.; Wang, K. The microRNA spectrum in 12 body fluids. *Clin. Chem.* **2010**, *56*, 1733–1741. [CrossRef] [PubMed]
36. de Gonzalo-Calvo, D.; Benítez, I.D.; Pinilla, L.; Carratalá, A.; Moncusí-Moix, A.; Gort-Paniello, C.; Molinero, M.; González, J.; Torres, G.; Bernal, M. Circulating microRNA profiles predict the severity of COVID-19 in hospitalized patients. *Transl. Res.* **2021**, *236*, 147–159. [CrossRef] [PubMed]
37. Qiu, X.-K.; Ma, J. Alteration in microRNA-155 level correspond to severity of coronary heart disease. *Scand. J. Clin. Lab. Investig.* **2018**, *78*, 219–223. [CrossRef] [PubMed]
38. Procyk, G.; Klimczak-Tomaniak, D.; Sygitowicz, G.; Tomaniak, M. Circulating and Platelet MicroRNAs in Cardiovascular Risk Assessment and Antiplatelet Therapy Monitoring. *J. Clin. Med.* **2022**, *11*, 1763. [CrossRef]
39. Baumann, V.; Winkler, J. miRNA-based therapies: Strategies and delivery platforms for oligonucleotide and non-oligonucleotide agents. *Future Med. Chem.* **2014**, *6*, 1967–1984. [CrossRef]
40. Page, M.J.; McKenzie, J.E.; Bossuyt, P.M.; Boutron, I.; Hoffmann, T.C.; Mulrow, C.D.; Shamseer, L.; Tetzlaff, J.M.; Akl, E.A.; Brennan, S.E. The PRISMA 2020 statement: An updated guideline for reporting systematic reviews. *BMJ* **2021**, *372*, n71. [CrossRef]
41. Lauretti, E.; Dincer, O.; Praticò, D. Regional and temporal miRNAs expression profile in a transgenic mouse model of tauopathy: Implication for its pathogenesis. *Mol. Psychiatry* **2021**, *26*, 7020–7028. [CrossRef] [PubMed]
42. Smith, P.Y.; Delay, C.; Girard, J.; Papon, M.-A.; Planel, E.; Sergeant, N.; Buée, L.; Hébert, S.S. MicroRNA-132 loss is associated with tau exon 10 inclusion in progressive supranuclear palsy. *Hum. Mol. Genet.* **2011**, *20*, 4016–4024. [CrossRef]
43. Tatura, R.; Buchholz, M.; Dickson, D.W.; van Swieten, J.; McLean, C.; Höglinger, G.; Müller, U. microRNA profiling: Increased expression of miR-147a and miR-518e in progressive supranuclear palsy (PSP). *Neurogenetics* **2016**, *17*, 165–171. [CrossRef] [PubMed]
44. Starhof, C.; Hejl, A.-M.; Heegaard, N.H.H.; Carlsen, A.L.; Burton, M.; Lilje, B.; Winge, K. The biomarker potential of cell-free microRNA from cerebrospinal fluid in Parkinsonian Syndromes. *Mov. Disord.* **2019**, *34*, 246–254. [CrossRef] [PubMed]
45. Manna, I.; Quattrone, A.; De Benedittis, S.; Vescio, B.; Iaccino, E.; Quattrone, A. Exosomal miRNA as peripheral biomarkers in Parkinson's disease and progressive supranuclear palsy: A pilot study. *Park. Relat. Disord.* **2021**, *93*, 77–84. [CrossRef]
46. Nonaka, W.; Takata, T.; Iwama, H.; Komatsubara, S.; Kobara, H.; Kamada, M.; Deguchi, K.; Touge, T.; Miyamoto, O.; Nakamura, T.; et al. A cerebrospinal fluid microRNA analysis: Progressive supranuclear palsy. *Mol. Med. Rep.* **2022**, *25*, 88. [CrossRef] [PubMed]
47. Ramaswamy, P.; Christopher, R.; Kumar Pal, P.; Debnath, M.; Yadav, R. Plasma microRNAs as a Potential Biomarker for Identification of Progressive Supranuclear Palsy. *Diagnostics* **2022**, *12*, 1204. [CrossRef]
48. Simoes, F.A.; Joilin, G.; Peters, O.; Schneider, L.-S.; Priller, J.; Spruth, E.J.; Vogt, I.; Kimmich, O.; Spottke, A.; Hoffmann, D.C.; et al. Potential of Non-Coding RNA as Biomarkers for Progressive Supranuclear Palsy. *Int. J. Mol. Sci.* **2022**, *23*, 14554. [CrossRef]
49. Pavelka, L.; Rauschenberger, A.; Hemedan, A.; Ostaszewski, M.; Glaab, E.; Krüger, R. Converging peripheral blood microRNA profiles in Parkinson's disease and progressive supranuclear palsy. *Brain Commun.* **2024**, *6*, fcae187. [CrossRef]
50. Zhu, Q.-B.; Unmehopa, U.; Bossers, K.; Hu, Y.-T.; Verwer, R.; Balesar, R.; Zhao, J.; Bao, A.-M.; Swaab, D. MicroRNA-132 and early growth response-1 in nucleus basalis of Meynert during the course of Alzheimer's disease. *Brain* **2016**, *139*, 908–921. [CrossRef]

International Journal of *Molecular Sciences*

MDPI

Article

TLR2 and TLR4 Are Expressed in Epiretinal Membranes: Possible Links with Vitreous Levels of Complement Fragments and DAMP-Related Proteins

Lucia Dinice [1,†], Graziana Esposito [1,†], Andrea Cacciamani [2,‡], Bijorn Omar Balzamino [1], Pamela Cosimi [2], Concetta Cafiero [3], Guido Ripandelli [2] and Alessandra Micera [1,*,‡]

1 Research and Development Laboratory for Biochemical, Molecular and Cellular Applications in Ophthalmological Science, IRCCS—Fondazione Bietti, 00184 Rome, Italy; lucia.dinice@fondazionebietti.it (L.D.); graziana.esposito@fondazionebietti.it (G.E.); bijorn.balzamino@fondazionebietti.it (B.O.B.)

2 Surgical Retina Research Unit, IRCCS—Fondazione Bietti, 00184 Rome, Italy; andrea_cacciamani@hotmail.com (A.C.); pcosimi@gmail.com (P.C.); guido.ripandelli@fondazionebietti.it (G.R.)

3 Anatomic Pathology Unit, Fabrizio Spaziani Hospital, 03100 Frosinone, Italy; concettacafiero@gmail.com

* Correspondence: alessandra.micera@fondazionebietti.it

† These authors contributed equally to this work.

‡ These authors contributed equally to this work and share last authorship.

Abstract: Previous studies reported the expression of toll-like receptors (TLRs), merely TLR2 and TLR4, and complement fragments (C3a, C5b9) in vitreoretinal disorders. Other than pathogens, TLRs can recognize endogenous products of tissue remodeling as damage-associated molecular pattern (DAMPs). The aim of this study was to confirm the expression of TLR2 and TLR4 in the fibrocellular membranes and vitreal fluids (soluble TLRs) of patients suffering of epiretinal membranes (ERMs) and assess their association with disease severity, complement fragments and inflammatory profiles. Twenty (n = 20) ERMs and twelve (n = 12) vitreous samples were collected at the time of the vitrectomy. Different severity-staged ERMs were processed for: immunolocalization (IF), transcriptomic (RT-PCR) and proteomics (ELISA, IP/WB, Protein Chip Array) analysis. The investigation of targets included TLR2, TLR4, C3a, C5b9, a few selected inflammatory biomarkers (Eotaxin-2, Rantes, Vascular Endothelial Growth Factor (VEGFA), Vascular Endothelial Growth Factor receptor (VEGFR2), Interferon-γ (IFNγ), Interleukin (IL1β, IL12p40/p70)) and a restricted panel of matrix enzymes (Matrix metalloproteinases (MMPs)/Tissue Inhibitor of Metallo-Proteinases (TIMPs)). A reduced cellularity was observed as function of ERM severity. TLR2, TLR4 and myD88 transcripts/proteins were detected in membranes and decreased upon disease severity. The levels of soluble TLR2 and TLR4, as well as C3a, C5b9, Eotaxin-2, Rantes, VEGFA, VEGFR2, IFNγ, IL1β, IL12p40/p70, MMP7 and TIMP2 levels were changed in vitreal samples. Significant correlations were observed between TLRs and complement fragments and between TLRs and some inflammatory mediators. Our findings pointed at TLR2 and TLR4 over-expression at early stages of ERM formation, suggesting the participation of the local immune response in the severity of disease. These activations at the early-stage of ERM formation suggest a potential persistence of innate immune response in the early phases of fibrocellular membrane formation.

Keywords: toll-like receptors (TLRs); epiretinal membranes (ERMs); vitreous; complement fragments; extra-cellular matrix (ECM); innate immunity; inflammation; tissue remodeling; vitreoretinal diseases

Citation: Dinice, L.; Esposito, G.; Cacciamani, A.; Balzamino, B.O.; Cosimi, P.; Cafiero, C.; Ripandelli, G.; Micera, A. TLR2 and TLR4 Are Expressed in Epiretinal Membranes: Possible Links with Vitreous Levels of Complement Fragments and DAMP-Related Proteins. *Int. J. Mol. Sci.* **2024**, *25*, 7732. https://doi.org/10.3390/ijms25147732

Academic Editor: Andrea Bernini

Received: 24 May 2024
Revised: 4 July 2024
Accepted: 8 July 2024
Published: 15 July 2024

1. Introduction

The epiretinal membranes (ERMs), also known as macular pucker or cellophane maculopathy, frequently develop with ageing and can be associated with diabetes, cardio-vascular disorders and metabolic syndrome [1]. Structurally, these avascular fibrocellular membranes are layered over the retina, in the macular zone, and contract the vitreoretinal interface, depending on the profibrogenic activity [1].

Several extra-retinal cells (T/B lymphocytes ($CD4^+$, $CD8^+$, $CD22^+$), macrophages ($CD68^+$), neutrophils and several MHC class II cell surface receptor (HLA-DR)-expressing cells [2,3] and a few retinal cell types (Müller cells, astrocytes, hyalocytes, retinal pigment epithelium cells and fibroblasts/myofibroblasts) were observed in whole flattened ERMs [4]. The contraction properties, depending on cell recruitment and extracellular matrix (ECM) synthesis/deposition (matrix remodeling), are regulated by local cell-to-ECM interactions and a plethora of proteins (growth factors, neuromediators and matrix-linked enzymes) [5]. This microenvironment triggers a tractional force between the membrane and retina (vitreoretinal interface), as enriched of the main profibrogenic factors, such as transforming growth factor β1 (TGFβ1), glial cell-derived growth factor (GDNF), vascular endothelial growth factor (VEGF) and neurotrophins (nerve growth factor (NGF), brain-derived neurotrophic factor (BDNF)) [4]. This microenvironment is the result of different cell types (activated Müller cells, contractile myofibroblasts (myoFBs) and reactive microglia [6,7]) and myofibroblast-like cells that participate in the synthesis of ECM and the release of growth factors/cytokines, influencing the contractile abilities of the membrane and contributing to disease pathogenesis (inflammation and angiogenesis) [4,8]. Müller cells, hyalocytes and retinal pigment epithelium-derived cells have been identified as potential myofibroblast precursors [4]. We previously analyzed the protein profile of vitreous and vitreal reflux collected, respectively, at the time of the vitrectomy and intravitreal anti-VEGF injection, showing that information about vitreal protein profiles might mirror of health of underneath retina, retinal ganglion cell (RGC) and/or photoreceptors [7–9].

The long-lasting ECM makeover (tissue remodeling) seems frequently associated with the production, release and accumulation of molecules associated with endogenous cellular and tissue stress signals (the Damage-Associated Molecular Patterns (DAMPs) and alarmins) [10]. Tissue-injury molecules (collagen and hyaluronan fragments) can be generated inside the fibrocellular membranes upon matrix retraction [11]. Accumulating inside tissues or being released in biological fluids, DAMPs might trigger a potent inflammatory response during non-infection-driven inflammation [10]. ECM fragments are sensed by toll-like receptors (TLRs) expressed by epithelial cells, innate-immune cells, antigen-presenting cells [12]. By allowing a specific recognition of Pathogen Associated Molecular Patterns, (PAMPs) and endogenous molecules (DAMPs), such as death-cell debris, intact or fragmented collagen and hyaluronic acid, inflammatory products and oxidative-modified lipids, TLRs prompt a local innate immune response and guarantee immune surveillance, allowing tissue homeostasis [12–15]. DAMPs preferentially interact with TLR2, TLR4 and complement cleaved fragments (C3a, C5b), in both membrane-bound or soluble forms [16–22]. Soluble TLRs and complement fragments accumulate in blood, urine, saliva, tears and aqueous and vitreous humors, as observed in infections and chronic autoimmune disorders [23]. Signals from innate immune response activation have been described in diabetic retinopathy (DR), age-related macular degeneration (AMD) and macular pucker (ERMs) [2,24]. The concept of functional cross-talk between TLRs and complement fragments has been accurately discussed in a recent revision of literature [23].

Therefore, the aim of the present study was to verify the expression of TLR2 and TLR4 in all the membranogenic phases of ERM disease, by means of immunolocalization (ERMs) and biomolecular analysis (ERMs/vitreous). The association between a few selected complement fragments (C3a, C5b-9) and inflammatory/remodeling mediators potentially expressed by these fibrocellular membranes was also investigated.

2. Results

In this observational single-point study, all ERMs (n = 20) and paired vitreal fluids were categorized as stage 2, stage 3 and stage 4 of the disease. The categorization was carried out by the surgeon at the time of clinical assessment before vitreoretinal surgery (Optical coherence tomography (OCT) bioinstrumental evaluation; Govetto's nomenclature) [25,26].

The characteristics of the study population are shown in Table 1. Due to the features of the diseases and the absence of controls, stage 1 was not included in this analysis, and stage 2 was used as the referring control for comparative studies, depending on the analysis.

Table 1. Study population and biosamples.

ID	Gender (M/F)	Age (Years)	Vitreous (Y/N)	ERM (Y/N)	Staging (Grades)	Comorbidities (List)
1	F	71	Y	Y	4	Hypertension, Cardiopathy
2	M	54	Y	Y	2	Hypertension, Cardiopathy, Type 2 Diabetes
3	F	63	N	Y	2	Hypercholesterolemia, Keratoconus
4	M	78	Y	Y	4	Hypertension, Hypercholesterolemia
5	F	69	N	Y	4	Hypercholesterolemia
6	F	66	N	Y	3	None
7	F	72	N	Y	2	None
8	F	70	Y	Y	3	Hypertension, Hypothyroidism
9	F	69	N	Y	3	Asthma, Thyroidectomy
10	F	65	Y	Y	4	Hypercholesterolemia, Type 2 Diabetes
11	F	83	Y	Y	2	Hypercholesterolemia
12	F	82	N	Y	4	Hypertension, Hypercholesterolemia
13	F	80	Y	Y	2	Hypertension, Cardiopathy, Type 2 Diabetes
14	F	69	Y	Y	2	Hypertension, Type 2 Diabetes, Hyperuricemia
15	F	81	Y	Y	3	Hypertension, Asthmatic Bronchitis
16	M	66	Y	Y	3	Type 2 Diabetes
17	F	71	Y	Y	3	Hypertension, Hypercholesterolemia
18	F	80	N	Y	2	Muscular Dystrophy
19	M	71	Y	Y	2	Hypertension, Asthma
20	M	69	N	Y	4	None

The chart summarizes demographic aspects of study population, sampling aspects (paired and unpaired aspects of ERM/vitreous samples) and comorbidities. Complete ophthalmic examination and ERM grading were assessed by two ophthalmologists (A.C. or G.R.). Visual acuity progressively declined from stage 1 to stage 4 [25]. Legend: ID, patients' progressive code; gender (M/F, male/female); age (years); vitreous (Y/N, yes/no); ERM (Y/N, yes/no); staging (grades); comorbidities, as depicted from initial anamnesis (list).

This small study population displayed different comorbidities, in line with age and disease features, distributed as follows: 50% (10/20) comorbidities with hypertension and 35% (7/20) comorbidities without hypertension. A total of 15% (3/20) declared no comorbidities of interest for this study (none, apparent good health).

2.1. TLR2 and TLR4 Immunoreactivity in ERMs

A widespread nuclear staining (DAPI/blue) was depicted in all ERMs, indicating the presence of several cell subsets embedded in the fibrous matrix (fibrocellular scaffolds). As shown in Figure 1, specific TLR2 (green/cy2) and TLR4 (red/cy3) immunoreactive cells were observed in ERMs at all stages (×400). Particularly, high TLR2 and moderate TLR4 expression characterize stage 2 (Figure 1A) with respect to stage 3 (Figure 1B) and stage 4 (Figure 1C). This decreasing trend was confirmed by imaging quantification. As shown in Figure 1D, the fluorescent analysis (IntDen) specific for each target showed higher TLR2 and TLR4 expression at early stages of the disease, with TLR2 IntDen being higher than TLR4. Of interest, ERM cellularity decreased with increasing severity, along with matrix deposition. Figure 1E shows the reduction in cellularity with increasing ERM severity, implying that the ERMs at stage 4 had less cellularity than at stage 3 and stage 2.

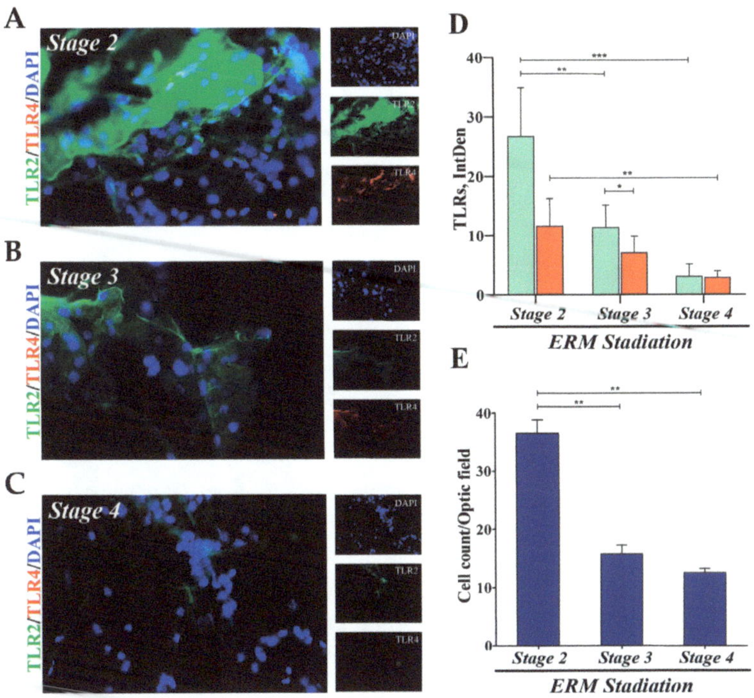

Figure 1. TLR2 and TLR4 protein expression in ERMs as a function of disease severity. A total of six membranes (N = 2 for stage 2, 1F/1M; N = 2 for stage 3, 1F/1M; N = 2 for stage 4, 1F/1M) were collected during pars plana vitrectomy and processed for Epifluorescent analysis (**A**–**C**), coupled to Integrated Density (IntDen; (**D**)) evaluation. Representative merged (main) and single (right side) panels of TLR2 (green) and TLR4 (red) in ERMs at stage 2 (**A**), stage 3 (**B**) and stage 4 (**C**). Membranes were counterstained with nuclear DAPI (blue) to better visualize the cells. Note the significant decrease in immunoreactivity (**D**) and cellularity (**E**), depending on severity ($p \leq 0.05$). Data represent mean ± SEM and values of fluorescent intensity (IntDen; ImageJ) are expressed in arbitrary units. Magnifications ×400 (bar size = 50 μm). Significances are shown in the panels (* $p \leq 0.05$; ** $p \leq 0.01$; *** $p \leq 0.005$), as calculated using one-way ANOVA followed by a Tukey–Kramer post hoc test (mean ± SEM).

2.2. TLR2 and TLR4 Immunoreactivity: Alpha-Smooth Muscle Actin (α-SMA)-Fibroblast like Cells, Glial Fibrillary Acidic Protein (GFAP)-Activated Müller Cells and Ionized Calcium-Binding Adapter Molecule 1 (Iba1)-Bearing Ameboid Microglia

The nuclear staining (DAPI/blue) of peeled-off ERM tissues highlighted the presence of several cell subsets inside the fibrocellular-matrix compartment of ERMs, as reported in previous studies. A specific TLR2 (Figure 2A,C,E) and TLR4 (Figure 2B,D,F) expression was observed in α-SMA positive myoFBs-like cells populating the whole flattened ERMs at stage 2 (Figure 2A,B), at stage 3 (Figure 2C,D) and at stage 4 (Figure 2E,F).

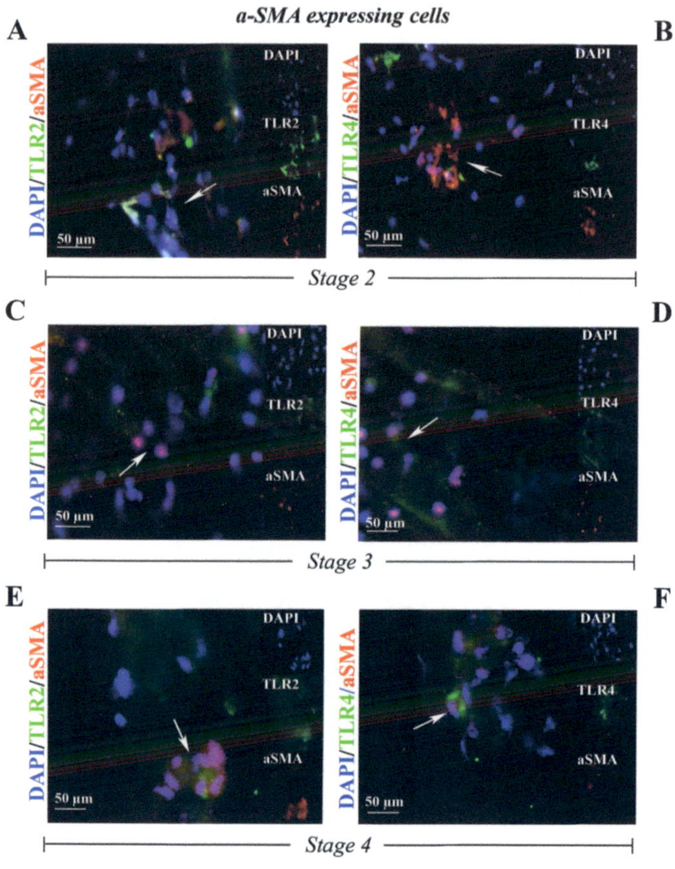

Figure 2. TLR2 and TLR4 immunoreactivity in myoFBs cells populating the ERM at stage 2 ((**A**,**B**) N = 2, 1F/1M), at stage 3 ((**C**,**D**) N = 2, 2F) and at stage 4 ((**E**,**F**) N = 2, 2F). Representative fluorescent panels of TLR2/green (**A**,**C**,**E**) or TLR4/green (**B**,**D**,**F**) and α-SMA/red from double-immunostained and nuclear counterstained (DAPI/blue) ERMs. Note the TLR2 and TLR4 immunoreactivity in, respectively, α-SMA positive myoFBs. Citofixed and whole-mounted membranes were used. Merged and single-channel panels are shown. Magnifications ×400 (bar size = 50 µm). Arrows indicate coexpression of targets.

Immunoreactivity for TLR2 and TLR4 was observed in the Iba1-positive cells (respectively, Figure 3A,C,E and Figure 3B,D,F) in ERMs at stage 2 (Figure 3A,B), at stage 3 (Figure 3C,D) and at stage 4 (Figure 3E,F).

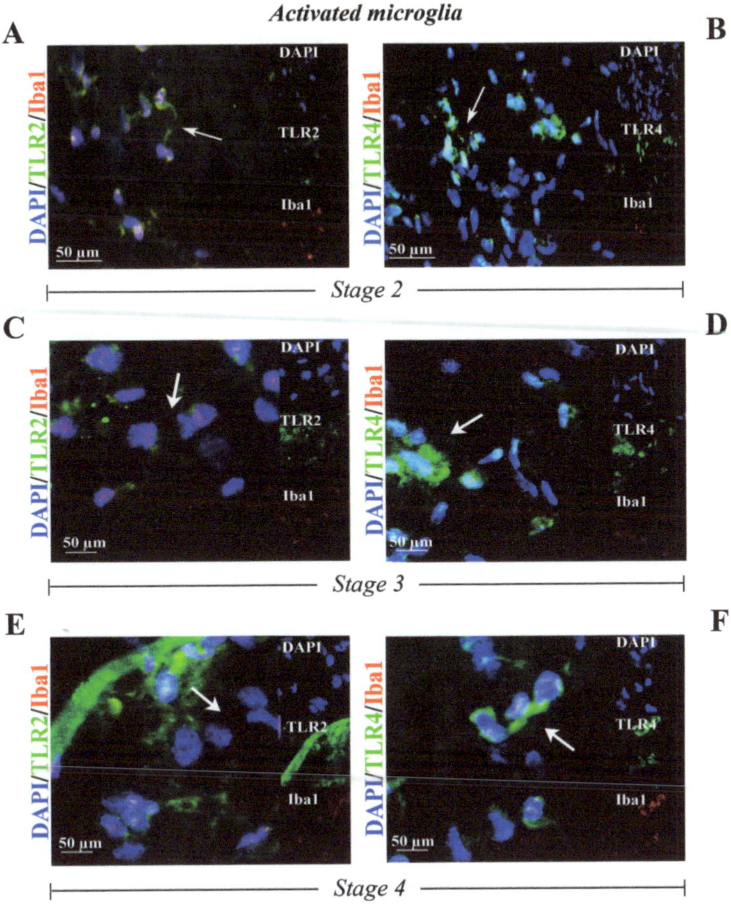

Figure 3. TLR2 and TLR4 immunoreactivity in reactive microglia cells populating the ERM at stage 2 ((**A**,**B**) N = 2, 2F), at stage 3 ((**C**,**D**) N = 1, 1F) and at stage 4 ((**E**,**F**) N = 1, 1F). Representative fluorescent panels of TLR2/green (**A**,**C**,**E**) or TLR4/green (**B**,**D**,**F**) and Iba1/red from double-immunostained and nuclear counterstained (DAPI/blue) ERMs. Note the TLR2 and TLR4 immunoreactivity in, respectively, some Iba1 immunoreactive cells. Citofixed and whole-mounted membranes were used. Merged and single-channel panels are shown. Magnifications ×400 (bar size = 50 µm). Arrows indicate coexpression of targets.

Finally, a specific TLR2 (Figure 4A,C,E) and TLR4 (Figure 4B,D,F) expression was observed in GFAP-positive cells inside the whole flattened ERMs at stage 2 (Figure 4A,B), at stage 3 (Figure 4C,D) and at stage 4 (Figure 4E,F).

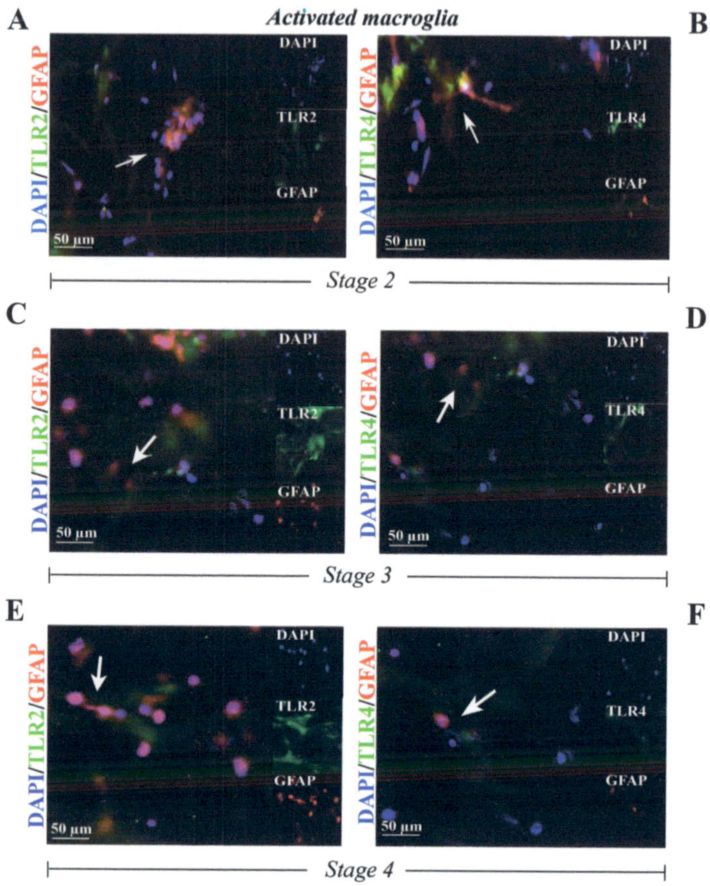

Figure 4. TLR2 and TLR4 immunoreactivity in activated Müller cells populating the ERMs at stage 2 ((**A**,**B**) N = 2, 2F), at stage 3 ((**C**,**D**) N = 1, 1F) and at stage 4 ((**E**,**F**) N = 1, 1F). Representative fluorescent panels of TLR2/green (**A**,**C**,**E**) or TLR4/green (**B**,**D**,**F**) and GFAP/red and nuclear counterstained (DAPI/blue) ERMs. Note the TLR2 and TLR4 immunoreactivity in, respectively, some cellular compartments of GFAP-positive Müller cells. Citofixed and whole-mounted membranes were used. Merged and single-channel panels are shown. Magnifications ×400 (bar size = 50 μm). Arrows indicate coexpression of targets.

2.3. *TLR2/TLR4/myD88 Transcripts Are Increased in ERMs as a Function of Disease Severity*

The transcription activity of *TLR2* and *TLR4* was verified by relative RT-PCR, considering the expression in stage 3 and stage 4 ERMs with respect to stage 2 (referred to as 1). As shown in Figure 5A, the relative analysis showed a significative upregulation of the *TLR2* transcript at stage 3 (4.59 ± 1.72 2logFC) and a trend toward an increase at stage 4 (0.65 ± 0.31 2logFC), with respect to stage 2. A slight upregulation of *TLR4* was detected at stage 3 (2.28 ± 1.41 2logFC) and a slight trend toward an increase was observed at stage 4 (0.96 ± 0.64 2logFC), as compared to stage 2. Regarding the specific adaptor molecule, a trend toward a decrease in the myD88 transcript expression was observed at stage 3 (−0.33 ± 0.36 2log-ratio; $p > 0.05$) and at stage 4 (−1.49 ± 0.39 2log-ratio; $p > 0.05$), with respect to stage 2 (referred to as 1) (Figure 5A).

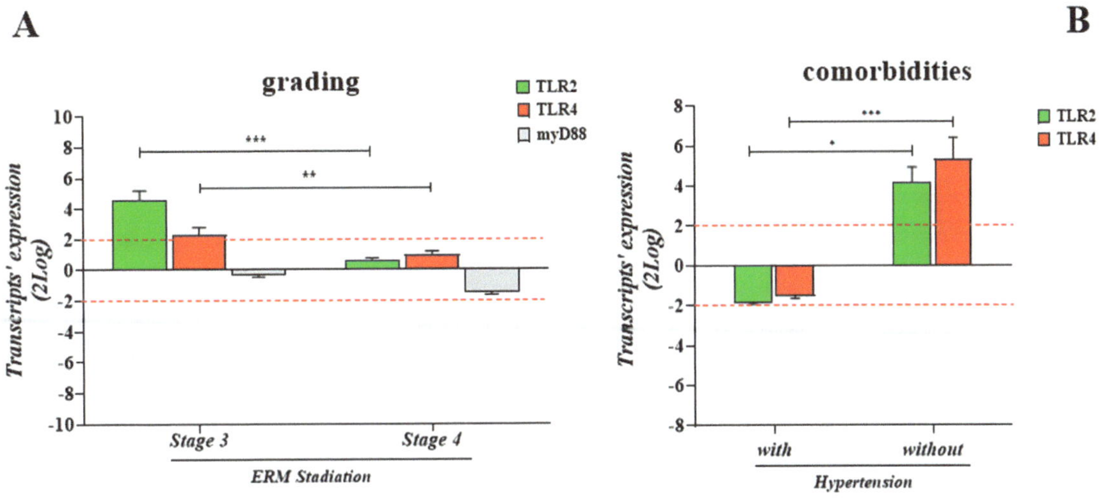

Figure 5. Transcripts' expression as a function of disease severity and comorbidities. (**A**) Histogram showing the significant upregulation of *TLR2*mRNA in ERM at stage 3 with respect to stage 4, as well as compared to stage 2. The same trend was observed for *TLR4*mRNA. An inverted trend was observed for the adaptor molecule *myD88*mRNA, as downregulation was observed in ERM at stage 4 with respect to stage 3, and no significant changes were observed with respect to stage 2. Values are relative expression ratios (fold-changes in log2-scale; mean ± SEM), as generated by REST-analysis, comparing stage 3 or stage 4 with respect to stage 2. (**B**) Transcriptomic analysis highlighted a decreased *TLR2* and *TLR4* expression in the presence of the major comorbidities associated with hypertension (10/20; ID: 1, 2, 4, 8, 12, 13, 14, 15, 17, 19), while an increased expression was detected in the absence of hypertension (7/20; ID: 3, 5, 9, 10, 11, 16, 18). Values are relative expression ratios (fold-changes in log2-scale; mean ± SEM), as generated by REST-analysis, comparing clusters vs. normal controls (3/20; IDs: 3, 5, 10). The *p*-values were calculated according to the REST-ANOVA Tukey–Kramer coupled analysis. Red dot lines are referred to 2log-FC PCR biological significance. Significances are shown in the panels (* $p \leq 0.05$; ** $p \leq 0.01$; *** $p \leq 0.005$), as calculated using one-way ANOVA followed by a Tukey–Kramer post hoc test (mean ± SEM).

For assessing the possibility that comorbidities might influence this transcription profile, a further clustering was carried out, considering comorbidities with and without hypertension, as shown in Figure 5B. Notably, *TLR2* and *TLR4* relative transcript expression was significantly higher in the ERMs from comorbidities without hypertension (+4.198 ± 0.735 FC 2log-ratio for *TLR2*; +5.331 ± 1.090 FC 2log-ratio for *TLR4*) and lower in comorbidities with hypertension (−1.869 ± 0.0735 FC 2log-ratio for *TLR2*; −1.528 ± 0.123 FC 2log-ratio for *TLR4*). Data were found with respect to same-stage ERMs obtained from patients without comorbidities (none).

2.4. The Soluble Forms of TLR2 and TLR4 Are Detected in Vitreous Samples

Immunoprecipitation, coupled to Western blot analysis (IP/WB), was used for quantifying the presence of soluble TLR2 and TLR4 proteins. The quantification of TLR2 (Figure 6A,C) and TLR4 (Figure 6B,D) in vitreous samples showed that both TLRs are less expressed at late stages and more highly expressed in the early stages of disease.

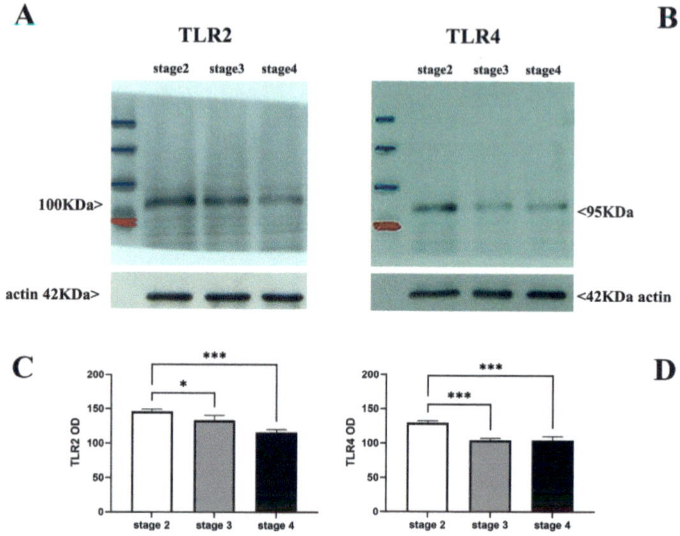

Figure 6. TLR2 and TLR4 levels in vitreous as a function of disease severity. A total of twelve specimens: N = 5 for stage 2 (ID: 2, 11, 13, 14, 19); N = 4 for stage 3 (ID: 8, 15, 16, 17); N = 3 for stage 4 (ID: 1, 4, 10) were processed for biochemical analysis. Briefly, the untouched vitreous samples were sonicated and cleared for Western blotting analysis. Representative vitreal levels of TLR2 and TLR4 (respectively, (**A**,**B**); IP/WB) are shown by gels, and densitometric IntDen analysis (**C**,**D**) was carried out considering all the samples. Note that TLR2 and TLR4 levels are comparable at early stages and greater than levels at late stages of the disease. Protein normalization was confirmed by actin expression. Data are mean ± SEM from optical density (OD) analysis (ImageJ) and *p*-values are shown by asterisks (see Section 4). Significant levels are shown in the panels (* $p \leq 0.05$; *** $p \leq 0.005$), as calculated using one-way ANOVA followed by a Tukey–Kramer post hoc test (mean ± SEM).

2.5. TLR2 and TLR4 Correlates with Biomarkers of Inflammation and Tissue Matrix Remodeling

An interesting aspect of the TLRs activation is the downstream signaling leading to the release of pro-inflammatory/angiogenic mediators (HLA-DR, p65-NFkB, Eotaxin-2, Rantes and VEGFA/VEGFR2) and matrix remodeling enzymes (MMPs/TIMPs), modulating antibacterial, antimicrobial and immune adaptive and apoptotic responses [27–30].

Figure 7A highlights the upregulations of *HLA-DR* and nuclear factor kappa-light-chain-enhancer of activated B cells (*p65-NFkB*) transcripts at stage 3 in ERMs, and a positive correlation between HLA-DR and respectively *TLR2* (rho = 0.978, $p < 0.04$) and *TLR4* (rho = 0.979, $p < 0.005$) was also observed. In these ERMs, as expected, the increased *p65-NFkB* transcripts expression observed at stage 3 were decreased at stage 4 (Figure 7A). Both *Eotaxin-2* and *Rantes* were decreased, while *VEGFA* and *VEGFR2* increased at stage 3 but were not retained at stage 4 in vitreous samples (Figure 7B). Regarding the vitreal Th1/Th2 profile, while *IFNγ* was reduced with ERM severity, *IL1β* and *IL12p70* increased depending on severity (Figure 7B).

The increased Eotaxin-2 levels detected in vitreal fluids, as confirmed by immunoprecipitation and immunoblotting (IP/WB) analysis, were not consistent with those detected in ERMs (Figure 7C). GFAP protein levels in both vitreous and membranes were increased, depending on ERM stadiation (Figure 7D), confirming the presence of active gliosis both locally and circulating the vitreal chamber.

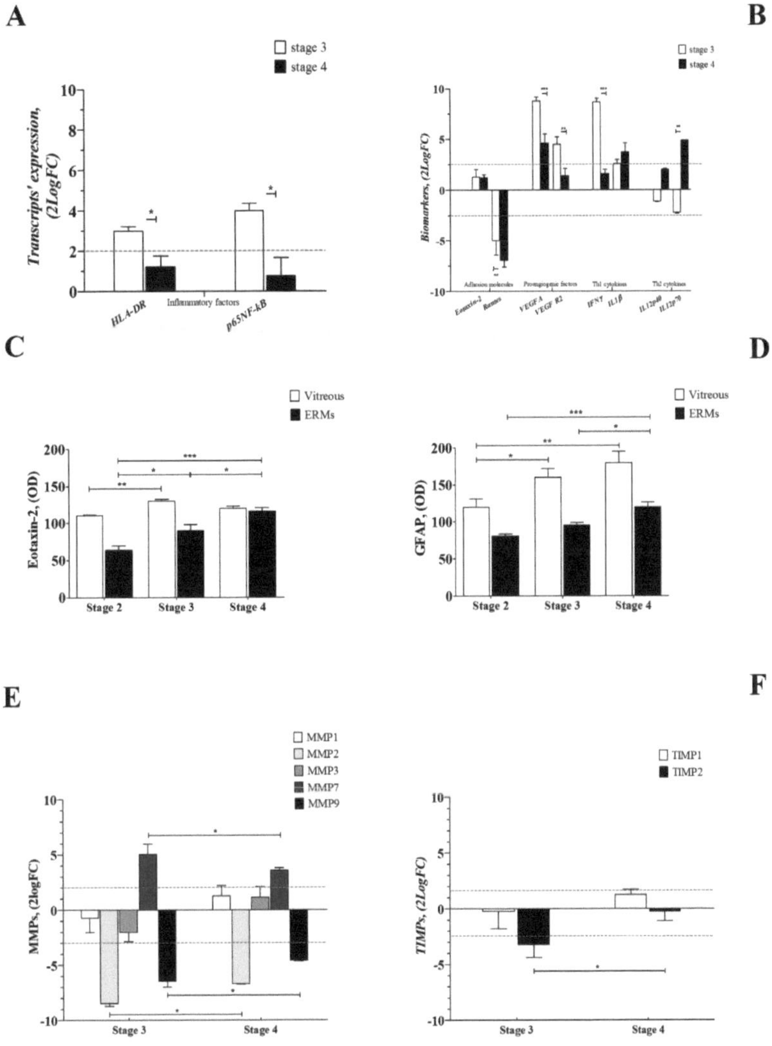

Figure 7. Inflammatory and remodeling profiles. (**A**,**B**) Inflammatory (**A**) and adhesion, proangiogenic and Th1/Th2 (**B**) patterns were analyzed, respectively, in ERMs and vitreous samples. (**A**) Note the *HLA-DR* and *p65NFkB* transcripts' upregulation in ERMs at stage 3. (**B**) Adhesion molecules appear deregulated, *VEGFA* and *VEGFR2* were upregulated at stage 3, and only *IL1β* and *IL12p70* were increased at stage 4, all in vitreous samples. (**C**,**D**) Eotaxin-2 (**C**) and GFAP (**D**) expression in vitreal samples and coupled ERMs, as assayed by IP/WB analysis. Note that Eotaxin-2 and GFAP were increased in both membrane formation and coupled vitreous, in a manner related to disease severity. All diseases stages are shown. (**E**,**F**) Histogram depicting the relative transcript expression of *MMPs* (**E**) and *TIMPs* (**F**) in ERM extracts as function disease severity. Note the *MMP2* and *MMP9* transcript downregulation and the *MMP7* transcript upregulation as early as stage 3. Whenever required, fold changes were calculated with respect to stage 2, and comparisons were carried out between stages. Data in the bar graph are shown as mean ± SEM (2log fold-changes) and *p*-values are shown by asterisks (see Section 4). Red dot lines are referred to 2log-FC PCR biological significance and significant levels are shown in the panels (* $p \leq 0.05$; ** $p \leq 0.01$; *** $p \leq 0.005$), as calculated using one-way ANOVA followed by a Tukey–Kramer post hoc test (mean ± SEM).

A transcriptomic analysis specific for *MMP1, MMP2, MMP3, MMP7, MMP9, TIMP1* and *TIMP2* in ERM extracts showed that only *MMP7* was significantly changed at stage 3 and stage 4 (Figure 7E). On the contrary, the *MMP2* and *MMP9* transcripts were significantly deregulated at stage 3, while the *MMP1* and *MMP3* transcripts were unchanged (Figure 7E). With respect to their tissue regulators, only *TIMP2* transcripts were decreased at stage 3 (Figure 7F). A strong positive correlation was detected between *MMP7* and *TLR2* at stage 3 (rho = 0.9948, $p < 0.05$).

2.6. TLR2 and TLR4 Correlate with Complement Fragments

Both the TLR family and complement system are main branches of innate immunity [31]. As shown in Figure 8A, increased levels of the soluble forms of C3a were quantified in the vitreous, as function of ERM severity. Higher C3a levels were observed at stage 4. By contrast, no significant changes were monitored for the vitreal C5b9 fragment. Unexpectedly, a positive correlation was observed between C3a and C5b9 markers (Figure 8B).

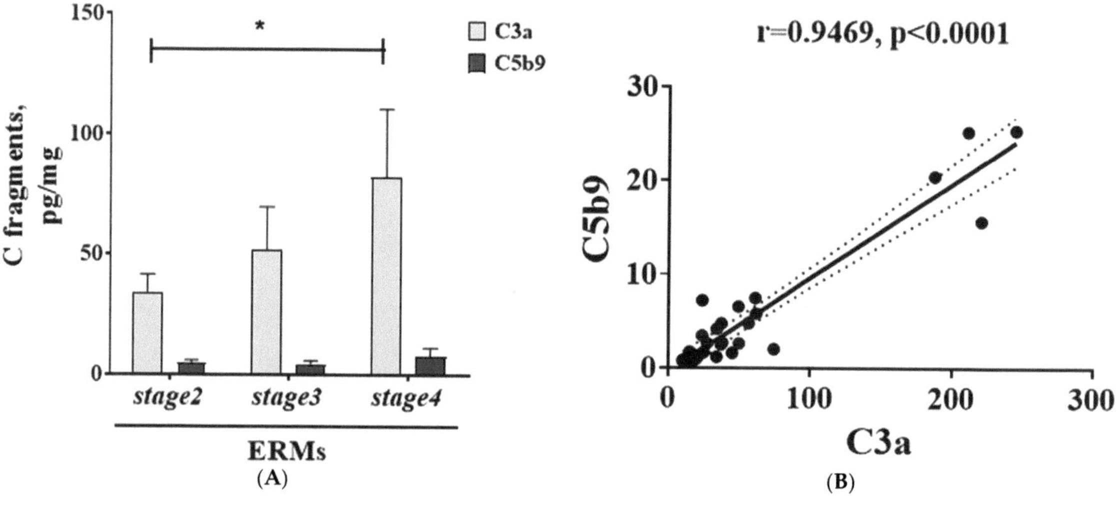

Figure 8. Complement fragments in vitreal fluids as function of ERM severity. Vitreal levels of C3a and C5b9 (ELISA; (**A**)); note that C3a levels were higher than C5b9 levels. (**B**) Scatterplot displaying the positive relationship between C3a and C5b9 markers. Pearson correlation coefficient (r) and significance are shown in the panel. Data are mean ± SEM, as detected by ELISA assay. Significant levels are shown in the panels (* $p \leq 0.05$), as calculated using one-way ANOVA followed by a Tukey–Kramer post hoc test (mean ± SEM).

By correlating the *TLRs* transcripts and vitreal complement values, positive correlations were detected between *TLR2*-C3a (Figure 9A,B), *TLR2*-C5b9 (Figure 9C,D), *TLR4*-C3a (Figure 9E,F) and *TLR4*-C5b9 (Figure 9G,H) at stage 3 and stage 4.

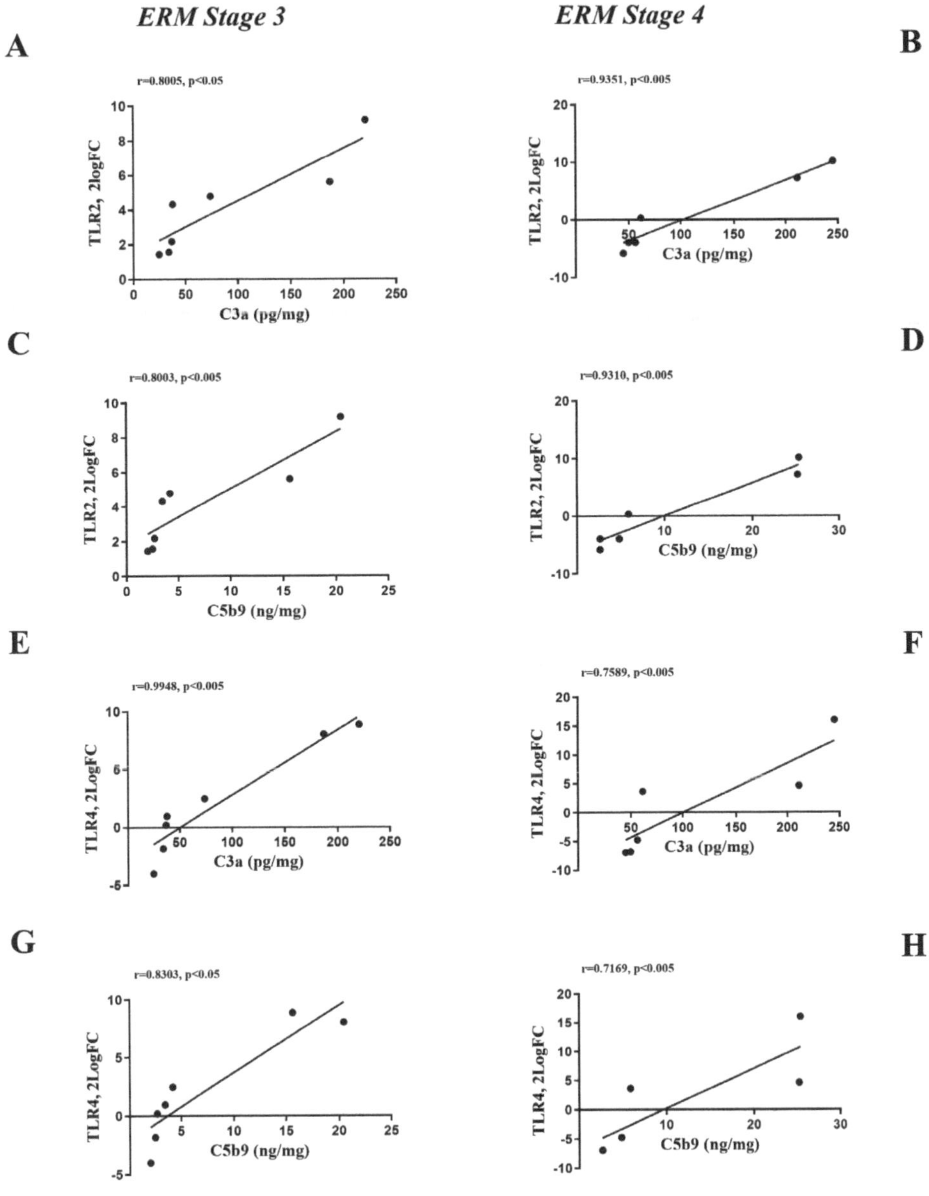

Figure 9. Correlation analysis for *TLRs* (transcripts) and complement fragments (vitreal levels) depending on ERM severity. Scatterplot specific for *TLR2*/C3a at stage 3 (**A**) and stage 4 (**B**); *TLR2*/C5b9 at stage 3 (**C**) and stage 4 (**D**); for *TLR4*/C3a at stage 3 (**E**) and stage 4 (**F**); for *TLR4*/C5b9 at stage 3 (**G**) and stage 4 (**H**). Note the positive correlation between complement fragments and *TLR2* and *TLR4*. The result of Pearson correlation analysis is shown inside each panel (correlation coefficient, r, and significance).

3. Discussion

Herein, we show the specific immunoreactivity of TLR2 and TLR4 in ERMs and their immunolocalization to GFAP-, Iba1- and α-SMA-expressing cells. TLR2 and TLR4 also correlated with C3b and C5b9 and some inflammatory mediators. The rationale behind TLR2 and TLR4 expression is discussed below.

Previous studies emphasized the involvement of membrane-associated and soluble TLRs in retinal degeneration diseases, as shown by in vitro/in vivo and human ex vivo studies (iatrogenic, diabetic, vascular and age-related vitreoretinal diseases) [32–37]. Particularly, TLR2 and TLR4 were associated with innate immune activation of complement fragments, promoting the phagocytic activities of mononuclear cells and oxidative stress-mediated responses [31,38,39]. This study extended the previous knowledge on TLR2 and TLR4 on ERMs by hypothesizing their expression in these fibrocellular membranes (ERMs) and coupled vitreal fluids, exploring the TLR2-TLR4's contribution to ERMs' severity. This hypothesis is sustained by the following: i. the presence of some cells of immune-derivation (Müller, astrocytes and microglia) inside the membrane and ii. the ability of TLRs to recognize endogenous danger molecules (DAMPs) released from the extracellular and intracellular space of damaged tissue or dead cells (collagen and hyaluronic acid fragments, hyaluronan, heat shock protein (HSP60), fibronectin and other tissue products) [32]. Activated DAMPs might trigger an uncontrolled and unbalanced innate immune response, contributing to and/or exacerbating the ERM-mediated retraction of the underneath retina [40].

First, the TLR2 and TLR4 immunofluorescence was inversely proportional to disease severity (ERM grading), and specific immunoreactivity for TLR2 and TLR4 was restricted to GFAP-reactive Müller cells, Iba1-bearing ameboid microglial cells and α-SMA-expressing cells (myofibroblast-like cells). These cells are known to recognize DAMPs exacerbating the local inflammation in the presence of matrix fragments/debris [41–47]. Previous studies also referred to activated Müller cells and reactive microglia as major cell sensor patterns in ERMs, and the myofibroblast-like cells' contribution was related to long-lasting ERM retraction [48]. The differential expression of TLR2 and TLR4 alongside the ERM phases of severity might be linked to the fact that Müller cells seem to play a primary role in the events leading to ERM formation (known as vertical gliosis) [48]. The impact of vertical traction on Müller cells might drive matrix synthesis and scaffold creation, paving the way for migration and cell organization of ERMs. In addition, these fibrocellular membranes confer resistance to the mechanical stimulation of the retina [48]. Our observation of TLR4 immunoreactive microglia is in line with previous studies, and no other TLRs, except for TLR2 and TLR4, have been detected on retinal microglia [49–51]. On the contrary, the observation of TLR2 and TLR4 immunoreactivity in activated GFAP-expressing Müller cells is in line with Kumar and Shamsuddin's findings, reporting the presence of TLR2 and TLR4 in retinal Müller cells [46,52]. Since some in vitro simulations highlighted the ability of TLR2 and TLR4 agonists to trigger the release of IL-6 and MIP-2/C-X-C Motif Chemokine Ligand (CXCL2) chemokines by reactive Müller cells, our observations in ERMs might suggest that TLRs expressing Müller cells can drive immunoregulatory functions in vitreoretinal diseases through DAMP products, in addition to the known role in anti-microbial surveillance (endophthalmitis) [46]. More interestingly, the presence of TLR-bearing myofibroblast-like cells might suggest a possible influence of DAMP-related TLRs on the contractile activities of these fibrocellular-matrix formations [45–47]. Collagen fragments might exacerbate the force generated by the fibrocellular membranes on the underneath retina. Precisely, type-II collagen can allow the jelly-like vitreous to adhere to the underneath retina, although ageing or traumatic/inflammatory events can trigger a gradual change of this jelly fluid, inducing its separation from retina (vitreous detachment) [53].

Our previous studies highlighted the existence of specific "ERM-vitreal protein-print" depending on the severity and origin of vitreoretinal damage, highlighting the possibility to use this protein signature for individualized therapy purposes [6]. A significant increase in TLR2 and TLR4 expression suggests the following: i. the activation of intracellular signaling, with p65NFkB nuclear translocation; ii. a huge cell-mediated response driven by professional cells (Antigen-Presenting Cell (APCs), phagocytes), accelerating the local inflammation by the adhesion of leukocytes and the maturation/migration of immune cells; and iii. RGC damage, with an exacerbation of local inflammation and impairments in the retinal network [54]. Herein, the observation of myD88 deregulation clearly indicates the presence of a less-activated pathway related to pro-inflammatory mediators (IFNγ, IL6, IL8), as reported by several in vivo and in vitro myD88 models [55].

Since a decreased cellularity occurs in late stages of the disease (ERM severity), another aspect of TLR2 and TLR4 signaling might be to drive the death of subsequent necrotic cells and APCs by means of NFκB signaling in a paracrine/autocrine fashion, as reported by in vitro simulations [56]. Under physiological processes, the removal of debris and dead cells is essential for protecting retinal networking, as documented by several experimental and ex vivo studies [57]. The apoptosis of myofibroblast-like cells might also occur in the later stages of ERM activity, supporting the reduced retraction intensity observed at stage 4, which often does not require ERM peel-off [25]. Our data on TLRs and HLA-DR expression are supported by previous studies reporting the specific cooperation of HLA-DR with TLR2 and TLR4 [58].

An interesting aspect of the TLR activation is the downstream signaling, leading to the release of pro-inflammatory/angiogenic and matrix enzymes able to modulate antibacterial, antimicrobial, as well as adaptive response and apoptotic response [27–30].

Our investigations in vitreal-ERM paired samples also highlighted an increase in Eotaxin-2 proteins in ERMs and vitreal fluids, in line with previous studies, anticipating the higher Eotaxin-2 levels in vitreous collected from retinal detachment and ERMs [59–61]. Of interest, the quantification of GFAP protein in these samples might corroborate the findings on activated Müller cells and their possible loss in the severe forms of the disease [62].

A balanced regulation between MMPs and TIMPs is essential in controlling the innate host defense and tissue repair/homeostasis [63]. Under a chronic inflammatory phenotype, TLRs can drive the matrix makeover via the modulation of specific enzymes (MMPs and TIMPs) [64]. Our transcriptomic analysis for MMPs/TIMPs enzymes showed that MMP7 and TIMP2 were significantly modulated in ERMs. A possible connection between MMP and TLRs could be envisaged, given that a strong positive correlation was detected between MMP7 and TLR2.

The Pattern Recognition Receptors (PRRs), including toll-like receptors (TLRs) and complement receptors (CRs), regulate the processes of recognition of pathogens/PAMPs and their clearance [65]. A strong interaction between complement fragments and TLR signaling has been reported in previous in vivo studies, suggesting a novel mechanism by which the complement system might promote inflammation and modulate adaptive immunity in tissues [31]. Activated complement fragments are crucial mediators of inflammation, and the complement cascade activation can actively mediate the inflammatory process working on leucocytes' recruitment and activation [66–68]. The presence of C5b9 complexes, some complement regulators, and abundant cytokines has been discussed in retinal diseases [69]. The innate response is highly influenced by some complement components (C3, C4, C5 and C9) and their membrane regulators (decay-accelerating factor, membrane cofactor protein (DAF9, and CD59) [70]. Herein, C3a and C5b9 were assayed in vitreous, and their expression was correlated with ERM severity. TLR2 was found to induce complement factors C3 [35]. TLR2 deficiency preserves tight junction expression and promotes RPE resistance to fragmentation [35]. The neutralization of TLR2 reduces the opsonizing fragments of C3 in the outer retina and protects the photoreceptor neurons from oxidative stress-induced degeneration [35]. The oxidative stress-induced formation of the terminal complement membrane attack complex and the infiltration of Iba1-positive cells are strikingly inhibited in TLR2-deficient retinas [35]. In previous studies, TLR2 was reported as a mediator of retinal degeneration in response to oxidative stress and was hypothesized to act as a "bridge" between oxidative damage and complement-mediated retinal pathology.

The limitations of this study are represented by the following: i. the absence of data on the quantification of soluble TLR2 and TLR4 forms and the detection of C5aR in ERMs, ii. a limited number of specimens for chip array analysis and correlation studies and finally, iii. the presence of some risk factors that might influence innate immune response, such as gender, ageing and some major metabolic comorbidities [71–75]. The small study population is in line with the observation that 2% of middle-aged individuals (over age 50) and 20% of elderly individuals (over age 75) show evidence of ERMs, although they do not need treatment [1]. The gender bias might also represent a limitation of this study, although age/sex matched cases and controls were compared. Although recognized as an immune privilege site, the eye might respond to changes in the sexual hormones and ageing. The hormonal levels can influence/modulate the immune response locally, and particularly females can display a higher immune response than age-matched males [73]. On the other hand, ageing is also known to influence circulating and tissue-specific hormonal levels [74]. In fact, immune parameters decline at a slower rate in aged women, although genome activity remains higher for acquired immunity than for innate immunity. On the other hand, men exhibit higher gene activity in innate immunity [74]. Certainly, our study populations included biosamples collected from patients having risk factors related to vascular impairments and/or metabolic, autoimmune and allergic disorders, while a small population of patients presented with an apparent good health condition. Previous studies suggest that the ERMs are relatively common among the aged population, and the potential risk factors include few but major metabolic comorbidities, such as hypertension, type 2 diabetes and hypercholesterolemia/dyslipidemia, in addition to age and sex [75]. The association of TLR2 and TLR4 with allergies is in line with previous studies, so we cannot exclude such influences on the modulation of TLRs in ERM severity [74]. Our preliminary analysis on the influence of the major comorbidities (clustered as with or without hypertension) on TLR expression suggests the need for a more comprehensive analysis related to ERM formation or recurrence, considering these metabolic influences. While it is well established that TLRs participate in the inflammatory process aimed at PAMPs' and/or DAMPs' removal, TLRs' involvement at the beginning and progression of chronic pathologies appears to be just a possibility [75].

4. Materials and Methods

This study was approved by the intramural ethical committee (IFO-Bietti, Rome, Italy) and performed in accordance with the ethical standards stated in the Declaration of Helsinki. The patients approved the experimental approach and signed the consent forms for specimen collection, handling and analysis.

4.1. Study Population, SD-OCT Classification and Subgrouping

ERM (20 peeled-off membranes) and vitreal (12 collected samples) specimens were sampled from 20 patients (16F/4M; 71.45 ± 7.36 years old) and underwent vitrectomy with peeling of epiretinal membranes. Before surgery, a full ophthalmic examination was carried out, including anamnesis, funduscopic evaluation, and Spectral Domain–Optical Coherence Tomography (Spectralis SD-OCT ver.1.5.12.0; Heidelberg Engineering, Heidelberg, Germany) [25,26]. The ERM classification was as follows: stage 1, mild and thin ERMs with presence of foveal depression; stage 2, ERMs associated with widening of nuclear layer and loss of foveal depression; stage 3, ERMs associated with continuous ectopic inner foveal layers crossing the entire foveal area; and stage 4, thick ERMs associated with continuous ectopic inner foveal layers and severe disruption of retinal layers [25]. Inclusion criteria were stage 2, stage 3 and stage 4 (subgrouping). Exclusion criteria comprised stage 1 ERMs (patients with anatomical and morphological retinal status do not need surgery), macular holes, proliferative diabetic retinopathy, age-related macular degeneration, retinal vascular occlusion, aphakia, high myopia (≥28 diopters), uncontrolled glaucoma. Additional exclusion criteria were prior vitrectomy, intravitreal injections and retinal laser photocoagulation, as well as opacity of optical media. The characteristics of the study population and specimen collection are summarized in Table 1. The chart also shows paired vitreal samples whenever obtained.

4.2. Sampling Mode and Pre-Analytics

The procedure for sampling included the vitreous collection at the beginning of standard 23G pars plana vitrectomy, just before opening the infusion port, and subsequent ERM peel-off. Untouched vitreous was delivered to the laboratory and processed as described below.

ERMs were placed as whole-mounted tissues on pretreated glass slides (BDH, Milan, Italy), postfixed (ThinPrep® PreservCyt solution, Hologic, Inc., Marlborough, MA, USA) and probed for immunofluorescent analysis. For Real-Time PCR and IP/WB analyses, ERMs were extracted in 300 µL of lysis solution (MirVana Paris™ Invitrogen; Thermo Fisher Scientific, Waltham, MA, USA) and native protein, as well as total RNA, were obtained from each tissue/section.

4.3. Double Immunofluorescent Analysis and Digital Acquisitions

A total of 20 ERMs (N = 8 for stage 2, 6F/2M; N = 6 for stage 3, 5F/1M; N = 6 for stage 4, 5F/1M) were used for immunofluorescent analysis. Briefly, whole-mounted ERMs were briefly rehydrated in PBS (10 mM Phosphate Buffer (PB) and 137 mM NaCl; pH 7.5). A blocking/permeabilizing step (0.1% BSA and 0.3% Triton X100 in PBS) was carried out before antibody incubation: anti-human TLR2 antibody (mouse; 1:200; Santa Cruz, CA, USA), anti-human TLR4 antibody (rabbit; 1:200; Santa Cruz), anti-human α-SMA antibody (mouse; 1:500; Sigma-Aldrich, Saint Louis, MO, USA), anti- Iba1 antibody (mouse; 1:500; Santa Cruz) and Anti- GFAP antibody (mouse; 1:500; Cell Signaling, Denver, MA, USA). Specific bindings were detected using secondary AlexaFluor-555 or AlexaFluor-488-coupled anti-rabbit (for TLR4) or anti-mouse (for TLR2, α-SMA, Iba1, GFAP) specie-specific F(ab)2 antibodies diluted (1:500) in 0.05% Tween20-PBS and incubated for 45 min on benchtop (Immunological Sciences, Rome, Italy). After nuclear counterstaining (blue/DAPI; Invitrogen Molecular Probes, MA, USA), the slides were mounted with an anti-fading supplemented Vectashield (Vector Laboratories, Inc., Burlingame, CA, USA) and examined under an epifluorescent direct microscope (Ni-Eclipse) equipped with UV lamp, digital camera (Axiocam) and NIS-Elements software F 4.00.00 for digital assets (8-bit TIFF format; Nikon, Konan, Minato-ku, Tokyo). Acquisitions were carried out with ×200 and ×400 objectives, both in single and merged image sets.

4.4. Molecular Analysis: Total RNA, cDNA and Relative Real-Time PCR

Twenty ERMs previously immunostained were used for molecular analyses. Briefly, total RNAs were extracted according to the MirVana Paris procedure with minor modifications, rehydrated in 11 µL RNase-free water (DEPC-treated/autoclaved MilliQ water, Sigma-Aldrich) and spectrophotometrically checked for RNA quantity/purity (>1.8; A280 program, Nanodrop; Celbio, Euroclone S.p.A, Milano, Italy). For cDNA synthesis, 100 ng of total RNA was retro-transcribed using the GoScript RT mix and random hexamers (Promega, Madison, WI, USA) in a LifePro Thermal Cycler (Euroclone, Rome, Italy). For amplification, 3 µL (target gene) and 1 µL (referring gene) cDNA were amplified with the SYBR-green hot-start PCR master mix (Hydra Taq; Biolab; Biocell, Rome, Italy) using the Eco™ Real-Time PCR platform (Illumina, San Diego, CA, USA).

Amplifications were carried out for both target/referring genes per sample and in parallel with negative/positive controls. PCR products (100–200 bps) were separated in 2.5% agarose gel (mini horizontal apparatus; Bio-Rad, Hercules, CA, USA), and bands were observed/acquired inside the UVP station (TIFF-format images). Cycle threshold values (Cq) from normalized samples with one melting curve were used for REST analysis [76]. Changes in target gene expression observed at stage 3 and stage 4 were provided as log2 expression ratio with respect to stage 2 (herein considered the reference subgroup), considering H3 as the reference gene. Primer pairs were synthesized by Eurofin MWG Genomics (www.eurofinsgenomics.eu, accessed on 29 march 2023), with at least one intron spanning primer (Primer3 Input (version 0.4.0)), and the sequences are reported in Supplementary Table S1.

4.5. Biochemical Analysis: IP/WB, ELISA and Chip-Based Protein Array

A total of 12 vitreous (N = 5 for stage 2, 3F/2M; N = 4 for stage 3, 3F/1M; N = 3 for stage 4, 2F/1M) were used for biochemical analyses. Briefly, the samples (250–500 μL) were quickly stabilized with protease inhibitors (1 μL/sample; Pierce, Thermo Fisher Scientific) and split into two aliquots that were allocated to IP/WB—ELISA and chip array analysis, as described below. Aliquots for IP/WB—ELISA were sonicated (VibraCell; Sonics, Newtown, CT, USA) and centrifuged to remove debris (13,000 rpm/7 min; Sigma 1–14 microfuge), while sister aliquots for chip array were centrifuged (2000 rpm for 7 min; Sigma 1–14 microfuge) to separate floating cells. The clarified supernatants were further sonicated (VibraCell, Newtown, CT, USA) and centrifuged to remove debris (13,000 rpm/ 7 min). A spectrophotometer analysis (3 μL) was carried out (N1000, Nanodrop) before producing aliquots.

IP/WB. The captured antibodies of interest (Eotaxin, Santa Cruz; GFAP, Immunological Sciences; TLR2, Santa Cruz and TLR4, Santa Cruz) were first immobilized onto the Pure Proteome Protein G Magnetic beads (15 μL, Millipore, Burlington, MA, USA) to generate the antibody–beads complex. The beads-bound antibodies were then added to the normalized samples (100 μg total protein) and after 2 h of incubation, the captured proteins were washed and eluted in denaturing loading buffer. All steps were performed under orbital shaking (Certomat II, Sartorius, D-72393 Burladingen, Deutschland). LB samples were preheated at 90 °C for 10 min and loaded on 4–12% precasted SDS-PAGE gels (Bio-Rad Laboratories Inc., Hercules, CA, USA), and electrophoresis was performed in a MiniProtean3 apparatus (Bio-Rad) under reducing conditions (120 V/frontline).

Electrophoresed bands were transferred to 0.22 μm membranes (Hybond, GE Healthcare, Buckinghamshire, UK) at 12 V/40 min in a semidry Trans-Blotting apparatus (Bio-Rad). The membranes were stained with the high-sensible Sypro Ruby protein blot stainer (Invitrogen, Waltham, MA, USA), according to the standard procedure, to visualize and acquire the separations. Immunoblotting, followed by chemiluminescent detection, was additionally performed to better visualize the proteins of interest. In both cases, optical density (OD) was performed using the freely available ImageJ software (software 1.54j, National Institutes of Health, Bethesda, MD, USA). Data were saved as 8-bit TIFF files and exported for figure assembly using the Adobe Photoshop 2024 program (Release 25.5.1, Adobe Systems, Inc., Mountain View, CA, USA). The loading of normalized samples was verified by checking the β-actin content (predicted molecular weight: 40–42 kDa; sc-47778, Santa Cruz).

ELISA. Aliquots (50 μL/sample) were 1:2 diluted in a sample dilution buffer provided by commercially available C3a and C5b-9 ELISA kits (Cloud-Clone Corp., Houston, TX, USA), including precoated 96-well plates and ready-to-use solutions. All incubation and washing procedures were performed according to the manufacturers' suggestions, with minor modifications. Absorbance (optical density—OD) values were recorded after plate reading (λ450–λ570 nm) and concentrations were produced according to standard curves (assay range: 5000 pg/mL–78 pg/mL for C3a and 80 ng/mL–1.25 ng/mL for C5b-9).

Chip-Based Protein Array. Sister aliquots were analyzed in customized G-series glass slides (14 identical subarrays/slide with 57 targets; RayBiotech, Norcross, CA, USA). Array chips were incubated with prediluted vitreous samples and labelled with a biotin-conjugated cocktail of antibodies, followed by a Cy3-conjugated streptavidin complex, according to the manufacturer's instructions. As a final point, the glass slides were washed once with MilliQ water, spin-dried (700 g/1 min) and scanned using a GenePix 4400 Microarray scanner (Molecular Devices LLC, Sunnyvale, Silicon Valley, CA, USA). The fluorescence intensity data (FI) were generated by GenePix Pro 6.0 software (Axon Instruments, Foster City, CA, USA) and provided as background-subtracted FI data (F532-B532, N factor) for each spot/volume. Values were expressed as ratio (pathological vs. reference signal). Inter-assay normalization was guaranteed by the presence of multiple internal controls for each subarray. To obtain appropriate Cy5 (background signal) and Cy3 (specific signal) images, the slides were scanned over previously validated acquisition parameters and the images/arrays (blocks) were uniformly adjusted for size, brightness and contrast at the time of acquisition.

4.6. Statistics

Data were first analyzed using the Kolmogorov–Smirnov and the Shapiro–Wilk tests to satisfy the assumption of values coming from a normally distributed population (Prism10.0; GraphPad Software Inc., San Diego, CA, USA). The one-way ANOVA analysis was used to compare protein expression between the subgroups. The REST–ANOVA coupled analysis was carried out for identifying significant changes in Real-Time PCR experiments. Correlations were assessed between subgroups using the Pearson rho correlation test. Some statistical analyses were confirmed with RStudio (Version: 2024.04.2+764, open-source platform). The two-sided unpaired *t*-test comparisons or the Wilcoxon–Mann–Whitney test followed by the Benjamini–Hochberg procedure were used for multiple test corrections. Significance levels were * $p < 0.05$, ** $p < 0.01$, *** $p < 0.001$ and **** $p < 0.0001$.

Continuous variables were reported as mean values and standard deviations, and categorical variables were represented as numbers and percentages. The statistical tests used in this study were the independent-sample *t*-test, Chi-squared test, Mann–Whitney U-test and Spearman correlation, as appropriate.

5. Conclusions

Corroborating studies suggest that the presence of innate immune cells and chronic low-grade inflammation might influence the development, function and evolution of ERMs and even their reappearance [77]. TLRs have the primary role to detect 'non-self' pathogen molecules, sensor endogenous tissue-damaged host-derived molecules (DAMPs) and guarantee tissue homeostasis. TLRs work in combination with complement fragments, ameliorating some inflammatory states linked to DAMPs' recognition, although exacerbations of the local inflammation and senescence have also been reported [31,78,79]. Our findings on the presence of TLR2 and TLR4 in these ERMs and their specific expression by contractile α-SMA-expressing cells, Iba1-immunoreactive microglial cells and GFAP-positive Müller cells would suggest the participation of innate response in the evolution and contraction abilities of ERMs [41–47]. Since the possibility of impaired TLR/DAMP signaling cannot be excluded, the TLR/complement system and its modulation might be, in the near future, an additional intravitreal target, especially to counteract recidivism. The ERM system might be an interesting "ex vivo" model to assess the inflammatory status and the cross-talk between the innate and adaptive response. Although vitrectomy and ERM peel-off continue to be the first-choice therapy for vitreoretinal diseases, much remains to be investigated regarding the development of alternative "drug" therapies for the prevention and early treatment of vitreous–retinal interface syndromes that are not associated with traumatic events. Further studies are underway to better elucidate these aspects, which could be useful in the near future for individualized therapies (personalized medicine).

Supplementary Materials: The supporting information can be downloaded at: https://www.mdpi.com/article/10.3390/ijms25147732/s1.

Author Contributions: Conceptualization, L.D., G.E., A.C. and A.M.; formal analysis, L.D., G.E., B.O.B., A.C. and P.C.; funding acquisition, A.C., G.R. and A.M.; investigation, L.D., G.E., A.C., P.C. and G.E.; methodology, L.D., G.E., A.C., P.C., C.C. and A.M.; project administration, A.C. and A.M.; resources, G.R. and A.M.; supervision, A.C. and A.M.; validation, L.D., G.E., B.O.B., A.C., P.C., C.C., G.R. and A.M.; writing—original draft, L.D., G.E., A.C. and A.M.; writing—review and editing, L.D., G.E., A.C., B.O.B. and A.M. All authors gave their final approval and agreed to be accountable for all aspects of the work. All authors have read and agreed to the published version of the manuscript.

Funding: This research was funded by the Italian Ministry of Health (RC2788872) and Fondazione Roma (Italy). The sponsors had no role in the design of this study nor in the collection, analysis or even interpretation of the data.

Institutional Review Board Statement: All procedures were designed for human participants and approved by the intramural IFO/Bietti Ethical Committee and were in accordance with the Helsinki Declaration (1964) and later amendments, in line with ethical standards.

Informed Consent Statement: Written consent for this study was provided by the participants, and signed informed consent was acquired before enrollment.

Data Availability Statement: All data generated or analyzed during this study are included in this published article.

Acknowledgments: This study was partially supported by the Italian Ministry of Health, 5xMille 2016 to Fondazione Bietti and Fondazione Roma (Italy). Many thanks to Giuliana Facciolo (Retinal Surgery Research Unit, IRCCS-Fondazione Bietti, Rome, Italy) for her helpful contribution in sampling and managing tissues at the time of the vitrectomy.

Conflicts of Interest: All authors declare no affiliations with or involvement in any organization or entity with any financial or non-financial interest in the subject matter or materials discussed in this manuscript.

References

1. Tsotridou, E.; Loukovitis, E.; Zapsalis, K.; Pentara, I.; Asteriadis, S.; Tranos, P.; Zachariadis, Z.; Anogeianakis, G. A Review of Last Decade Developments on Epiretinal Membrane Pathogenesis. *Med. Hypothesis Discov. Innov. Ophthalmol.* **2020**, *9*, 91–110.
2. Shi, Y.; Hui, Y.; Ma, J. An immunohistochemical study of IgG, complement C3, collagen type III and macrophage-marker Ki-M7 in epiretinal membranes. *Yan Ke Xue Bao* **1996**, *12*, 10–14. [PubMed]
3. Tang, S.; Scheiffarth, O.F.; Wildner, G.; Thurau, S.R.; Lund, O.E. Lymphocytes, macrophages and HLA-DR expression in vitreal and epiretinal membranes of proliferative vitreoretinopathy. An immunohistochemical study. *Ger. J. Ophthalmol.* **1992**, *1*, 176–179. [PubMed]
4. Bianchi, L.; Altera, A.; Barone, V.; Bonente, D.; Bacci, T.; De Benedetto, E.; Bini, L.; Tosi, G.M.; Galvagni, F.; Bertelli, E. Untangling the Extracellular Matrix of Idiopathic Epiretinal Membrane: A Path Winding among Structure, Interactomics and Translational Medicine. *Cells* **2022**, *15*, 2531. [CrossRef] [PubMed]
5. Hu, M.; Ling, Z.; Ren, X. Extracellular matrix dynamics: Tracking in biological systems and their implications. *J. Biol. Eng.* **2022**, *16*, 13. [CrossRef] [PubMed]
6. Cacciamani, A.; Parravano, M.; Scarinci, F.; Esposito, G.; Varano, M.; Micera, A. A Simple Spontaneous Vitreal Reflux Collecting Procedure During Intravitreal Injection: Set-Up and Validation Studies. *Curr. Eye Res.* **2016**, *41*, 971–976. [CrossRef] [PubMed]
7. Dinice, L.; Cacciamani, A.; Esposito, G.; Taurone, S.; Carletti, R.; Ripandelli, G.; Artico, M.; Micera, A. Osteopontin in vitreous and idiopathic epiretinal membranes. *Graefes Arch. Clin. Exp. Ophthalmol.* **2020**, *258*, 1503–1513. [CrossRef] [PubMed]
8. Da Silva, R.A.; Roda, V.M.P.; Matsuda, M.; Siqueira, P.V.; Lustoza-Costa, G.J.; Wu, D.C.; Hamassaki, D.E. Cellular components of the idiopathic epiretinal membrane. *Graefes Arch. Clin. Exp. Ophthalmol.* **2022**, *260*, 1435–1444. [CrossRef]
9. El-Asrar, A.M.A.; Nawaz, M.I.; Kangave, D.; Siddiquei, M.M.; Geboes, K. Osteopontin and other regulators of angiogenesis and fibrogenesis in the vitreous from patients with proliferative vitreoretinal disorders. *Mediat. Inflamm.* **2012**, *2012*, 493043. [CrossRef]
10. Bianchi, M.E. DAMPs, PAMPs and alarmins: All we need to know about danger. *J. Leukoc. Biol.* **2007**, *81*, 1–5. [CrossRef]
11. Gaudet, A.D.; Popovich, P.G. Extracellular matrix regulation of inflammation in the healthy and injured spinal cord. *Exp. Neurol.* **2014**, *258*, 24–34. [CrossRef] [PubMed]
12. Frevert, C.W.; Felgenhauer, J.; Wygrecka, M.; Nastase, M.V.; Schaefer, L. Danger-Associated Molecular Patterns Derived From the Extracellular Matrix Provide Temporal Control of Innate Immunity. *J. Histochem. Cytochem.* **2018**, *66*, 213–227. [CrossRef] [PubMed]
13. Clett, E. Endogenous ligands of TLR2 and TLR4: Agonists or assistants? *J. Leukoc. Biol.* **2010**, *87*, 989–999.
14. Sirisinha, S. Insight into the mechanisms regulating immune homeostasis in health and disease. *Asian Pac. J. Allergy Immunol.* **2011**, *29*, 1–14. [PubMed]
15. Micera, A.; Balzamino, B.O.; Di Zazzo, A.; Biamonte, F.; Sica, G.; Bonini, S. Toll-Like Receptors and Tissue Remodeling: The Pro/Cons Recent Findings. *J. Cell Physiol.* **2016**, *231*, 531–544. [CrossRef] [PubMed]
16. Medzhitov, R. Toll-like receptors and innate immunity. *Nat. Rev. Immunol.* **2001**, *1*, 135–145. [CrossRef] [PubMed]
17. Tukhvatulin, A.I.; Logunov, D.Y.; Shcherbinin, D.N.; Shmarov, M.M.; Naroditsky, B.S.; Gudkov, A.V.; Gintsburget, A.L. Toll-Like Receptors and Their Adapter Molecules. *Biochemistry* **2010**, *75*, 1098–1114. [CrossRef] [PubMed]
18. Micera, A.; Stampachiacchiere, B.; Aronni, S.; Serapiao dos Santos, M.; Lambiase, A. Toll-like receptors and the eye. *Curr. Opin. Allergy Clin. Immunol.* **2005**, *5*, 451–458. [CrossRef] [PubMed]
19. Uematsu, S.; Akira, S. Toll-Like receptors (TLRs) and their ligands. *Handb. Exp. Pharmacol.* **2008**, *183*, 1–20.
20. Stepp, M.A.; Menko, A.S. Immune responses to injury and their links to eye disease. *Transl. Res.* **2021**, *236*, 52–71. [CrossRef]

21. Hossain, M.J.; Morandi, E.; Tanasescu, R.; Frakich, N.; Caldano, M.; Onion, D.; Faraj, T.A.; Clett, E.; Gran, B. The Soluble Form of Toll-Like Receptor 2 Is Elevated in Serum of Multiple Sclerosis Patients: A Novel Potential Disease Biomarker. *Front Immunol.* **2018**, *9*, 457. [CrossRef] [PubMed]
22. Zunt, S.L.; Burton, L.V.; Goldblatt, L.I.; Dobbins, E.E.; Srinivasanet, M. Soluble forms of Toll-like receptor 4 are present in human saliva and modulate tumor necrosis factor-alpha secretion by macrophage-like cells. *Clin. Exp. Immunol.* **2009**, *156*, 285–293. [CrossRef] [PubMed]
23. Hajishengallis, G.; Lambris, J.D. Crosstalk pathways between Toll-like receptors and the complement system. *Trends Immunol.* **2010**, *31*, 154–163. [CrossRef]
24. Zhao, M.; Li, C.H.; Liu, Y.L. Toll-like receptor (TLR)-2/4 expression in retinal ganglion cells in a high-glucose environment and its implications. *Genet. Mol. Res.* **2016**, *25*, 2. [CrossRef] [PubMed]
25. Govetto, A.; Lalane, R.A.; Sarraf, D.; Figueroa, M.S.; Hubschman, J.P. Insights Into Epiretinal Membranes: Presence of Ectopic Inner Foveal Layers and a New Optical Coherence Tomography Staging Scheme. *Am. J. Ophthalmol.* **2017**, *175*, 99–113. [CrossRef] [PubMed]
26. Cacciamani, A.; Cosimi, P.; Ripandelli, G.; Di Nicola, M.; Scarinci, F. Epiretinal Membrane Surgery: Structural Retinal Changes Correlate with the Improvement of Visual Function. *J. Clin. Med.* **2021**, *10*, 90. [CrossRef]
27. Zakeri, A.; Russo, M. Dual Role of Toll-like Receptors in Human and Experimental Asthma Models. *Front. Immunol.* **2018**, *9*, 1027. [CrossRef] [PubMed]
28. Isaza-Correa, J.M.; Liang, Z.; van den Berg, A.; Diepstra, A.; Visser, L. Toll-like receptors in the pathogenesis of human B cell malignancies. *J. Hematol. Oncol.* **2014**, *7*, 57. [CrossRef]
29. Zhang, G.; Ghosh, S. Toll-like receptor–mediated NF-κB activation: A phylogenetically conserved paradigm in innate immunity. *J. Clin. Investig.* **2001**, *107*, 13–19. [CrossRef]
30. Yu, W.-C.; Cong, J.-P.; Mi, L.-Y. Expressions of TOLL-like receptor 4 (TLR-4) and matrix metalloproteinase 9 (MMP-9)/Tissue inhibitor of metalloproteinase 1 (TIMP-1) in pulmonary blood vessels with chronic obstructive pulmonary diseases and their relationships with pulmonary vascular remodelling. *Rev. Assoc. Med. Bras.* **2018**, *64*, 361–367.
31. Hajishengallis, G.; Lambris, J.D. More than complementing Tolls: Complement–Toll-like receptor synergy and crosstalk in innate immunity and inflammation. *Immunol. Rev.* **2016**, *274*, 233–244. [CrossRef] [PubMed]
32. Mahaling, B.; Low, S.W.Y.; Beck, M.; Kumar, D.; Ahmed, S.; Connor, T.; Ahmad, B.; Chaurasia, S.S. Damage-Associated Molecular Patterns (DAMPs) in Retinal Disorders. *Int. J. Mol. Sci.* **2022**, *23*, 2591. [CrossRef] [PubMed]
33. Bolourani, S.; Brenner, M.; Wang, P. The interplay of DAMPs, TLR4, and proinflammatory cytokines in pulmonary fibrosis. *J. Mol. Med.* **2021**, *99*, 1373–1384. [CrossRef] [PubMed]
34. Lee, J.J.; Wang, P.W.; Yang, I.H.; Huang, H.-M.; Chang, C.-S.; Wu, C.-L.; Chuang, J.-H. High-fat diet induces toll-like receptor 4-dependent macrophage/microglial cell activation and retinal impairment. *Investig. Ophthalmol. Vis. Sci.* **2015**, *56*, 3041–3050. [CrossRef] [PubMed]
35. Mulfaul, K.; Ozaki, E.; Fernando, N.; Brennan, K.; Chirco, K.R.; Connolly, E.; Greene, C.; Maminishkis, A.; Salomon, R.G.; Linetsky, M.; et al. Toll-like Receptor 2 Facilitates Oxidative Damage-Induced Retinal Degeneration. *Cell. Rep.* **2020**, *30*, 2209–2224.e5. [CrossRef] [PubMed]
36. Klettner, A.; Roider, J. Retinal Pigment Epithelium Expressed Toll-like Receptors and Their Potential Role in Age-Related Macular Degeneration. *Int. J. Mol. Sci.* **2021**, *22*, 8387. [CrossRef] [PubMed]
37. Bayan, N.; Yazdanpanah, N.; Rezaei, N. Role of toll-like receptor 4 in diabetic retinopathy. *Pharmacol. Res.* **2022**, *175*, 105960. [CrossRef] [PubMed]
38. Lagos, L.S.; Luu, T.V.; De Haan, B.; Faas, M.; De Vos, P. TLR2 and TLR4 activity in monocytes and macrophages after exposure to amoxicillin, ciprofloxacin, doxycycline and erythromycin. *J. Antimicrob. Chemother.* **2022**, *77*, 2972–2983. [CrossRef]
39. El-Zayat, S.R.; Sibaii, H.; Mannaa, F.A. Toll-like receptors activation, signaling, and targeting: An overview. *Bull. Natl. Res. Cent.* **2019**, *43*, 187. [CrossRef]
40. Xu, H.; Chen, M. Diabetic retinopathy and dysregulated innate immunity. *Vision Res.* **2017**, *139*, 39–46. [CrossRef]
41. Yazici, A.T.; Alagoz, N.; Cxelik, H.U.; Alagöz, C.M.; Cakir, M.; Cekiç, O.; Yilmaz, O.F. Idiopathic and secondary epiretinal membranes: Do they differ in terms of morphology? An optical coherence tomography-based study. *Retina* **2011**, *31*, 779–784. [CrossRef] [PubMed]
42. Oberstein, S.Y.L.; Byun, J.; Herrera, D.; Chapin, E.A.; Fisher, S.K.; Lewis, G.P. Cell proliferation in human epiretinal membranes: Characterization of cell types and correlation with disease condition and duration. *Mol. Vis.* **2011**, *17*, 1794–1805. [PubMed]
43. Minchiotti, S.; Stampachiacchiere, B.; Micera, A.; Lambiase, A.; Ripandelli, G.; Billi, B.; Bonini, S. Human idiopathic epiretinal membranes express NGF and NGF receptors. *Retina* **2008**, *28*, 628–637. [CrossRef]
44. Guerin, C.J.; Wolfshagen, R.W.; Eifrig, D.E.; Anderson, D.H. Immunocytochemical identification of Muller's glia as a component of human epiretinal membranes. *Investig. Ophthalmol. Vis. Sci.* **1990**, *31*, 1483–1491.
45. He, C.; Lai, P.; Wang, J.; Zhou, T.; Huang, Z.; Zhou, L.; Liu, X. TLR2/4 deficiency prevents oxygen-induced vascular degeneration and promotes revascularization by downregulating IL-17 in the retina. *Sci. Rep.* **2016**, *6*, 27739. [CrossRef] [PubMed]
46. Lin, X.; Fang, D.; Zhou, H.; Su, S.B. The expression of Toll-like receptors in murine Müller cells, the glial cells in retina. *Neurol. Sci.* **2013**, *34*, 1339–1346. [CrossRef]

47. Holm, T.H.; Draeby, D.; Owens, T. Microglia are required for astroglial Toll-like receptor 4 response and for optimal TLR2 and TLR3 response. *Glia* **2012**, *60*, 630–638. [CrossRef] [PubMed]
48. Semeraro, F.; Morescalchi, F.; Duse, S.; Gambicorti, E.; Russo, A.; Costagliola, C. Current Trends about Inner Limiting Membrane Peeling in Surgery for Epiretinal Membranes. *J. Ophthalmol.* **2015**, *2015*, 671905. [CrossRef]
49. Xu, W.-Q.; Wang, Y.-S. The role of Toll-like receptors in retinal ischemic diseases. *Int. J. Ophthalmol.* **2016**, *9*, 1343–1351.
50. Ko, M.K.; Saraswathy, S.; Parikh, J.G.; Rao, N.A. The role of TLR4 activation in photoreceptor mitochondrial oxidative stress. *Investig. Ophthalmol. Vis. Sci.* **2011**, *52*, 5824–5835. [CrossRef]
51. Kochan, T.; Singla, A.; Tosi, J.; Kumar, A. Toll-like receptor 2 ligand pretreatment attenuates retinal microglial inflammatory response but enhances phagocytic activity toward Staphylococcus aureus. *Infect. Immun.* **2012**, *80*, 2076–2088. [CrossRef] [PubMed]
52. Kumar, A.; Shamsuddin, N. Retinal Müller glia initiate innate response to infectious stimuli via toll-like receptor signaling. *PLoS ONE* **2012**, *7*, e29830.
53. Bu, S.-C.; Kuijer, R.; Li, X.-R.; Hooymans, J.M.M.; Los, L.I. Idiopathic Epiretinal Membrane. *Retina* **2014**, *34*, 2317–2335. [CrossRef] [PubMed]
54. Owuraku, T.-L.; Imran, M.; Winfried, M.A. Toll-Like Receptor Signalling Pathways and the Pathogenesis of Retinal Diseases. *Front. Ophthalmol.* **2022**, *2*, 850394.
55. Garces, K.; Carmy, T.; Illiano, P.; Brambilla, R.; Hackam, A.S. Increased Neuroprotective Microglia and Photoreceptor Survival in the Retina from a Peptide Inhibitor of Myeloid Differentiation Factor 88 (MyD88). *J. Mol. Neurosci.* **2020**, *70*, 968–980. [CrossRef] [PubMed]
56. Li, M.; Carpio, D.F.; Zheng, Y.; Bruzzo, P.; Singh, V.; Ouaaz, F.; Medzhitov, R.M.; Beg, A.A. An essential role of the NF-kappa B/Toll-like receptor pathway in induction of inflammatory and tissue-repair gene expression by necrotic cells. *J Immunol.* **2001**, *166*, 7128–7135. [CrossRef]
57. Szczepan, M.; Llorián-Salvador, M.; Chen, M.; Xu, H. Immune Cells in Subretinal Wound Healing and Fibrosis. *Front. Cell. Neurosci.* **2022**, *16*, 916719. [CrossRef] [PubMed]
58. Shio, M.T.; Hassan, G.S.; Shah, W.A.; Nadiri, A.; El Fakhry, Y.; Li, H.; Mourad, W. Coexpression of TLR2 or TLR4 with HLA-DR potentiates the superantigenic activities of Mycoplasma arthritidis-derived mitogen. *J. Immunol.* **2014**, *192*, 2543–2550. [CrossRef] [PubMed]
59. Balogh, A.; Milibák, T.; Szabó, V.; Nagy, Z.Z.; Kaarniranta, K.; Resch, M.D. Immunological biomarkers of the vitreous responsible for proliferative alteration in the different forms of retinal detachment. *BMC Ophthalmol.* **2020**, *20*, 491. [CrossRef]
60. Tsai, C.-S.; Huang, C.-Y.; Chen, C.-H.; Lin, Y.-W.; Shih, C.-M.; Tsao, N.-W.; Chiang, K.-H.; Lee, C.-Y.; Jeng, H.; Lin, F.-Y. Eotaxin-2 increased toll-like receptor 4 expression in endothelial cells in vitro and exacerbates high-cholesterol diet-induced atherogenesis in vivo. *Am. J. Transl. Res.* **2016**, *8*, 5338–5353.
61. Jünemann, A.G.M.; Rejdak, R.; Huchzermeyer, C.; Maciejewski, R.; Grieb, P.; Kruse, F.E.; Zrenner, E.; Rejdak, K.; Petzold, A. Elevated vitreous body glial fibrillary acidic protein in retinal diseases. *Graefes Arch. Clin. Exp. Ophthalmol.* **2015**, *253*, 2181–2186. [CrossRef] [PubMed]
62. Ali, M.F.; Dasari, H.; Van Keulen, V.P.; Cornec, D.; Vasmatzis, G.; Peikert, T.; Carmona, E.M. Microbial Antigens Stimulate Metalloprotease-7 Secretion in Human B-Lymphocytes Using mTOR-Dependent and Independent Pathways. *Sci. Rep.* **2017**, *7*, 3869. [CrossRef]
63. Lisboa, R.A.; Andrade, M.V.; Cunha-Mel, J.R. Toll-like receptor activation and mechanical force stimulation promote the secretion of matrix metalloproteinases 1, 3 and 10 of human periodontal fibroblasts via p38, JNK and NF-kB. *Arch. Oral Biol.* **2013**, *58*, 731–739. [CrossRef]
64. Kumar, V. The complement system, toll-like receptors and inflammasomes in host defense: Three musketeers' one target. *Int. Rev. Immunol.* **2019**, *38*, 131–156. [CrossRef] [PubMed]
65. Mondino, B.J.; Sidikaro, Y.; Sumner, H. Anaphylatoxin levels in human vitreous humor. *Investig. Ophthalmol. Vis. Sci.* **1988**, *29*, 1195–1198.
66. Mondino, B.J.; Sumner, H. Anaphylatoxin levels in human aqueous humor. *Investig. Ophthalmol. Vis. Sci.* **1986**, *8*, 1288–1292.
67. Mondino, B.J.; Glovsky, M.M.; Ghekiere, L. Activated complement in inflamed aqueous humor. *Investig. Ophthalmol. Vis. Sci.* **1984**, *25*, 871–873.
68. Lueck, K.; Wasmuth, S.; Williams, J.; Hughes, T.R.; Morgan, B.P.; Lommatzsch, A.; Greenwood, J.; Moss, S.E.; Pauleikhoff, D. Sub-lytic C5b-9 induces functional changes in retinal pigment epithelial cells consistent with age-related macular degeneration. *Eye* **2011**, *25*, 1074–1082. [CrossRef] [PubMed]
69. Dunkelberger, J.R.; Song, W.-C. Complement and its role in innate and adaptive immune responses. *Cell Res.* **2010**, *20*, 34–50. [CrossRef]
70. Jaén, R.I.; Val-Blasco, A.; Prieto, P.; Gil-Fernández, M.; Smani, T.; López-Sendón, J.L.; Delgado, C.; Boscá, L.; Fernández-Velasco, M. Innate Immune Receptors, Key Actors in Cardiovascular Diseases. *JACC Basic Transl. Sci.* **2020**, *5*, 735–749. [CrossRef]
71. Wenger, M.; Grosse-Kathoefer, S.; Kraiem, A.; Pelamatti, E.; Nunes, N.; Pointner, L.; Aglas, L. When the allergy alarm bells toll: The role of Toll-like receptors in allergic diseases and treatment. *Front. Mol. Biosci.* **2023**, *10*, 1204025. [CrossRef] [PubMed]
72. Calabrò, A.; Accardi, G.; Aiello, A.; Caruso, C.; Candore, G. Sex and gender affect immune aging. *Front Aging.* **2023**, *28*, 1272118. [CrossRef] [PubMed]

73. Hägg, S.; Jylhävä, J. Sex differences in biological aging with a focus on human studies. *eLife* **2021**, *10*, e63425. [CrossRef] [PubMed]
74. Xiao, W.; Chen, X.; Yan, W.; Zhu, Z.; He, M. Prevalence and risk factors of epiretinal membranes: A systematic review and meta-analysis of population-based studies. *BMJ Open* **2017**, *7*, e014644. [CrossRef] [PubMed]
75. Kumar, V. Toll-like receptors in immunity and inflammatory diseases: Past, present, and future. *Int. Immunopharmacol.* **2018**, *59*, 391–412.
76. Pfaffl, M.W.; Horgan, G.W.; Dempfle, L. Relative Expression Software Tool (REST©) for group wise comparison and statistical analysis of relative expression results in real-time PCR. *Nucleic Acids Res.* **2002**, *30*, E36. [CrossRef] [PubMed]
77. Gong, T.; Liu, L.; Jiang, W.; Zhou, R. DAMP-sensing receptors in sterile inflammation and inflammatory diseases. *Nat. Rev. Immunol.* **2020**, *20*, 95–112. [CrossRef] [PubMed]
78. Lee, H.; Lee, S.; Cho, I.-H.; Lee, S.J. Toll-like receptors: Sensor molecules for detecting damage to the nervous system. *Curr. Protein Pept. Sci.* **2013**, *14*, 33–42. [CrossRef]
79. Barratt-Due, A.; Pischke, S.E.; Nilsson, P.H.; Espevik, T.; Mollnes, T.E. Dual inhibition of complement and Toll-like receptors as a novel approach to treat inflammatory diseases-C3 or C5 emerge together with CD14 as promising targets. *J. Leukoc. Biol.* **2017**, *101*, 193–204. [CrossRef]

International Journal of
Molecular Sciences

MDPI

Communication

A Plasma Pyrophosphate Cutoff Value for Diagnosing Pseudoxanthoma Elasticum

Isabelle Rubera [1], Laetitia Clotaire [1], Audrey Laurain [1], Alexandre Destere [2,3], Ludovic Martin [4], Christophe Duranton [1,*,†] and Georges Leftheriotis [1,3,†]

[1] University Côte d'Azur, CNRS, LP2M, Labex ICST, 06107 Nice, France; isabelle.rubera@univ-cotedazur.fr (I.R.); laetitia.clotaire@etu.univ-cotedazur.fr (L.C.); audrey.laurain@aphp.fr (A.L.); leftheriotis.g@chu-nice.fr (G.L.)
[2] Pharmacology Department, Nice University Hospital, 06000 Nice, France
[3] FAVA-MULTI South Competence Center for Rare Arterial Calcifying Disease, Nice University Hospital, 06000 Nice, France
[4] PXE Reference Center, MAGEC Nord, Angers University Hospital, 49000 Angers, France; lumartin@chu-angers.fr
* Correspondence: christophe.duranton@univ-cotedazur.fr; Tel.: +33-(0)489153550
† These authors contributed equally to this work.

Abstract: Pseudoxanthoma elasticum (PXE) is a rare inherited systemic disease responsible for a juvenile peripheral arterial calcification disease. The clinical diagnosis of PXE is only based on a complex multi-organ phenotypic score and/or genetical analysis. Reduced plasma inorganic pyrophosphate concentration [PPi]p has been linked to PXE. In this study, we used a novel and accurate method to measure [PPi]p in one of the largest cohorts of PXE patients, and we reported the valuable contribution of a cutoff value to PXE diagnosis. Plasma samples and clinical records from two French reference centers for PXE (PXE Consultation Center, Angers, and FAVA-MULTI South Competent Center, Nice) were assessed. Plasma PPi were measured in 153 PXE and 46 non-PXE patients. The PPi concentrations in the plasma samples were determined by a new method combining enzymatic and ion chromatography approaches. The best match between the sensitivity and specificity (Youden index) for diagnosing PXE was determined by ROC analysis. [PPi]p were lower in PXE patients (0.92 ± 0.30 μmol/L) than in non-PXE patients (1.61 ± 0.33 μmol/L, $p < 0.0001$), corresponding to a mean reduction of 43 ± 19% (SD). The PPi cutoff value for diagnosing PXE in all patients was 1.2 μmol/L, with a sensitivity of 83.3% and a specificity of 91.1% (AUC = 0.93), without sex differences. In patients aged <50 years (i.e., the age period for PXE diagnosis), the cutoff PPi was 1.2 μmol/L (sensitivity, specificity, and AUC of 93%, 96%, and 0.97, respectively). The [PPi]p shows high accuracy for diagnosing PXE; thus, quantifying plasma PPi represents the first blood assay for diagnosing PXE.

Keywords: diagnosis; rare disease; PXE; threshold; PPi; PPi quantification

Citation: Rubera, I.; Clotaire, L.; Laurain, A.; Destere, A.; Martin, L.; Duranton, C.; Leftheriotis, G. A Plasma Pyrophosphate Cutoff Value for Diagnosing Pseudoxanthoma Elasticum. *Int. J. Mol. Sci.* **2024**, *25*, 6502. https://doi.org/10.3390/ijms25126502

Academic Editor: Andrea Bernini

Received: 7 May 2024
Revised: 3 June 2024
Accepted: 9 June 2024
Published: 13 June 2024

1. Introduction

Pseudoxanthoma elasticum (PXE) is a chronic and disabling monogenic rare disease caused primarily by mutations in the *ABCC6* gene [1]. Its prevalence has been estimated between 1/25,000 and 1/100,000 in the general population, with female predominance [2]. PXE is characterized by abnormal ectopic calcifications in soft connective tissues such as the skin, retina and cardiovascular system. The clinical diagnosis of PXE [3,4] is currently based on clinical criteria (PHENODEX score [5]), which include the presence of 1 or 2 major ocular lesions (i.e., peau d'orange and/or angioid streaks) associated with 1 or 2 skin and/or mucosal lesions (i.e., pseudo-xanthomatous papules of the flexion folds) and/or a positive skin biopsy taken from the lesional skin, the neck or the flexural area in the absence of suggestive skin changes. Genetic diagnosis is a powerful method to detect

PXE [6] but may be unsuccessful since a small but significant level of PXE patients have no *ABCC6* mutations. PXE demonstrates considerable inter- and intrafamilial heterogeneity and, in some cases, overlaps with other calcifying diseases, such as generalized arterial calcification in infancy (GACI) [7], that is linked to heterozygous pathogenic mutations of the *ENPP1* gene [8]. Moreover, a PXE-like phenotype was also described due to pathogenic variants in the *GGCX* gene that encoded the gamma-glutamyl-carboxylase protein with alteration of anti-calcification processes [9]. Therefore, genetic diagnosis is presently reserved for atypical cases.

Due to the high phenotypic and clinical variability of PXE, there is a growing need to improve diagnostic accuracy due to prognostic implications and future development of therapeutic options for this disease.

The *ABCC6* gene involved in PXE encodes the ATP-binding cassette transporter ABCC6, the activity of which appears to be linked to the cellular outflow of ATP and the subsequent generation of inorganic pyrophosphate (PPi) [10]. Pyrophosphate is known as a physiological plasma anti-calcifying factor, and plasma PPi is chronically decreased in *Abcc6*$^{-/-}$ animal models [10–12] and by 25 to 60% in PXE patients [12–17]. To date, in the absence of standardized measurement methods to determine PPi concentration in the plasma, the diagnostic value of PPi for PXE has never been tested, and its relevance remains to be established.

Herein, we develop a new method combining enzymatic and ion chromatographic techniques to determine the plasma PPi concentration with accuracy, allowing to report a diagnostic value in one of the largest European cohorts of PXE patients.

2. Results

2.1. A Significant Decrease of Plasma PPi Concentration in a PXE Cohort

For this study, we have used a new method coupling two different technical approaches (enzymatic and ion chromatography), allowing a cross-control to accurately determine the PPi concentration in plasma samples.

In the non-PXE control cohort (46 healthy volunteers, (mean ± SD) age 45 ± 15 years; 55% women), the mean plasma PPi concentration was 1.61 ± 0.33 µmol/L, while in the PXE cohort (153 patients, 49 ± 15 years; 107 (71%) women), it was 0.93 ± 0.30 µmol/L (Figure 1). The PXE patients had a 43 ± 19% reduction in PPi concentration compared to non-PXE patients.

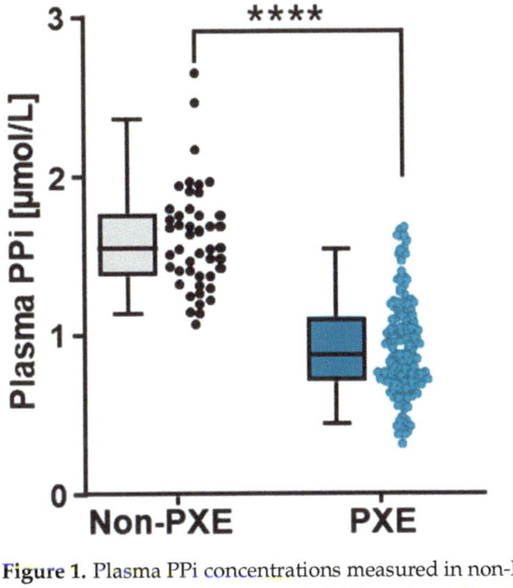

Figure 1. Plasma PPi concentrations measured in non-PXE (n = 46) and PXE patients (n = 153). Solid bars are the median values, and whiskers show the interquartiles (5/95). **** $p < 0.0001$.

2.2. Determination of a PPi Cutoff Value for PXE Diagnosis

The statistical analysis plan was used, and the main results of this study are presented in Figure 2. Briefly, two ROC analyses were performed with the training dataset, comprising all PXE and with a training dataset of <50 years old PXE patients (<50 y.o., Figure 3).

A PPi cutoff (Youden index) for diagnosing PXE in all patients was calculated at 1.2 µmol/L, with a sensitivity of 83.3%, a specificity of 91.1%, and an AUC of 0.93 (Figures 2 and 3A, Table 1).

With respect to the "simulated-training" dataset, the cutoff was 1.2 µmol/L, leading to a sensitivity of 83.0% and a specificity of 95.0% (Figure 2).

Analysis of patients aged <50 years old, the approximate age at which PXE should be diagnosed before the onset of complications, revealed that the PPi cutoff value was 1.2 µmol/L, with a sensitivity, specificity, and AUC of 92.6%, 95.8%, and 0.97, respectively (Figure 3B and Table 1). With respect to the "simulated-training" dataset, the cutoff was 1.2 µmol/L, with a sensitivity of 94.3% and a specificity of 97% (Figure 2).

The performance of the "test" dataset for all patients and patients aged <50 y.o. was (sensitivity/specificity) 92.3%/91.7% and 100%/100%, respectively (Figure 2).

The performances of the different thresholds (100% sensitivity threshold, Youden index, and 100% specificity threshold) in the overall population and patients aged <50 y.o. in the training and test dataset are presented in Table 1. The PPi cutoffs yielding 100% specificity (1.1 µM) or 100% sensitivity (1.7 µM) are identical in the overall population and in those aged <50 y.o. (Table 1).

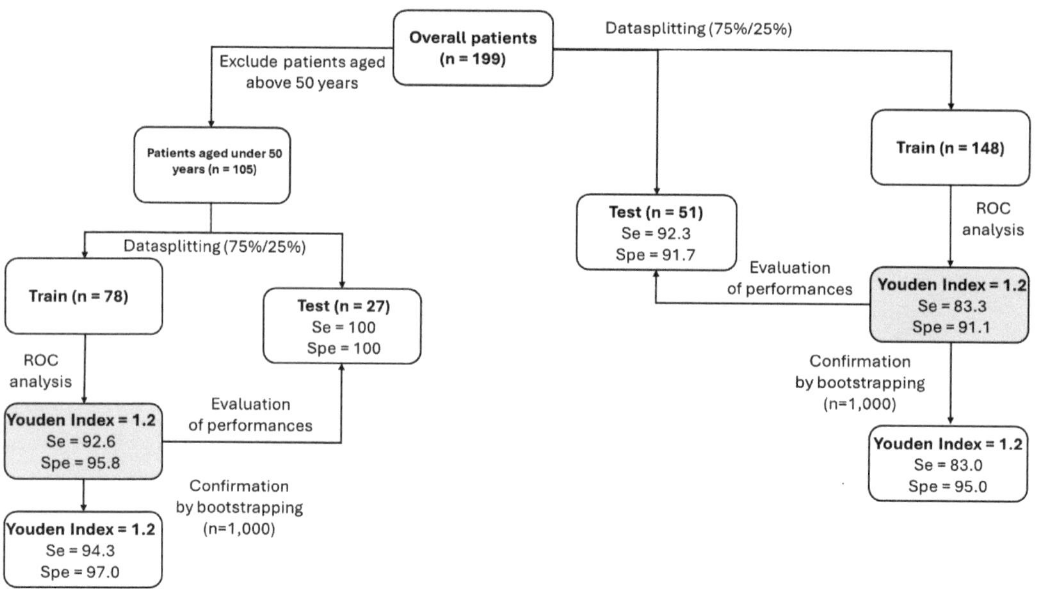

Figure 2. Proposed model for accurate determination of the PPi cutoff for PXE diagnosis. The overall population (n = 199) was split into training (75%, n = 148) and test (25%, n = 51) datasets. Cutoff values were determined by ROC analysis of the "training" dataset and of the "simulated-training" dataset (n = 1000, using a bootstrap analysis of the "training" dataset). The same strategy was used to determine the cutoff value in patients aged <50 y.o. Se: sensitivity; Spe: specificity.

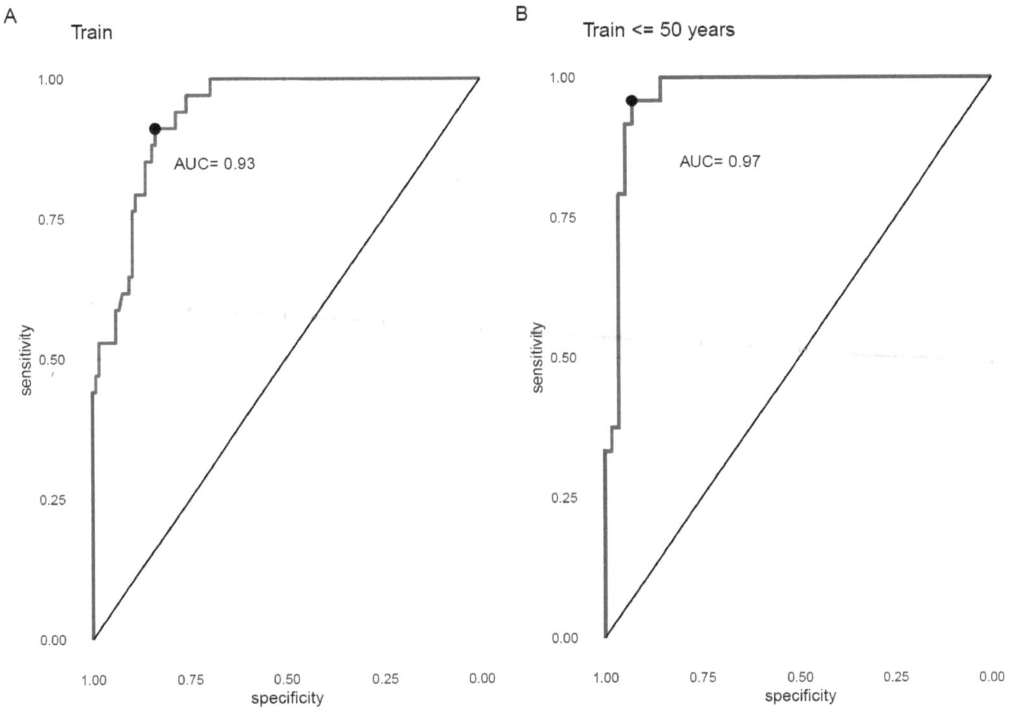

Figure 3. Determination of the best cutoff value of plasma inorganic pyrophosphate (PPi) for diagnosing PXE using receiver operating characteristic (ROC) curves in all patients (**A**) and in patients < 50 years old. (**B**). The black dots indicate the best compromise between sensitivity and specificity (Youden index). AUC: area under the curve.

Table 1. Performance of the determined thresholds (100% sensitivity, Youden index, and 100% specificity) in the "training" and "test" datasets for all patients and patients aged <50 y.o.

Threshold	Overall Patients				Patients Aged under 50 Years (<50 y.o.)			
	Train (n = 148)		Test (n = 51)		Train (n = 78)		Test (n = 27)	
	Sensitivity (%)	Specificity (%)	Sensitivity (%)	Specificity (%)	Sensitivity (%)	Specificity (%)	Sensitivity (%)	Specificity (%)
<1.70	100.0	44.1	100.0	33.3	100.0	33.3	100.0	37.5
<1.20	**83.3**	**91.1**	**92.3**	**91.7**	**92.6**	**95.8**	**100.0**	**100.0**
<1.10	69.3	100.0	79.5	100.0	85.2	100.0	89.5	100.0

2.3. Influence of Sex on the PPi Cutoff for PXE Diagnosis

The PPi cutoff value in the male and female "training" dataset aged <50 years old was 1.2 μmol/L for both men (sensitivity 94.4%, specificity 100%) and women (92.9% and 93.9%, respectively).

3. Discussion

Our study reports a cutoff plasma PPi value for diagnosing PXE, an inherited disease characterized by chronic disability due to extensive and progressive ectopic calcifications in connective tissues. Since 1994, the diagnosis of PXE has been based only on clinical symptoms, skin biopsy results, and family history [8]. Due to considerable inter- and intrafamilial phenotype heterogeneity, the involvement of different organ systems in this disease may greatly differ. The reasons for this phenotypic heterogeneity currently remain unknown, and attempts to establish genotype/phenotype correlations have failed to establish a consistent one [18] except in one study [19].

A low plasma PPi has been consistently reported in PXE patients with a reduction rate ranging from 25 to 60% [12–17] despite significant differences in the absolute values: plasma PPi levels in PXE patients versus healthy controls were reported in a Belgian cohort (mean value 0.497 vs. 0.985 μM [17]), in a French cohort (median value 0.77 vs. 1.53 μM [15]), in a Dutch cohort (0.53 vs. 1.13 μM [14]) and in a Spanish cohort (1.11 vs. 1.43 μM [16]). Altogether, comparisons of plasma PPi concentration between all these studies are somewhat difficult since it depends on (1) the type of blood collection tube (CTAD, citrate, heparin, EDTA, . . .), (2) the pre-analytic preparation of the plasma samples (i.e., ultrafiltration, platelet-depletion), and (3) the method used to measure PPi. A recent study [17] pointed out that PPi level does not correlate with the genotype or phenotype of PXE disease. However, we have already reported a strong correlation between age, arterial calcification, and disease severity [15], suggesting that time of exposure to a low plasma PPi concentration is a major determinant of arterial calcification and clinical severity in PXE.

According to our study, a plasma PPi cutoff value of 1.2 μmol/L detected almost 83% of PXE patients with high specificity (91%) and was not influenced by sex. Interestingly, when the population age was limited to <50 y.o. (i.e., the age period when PXE is usually diagnosed), the plasma PPi cutoff value was still 1.2 μmol/L, while the sensitivity and specificity increased to 93 and 96%, respectively. Therefore, PPi represents a promising biological blood marker that provides additional diagnostic criteria and reliable value for diagnosing PXE. To date, the use of plasma PPi for PXE diagnosis has never been established due to the lack of "gold standard" methods and technical limitations in precisely determining PPi concentrations, which are both overcome using our method.

Although the number of PXE patients included in this study is small, it remains among the largest since PXE is a rare disease. The threshold validation was performed on a small subset of the internal validation cohort, which can limit the performance of the external validation cohort. In addition, the use of the cutoff value is limited to our method, which combines enzymatic and ion chromatography techniques as a cross-control, which is not the case for the other tests usually carried out in laboratories. Therefore, access to this assay method in clinical practice remains to be developed.

In conclusion, our results show that the plasma PPi concentration may reliably contribute to PXE diagnosis. Furthermore, this blood test will significantly contribute to the development and validation of future treatments for this disease by clinical trials. A strategy currently being explored by the medical community is based on normalizing plasma PPi concentration in PXE patients [20]; therefore, it is essential to monitor PPi changes accurately over time during treatment.

4. Materials and Methods

4.1. Study Population

Medical records and biological samples collected between 2018 and 2023 from PXE patients were obtained from the biobank of the National PXE Reference Center (MAGEC Nord, Angers University Hospital, Angers, France) as part of the protocol for phenotyping the French PXE cohort (ClinicalTrials.gov Identifier: NCT01446380) and by the FAVA-MULTI South Competence Center for Rare Arterial Calcifying Disease, Nice University Hospital, Nice, France (ClinicalTrials.gov Identifier: NCT04868578). PXE diagnosis was performed by experienced clinicians from the two centers (LM and GL) on the basis of the actual diagnostic criteria for PXE [8,10]. Non-PXE patients (healthy volunteers) were recruited from routine clinical check-ups as controls. All PXE and non-PXE patients provided informed consent for the study.

4.2. Preanalytical Preparation of Blood Samples

Blood samples were collected in CTAD tubes on the morning after overnight fasting. These tubes were kept on ice until centrifugation (1000× *g*, 15 min, 4 °C). Plasma samples were collected and ultrafiltered at 4 °C using Amicon 0.5 mL Ultracel® filters (Sigma-Aldrich, Saint-Quentin Fallavier, France) to eliminate platelets and 95% of the proteins (including albumin). All the preanalytical steps were performed as quickly as possible to avoid PPi degradation or blood cell lysis.

4.3. Plasma PPi Concentration Measurement

The plasma PPi concentration was determined by a new method combining enzymatic and ion chromatography (IC) techniques. This method allows cross-control of the measured PPi values and increases the reliability of the result.

Enzymatic method: Quantification of the PPi concentration in ultrafiltered plasma was performed using a modified method based on a previously described enzyme assay [5]. PPi was converted to ATP using sulfurylase (Bio-Techne, Rennes, France) in the presence of APS (adenosine-5′-phosphosulfate; Sigma-Aldrich, Saint-Quentin Fallavier, France) after incubating for 40 min at 37 °C and subsequently for 10 min at 90 °C to inactivate sulfurylase. The generated ATP was quantified using an ATP Determination Kit (ATPlite; PerkinElmer, Courtabeuf, France). The luminescence was measured on a microplate reader (Synergy HT, BioTek, Shoreline, WA, USA). PPi values were obtained by subtracting the basal ATP level measured with ATP sulfurylase-inactivated solution in each sample.

Ion chromatography method: Patient plasma ultrafiltrates were deproteinized using acetonitrile (1/1). The samples were strongly mixed and centrifuged (12,000× *g*, 10 min, 4 °C). The same protocol (acetonitrile) was also used to quantify the calibration standard solutions (sodium pyrophosphate dibasic, Santa Cruz Technologies, Heidelberg, Germany). PPi detection by conductimetry was performed using an ion chromatographic Dionex ICS-5000 plus system (Thermo Scientific, Courtabeuf, France). The system was equipped with an eluent generator (EGC500KOH), a guard precolumn (AG11-HC), and an analytical column (IonPac AS-11-HC). PPi peak quantification was performed using "Chromeleon software version 7.2 SR4" (Thermo Scientific, Courtabeuf, France) by measuring the surface area, and the results were compared to a PPi standard curve.

4.4. Statistical Analysis

All the statistical analyses were performed in Prism 10.0 (GraphPad) or R software version 4.3.2 [21]. Data are presented as mean ± SD (standard deviation). Statistical differences between PXE patients and non-PXE patients were compared using the Mann–Whitney test.

The cutoff values for the plasma PPi concentration were determined using the receiver operating characteristic (ROC) curve method to determine the best compromise between sensitivity and specificity (Youden index) and also to explore the performance of other thresholds (100% sensitivity and 100% specificity). The ROC analysis was performed using the R package version 1.18.5 "pROC" [22].

The diagnostic performance of a given PPi cutoff value was tested after the overall population was split into "training" (75%, n = 148) and "test" (25%, n = 51) datasets. The cutoffs determined by the ROC analysis were subsequently compared to a cutoff determined using a bootstrap analysis issued from the "training" set (i.e., "simulated-training" dataset = 1000). Finally, the cutoff was tested in the "test" dataset. A similar analysis was performed to determine the cutoff value for patients <50 years old (<50 y.o.). The effect of sex on the cutoff was also determined in the "simulated-training" set and further applied to the test dataset to evaluate the diagnostic performance as a function of sex.

Author Contributions: I.R., C.D., L.M. and G.L. designed the research. G.L. and L.M. enrolled the patients and managed the PXE cohort. I.R., L.C., C.D. and A.L. performed the quantification of plasma inorganic pyrophosphate. A.D. analyzed the data and performed all statistical analysis. I.R., C.D. and G.L. wrote the manuscript. All authors have read and agreed to the published version of the manuscript.

Funding: The study was supported by the French Ministry of Public Health (PHRC-19-0402, Nice University Hospital. PROPHECI-PPI Study. NCT04868578), by the Angers University Hospital (49RC08-0018, NCT01446380), and by the "Agence Nationale de la Recherche" (ANR KI-PYRINA AAP CE17 grant).

Institutional Review Board Statement: The study was approved by the French ethical committee (references: 2008-A01372-53 (7 January 2010) and 2021-A03036-35, (7 February 2022)).

Informed Consent Statement: Informed consent was obtained from all subjects (PXE and non-PXE patients) involved in the study.

Data Availability Statement: The datasets generated during and/or analyzed during the current study are available from the corresponding author upon reasonable request. Biological samples for the study are stored and available in the Angers' biobank under the reference BB-0033-00038.

Acknowledgments: The authors would like to address their warm thanks to the patients who participated in the study and all the staff who contributed to its setup and logistics, including M. Bonnard, S. Bonnet, H. Humeau, S. Contu, and O. Kroselj.

Conflicts of Interest: The authors declare no conflicts of interest.

References

1. Le Saux, O.; Urban, Z.; Tschuch, C.; Csiszar, K.; Bacchelli, B.; Quaglino, D.; Pasquali-Ronchetti, I.; Pope, F.M.; Richards, A.; Terry, S.; et al. Mutations in a Gene Encoding an ABC Transporter Cause Pseudoxanthoma Elasticum. *Nat. Genet.* **2000**, *25*, 223–227. [CrossRef]
2. Germain, D.P. Pseudoxanthoma Elasticum. *Orphanet J. Rare Dis.* **2017**, *12*, 85. [CrossRef] [PubMed]
3. Lebwohl, M.; Neldner, K.; Pope, F.M.; De Paepe, A.; Christiano, A.M.; Boyd, C.D.; Uitto, J.; McKusick, V.A. Classification of Pseudoxanthoma Elasticum: Report of a Consensus Conference. *J. Am. Acad. Dermatol.* **1994**, *30*, 103–107. [CrossRef]
4. Uitto, J.; Jiang, Q.; Váradi, A.; Bercovitch, L.G.; Terry, S.F. Pseudoxanthoma Elasticum: Diagnostic features, classification, and treatment options. *Expert Opin. Orphan Drugs* **2014**, *2*, 567–577. [CrossRef] [PubMed]
5. Plomp, A.S.; Toonstra, J.; Bergen, A.A.; van Dijk, M.R.; de Jong, P.T. Proposal for Updating the Pseudoxanthoma Elasticum Classification System and a Review of the Clinical Findings. *Am. J. Med. Genet. A* **2010**, *152A*, 1049–1058. [CrossRef] [PubMed]
6. Vanakker, O.M.; Leroy, B.P.; Coucke, P.; Bercovitch, L.G.; Uitto, J.; Viljoen, D.; Terry, S.F.; Van Acker, P.; Matthys, D.; Loeys, B.; et al. Novel Clinico-Molecular Insights in Pseudoxanthoma Elasticum Provide an Efficient Molecular Screening Method and a Comprehensive Diagnostic Flowchart. *Hum. Mutat.* **2008**, *29*, 205. [CrossRef] [PubMed]
7. Ralph, D.; Nitschke, Y.; Levine, M.A.; Caffet, M.; Wurst, T.; Saeidian, A.H.; Youssefian, L.; Vahidnezhad, H.; Terry, S.F.; Rutsch, F.; et al. ENPP1 Variants in Patients with GACI and PXE Expand the Clinical and Genetic Heterogeneity of Heritable Disorders of Ectopic Calcification. *PLoS Genet.* **2022**, *18*, e1010192. [CrossRef]
8. Rutsch, F.; Ruf, N.; Vaingankar, S.; Toliat, M.R.; Suk, A.; Höhne, W.; Schauer, G.; Lehmann, M.; Roscioli, T.; Schnabel, D.; et al. Mutations in ENPP1 Are Associated with "idiopathic" Infantile Arterial Calcification. *Nat. Genet.* **2003**, *34*, 379–381. [CrossRef] [PubMed]
9. Vanakker, O.M.; Martin, L.; Gheduzzi, D.; Leroy, B.P.; Loeys, B.L.; Guerci, V.I.; Matthys, D.; Terry, S.F.; Coucke, P.J.; Pasquali-Ronchetti, I.; et al. Pseudoxanthoma Elasticum-like Phenotype with Cutis Laxa and Multiple Coagulation Factor Deficiency Represents a Separate Genetic Entity. *J. Investig. Dermatol.* **2007**, *127*, 581–587. [CrossRef]
10. Jansen, R.S.; Küçükosmanoglu, A.; de Haas, M.; Sapthu, S.; Otero, J.A.; Hegman, I.E.M.; Bergen, A.A.B.; Gorgels, T.G.M.F.; Borst, P.; van de Wetering, K. ABCC6 Prevents Ectopic Mineralization Seen in Pseudoxanthoma Elasticum by Inducing Cellular Nucleotide Release. *Proc. Natl. Acad. Sci. USA* **2013**, *110*, 20206–20211. [CrossRef] [PubMed]
11. Pomozi, V.; Brampton, C.; van de Wetering, K.; Zoll, J.; Calio, B.; Pham, K.; Owens, J.B.; Marh, J.; Moisyadi, S.; Varadi, A.; et al. Pyrophosphate Supplementation Prevents Chronic and Acute Calcification in ABCC6-Deficient Mice. *Am. J. Pathol.* **2017**, *187*, 1258–1272. [CrossRef]
12. Kauffenstein, G.; Yegutkin, G.G.; Khiati, S.; Pomozi, V.; Le Saux, O.; Leftheriotis, G.; Lenaers, G.; Henrion, D.; Martin, L. Alteration of Extracellular Nucleotide Metabolism in Pseudoxanthoma Elasticum. *J. Investig. Dermatol.* **2018**, *138*, 1862–1870. [CrossRef] [PubMed]
13. Jansen, R.S.; Duijst, S.; Mahakena, S.; Sommer, D.; Szeri, F.; Varadi, A.; Plomp, A.; Bergen, A.A.; Oude Elferink, R.P.; Borst, P.; et al. ABCC6-Mediated ATP Secretion by the Liver Is the Main Source of the Mineralization Inhibitor Inorganic Pyrophosphate in the Systemic Circulation-Brief Report. *Arterioscler. Thromb. Vasc. Biol.* **2014**, *34*, 1985–1989. [CrossRef]

14. Kozák, E.; Bartstra, J.W.; de Jong, P.A.; Mali, W.P.T.M.; Fülöp, K.; Tőkési, N.; Pomozi, V.; Risseeuw, S.; Norel, J.O.; van Leeuwen, R.; et al. Plasma Level of Pyrophosphate Is Low in Pseudoxanthoma Elasticum Owing to Mutations in the ABCC6 Gene, but It Does Not Correlate with ABCC6 Genotype. *J. Clin. Med.* **2023**, *12*, 1047. [CrossRef] [PubMed]
15. Leftheriotis, G.; Navasiolava, N.; Clotaire, L.; Duranton, C.; Le Saux, O.; Bendahhou, S.; Laurain, A.; Rubera, I.; Martin, L. Relationships between Plasma Pyrophosphate, Vascular Calcification and Clinical Severity in Patients Affected by Pseudoxanthoma Elasticum. *J. Clin. Med.* **2022**, *11*, 2588. [CrossRef]
16. Sánchez-Tévar, A.M.; García-Fernández, M.; Murcia-Casas, B.; Rioja-Villodres, J.; Carrillo, J.L.; Camacho, M.; Van Gils, M.; Sánchez-Chaparro, M.A.; Vanakker, O.; Valdivielso, P. Plasma Inorganic Pyrophosphate and Alkaline Phosphatase in Patients with Pseudoxanthoma Elasticum. *Ann. Transl. Med.* **2019**, *7*, 798. [CrossRef] [PubMed]
17. Van Gils, M.; Depauw, J.; Coucke, P.J.; Aerts, S.; Verschuere, S.; Nollet, L.; Vanakker, O.M. Inorganic Pyrophosphate Plasma Levels Are Decreased in Pseudoxanthoma Elasticum Patients and Heterozygous Carriers but Do Not Correlate with the Genotype or Phenotype. *J. Clin. Med.* **2023**, *12*, 1893. [CrossRef] [PubMed]
18. Bartstra, J.W.; Risseeuw, S.; de Jong, P.A.; van Os, B.; Kalsbeek, L.; Mol, C.; Baas, A.F.; Verschuere, S.; Vanakker, O.; Florijn, R.J.; et al. Genotype-Phenotype Correlation in Pseudoxanthoma Elasticum. *Atherosclerosis* **2021**, *324*, 18–26. [CrossRef] [PubMed]
19. Legrand, A.; Cornez, L.; Samkari, W.; Mazzella, J.-M.; Venisse, A.; Boccio, V.; Auribault, K.; Keren, B.; Benistan, K.; Germain, D.P.; et al. Mutation Spectrum in the ABCC6 Gene and Genotype-Phenotype Correlations in a French Cohort with Pseudoxanthoma Elasticum. *Genet. Med.* **2017**, *19*, 909–917. [CrossRef] [PubMed]
20. Centre Hospitalier Universitaire de Nice. Efficacy of PyROphosPHate Supplementation to Fight ECtopIc Calcification in PseudoXanthoma Elasticum—PROPHECI-PPI Study. 13 December 2022. Available online: https://classic.clinicaltrials.gov/ct2/show/NCT04868578 (accessed on 3 May 2021).
21. R Core Team. *R A Language and Environment for Statistical Computing*; R Foundation for Statistical Computing: Vienna, Austria, 2022. Available online: https://www.scirp.org/(S(lz5mqp453ed%20snp55rrgjct55))/reference/referencespapers.aspx?referenceid=3456808 (accessed on 30 October 2023).
22. Robin, X.; Turck, N.; Hainard, A.; Tiberti, N.; Lisacek, F.; Sanchez, J.-C.; Müller, M. pROC: An Open-Source Package for R and S+ to Analyze and Compare ROC Curves. *BMC Bioinform.* **2011**, *12*, 77. [CrossRef] [PubMed]

International Journal of
Molecular Sciences

MDPI

Communication

Comparative Single Vesicle Analysis of Aqueous Humor Extracellular Vesicles before and after Radiation in Uveal Melanoma Eyes

Shreya Sirivolu [1,2,†], Chen-Ching Peng [1,2,†], Paolo Neviani [3], Benjamin Y. Xu [2], Jesse L. Berry [1,2,4,5,*] and Liya Xu [1,2,*]

1 The Vision Center at Children's Hospital Los Angeles, Los Angeles, CA 90027, USA; sirivolu@usc.edu (S.S.); ppeng@chla.usc.edu (C.-C.P.)
2 USC Roski Eye Institute, Keck School of Medicine, University of Southern California, Los Angeles, CA 90033, USA
3 Extracellular Vesicle Core, Children's Hospital Los Angeles, Los Angeles, CA 90027, USA
4 The Saban Research Institute, Children's Hospital Los Angeles, Los Angeles, CA 90027, USA
5 Norris Comprehensive Cancer Center, Keck School of Medicine, University of Southern California, Los Angeles, CA 90033, USA
* Correspondence: jesse.berry@med.usc.edu (J.L.B.); lixu@chla.usc.edu (L.X.)
† These authors contributed equally to this work.

Abstract: Small extracellular vesicles (sEVs) have been shown to promote tumorigenesis, treatment resistance, and metastasis in multiple cancer types; however, sEVs in the aqueous humor (AH) of uveal melanoma (UM) patients have never previously been profiled. In this study, we used single particle analysis to characterize sEV subpopulations in the AH of UM patients by quantifying their size, concentration, and phenotypes based on cell surface markers, specifically the tetraspanin co-expression patterns of CD9, CD63, and CD81. sEVs were analyzed from paired pre- and post-treatment (brachytherapy, a form of radiation) AH samples collected from 19 UM patients. In post-brachytherapy samples, two subpopulations, CD63/81+ and CD9/63/81+ sEVs, were significantly increased. These trends existed even when stratified by tumor location and GEP class 1 and class 2 (albeit not significant for GEP class 2). In this initial report of single vesicle profiling of sEVs in the AH of UM patients, we demonstrated that sEVs can be detected in the AH. We further identified two subpopulations that were increased post-brachytherapy, which may suggest radiation-induced release of these particles, potentially from tumor cells. Further study of the cargo carried by these sEV subpopulations may uncover important biomarkers and insights into tumorigenesis for UM.

Keywords: extracellular vesicles (EVs); uveal melanoma; aqueous humor (AH); liquid biopsy; ocular cancer

Citation: Sirivolu, S.; Peng, C.-C.; Neviani, P.; Xu, B.Y.; Berry, J.L.; Xu, L. Comparative Single Vesicle Analysis of Aqueous Humor Extracellular Vesicles before and after Radiation in Uveal Melanoma Eyes. *Int. J. Mol. Sci.* **2024**, *25*, 6035. https://doi.org/10.3390/ijms25116035

Academic Editors: Ylenia Zambito and Andrea Bernini

Received: 10 April 2024
Revised: 26 May 2024
Accepted: 29 May 2024
Published: 30 May 2024

1. Introduction

Extracellular vesicles (EVs) are the collective term for various secreted membrane-enclosed nano-sized vesicles released by virtually every cell type [1]. EVs are classified by size, with small EVs (sEVs), ranging from 30 to 150 nm, being the predominant size found in intraocular biofluids [2]. Among EVs, sEVs are of the greatest interest as potential cancer biomarkers [3,4] with tumor-derived sEVs being shown to promote tumor formation, progression, resistance, immune response regulation, and metastases [4–7], resulting in the emerging role of sEVs in biomarker research [8]. EVs have also been reported for different malignances, such as colorectal cancer, to carry oncogenic factors that can trigger malignant transformation in target cells [9]. The surfaces of sEVs are highly enriched in tetraspanin, a protein superfamily that organize membrane microdomains by forming clusters and interacting with a large variety of transmembrane and cytosolic signaling proteins [10–13]. While they have been used for the phenotypic expression profiling of EVs, tetraspanins have a role beyond serving as cell surface markers that includes extracellular vesicle biogenesis, cargo selection, cell targeting, and cell uptake [14].

We previously published that a distinct eye-specific subpopulation of sEVs can be detected in pediatric aqueous humor (AH) in patients with several ocular diseases, which included congenital cataract, congenital glaucoma, pediatric retinal disease, and retinoblastoma [15]. We identified enrichment of the mono-CD63+ sEV subpopulation in the AH across all disease types [15]. In a study analyzing EVs in multiple cancers, CD63+ vesicles were not present in human plasma, serum, and bone marrow and were present in 10% or fewer samples of lymphatic and bile duct fluid [16]. This suggests that the mono-CD63+ sEV subpopulation may be specific to the eye and detectable in the AH. The specific role of CD63+ sEVs in ocular tumors is yet to be described. Previous reports suggest that CD63 is closely associated with lysosomal trafficking and is a key player in exosome formation and release by participating in the endosomal sorting complex required for transport (ESCART)-independent pathway [17]. In studying patients with retinoblastoma, a primary pediatric ocular malignancy, we observed a significantly dominant subpopulation of CD63/81+ sEVs [18]. This subpopulation was more enriched before treatment and in patients with more significant tumor burden, suggesting they are a tumor-derived subpopulation.

The composition of EVs in the AH of uveal melanoma (UM) patients has not been previously reported. Uveal melanoma is the most common primary ocular malignancy in adults [19]. It is a relatively rare disease (incidence: 5.1 cases per million per year), with tumors located in either the iris (4%), ciliary body (6%), or the choroid (90%) [19]. Plaque brachytherapy, a form of localized radiation therapy, is a standard-of-care procedure to preserve the eye in uveal melanoma patients. The procedure involves the placement of a radioactive plaque onto the scleral wall, followed by its subsequent removal several days later [20]. Gene expression profiling (GEP) is a widely used prospectively validated tool to stratify the risk of metastasis by assigning UM patients to two highly prognostic molecular classes: class 1 (low metastatic risk) and class 2 (high metastatic risk) [21]. Our aim is to use single particle analysis to characterize sEV subpopulations in the AH of UM patients by quantifying their size, concentration, and phenotypes based on tetraspanin expression patterns. Our analysis included paired pre- and post-brachytherapy AH samples, which were further stratified by GEP class and by tumor location.

2. Results

2.1. Patient Clinical Characteristics and Demographics

Nineteen UM patients were included in this study, with paired AH samples (pre- and post-brachytherapy collected from each patient). Based on a clinically validated and widely used 15-gene expression profile test performed on tumor biopsy samples by Castle Biosciences, 12 patients were GEP class 1, 4 patients were GEP class 2, and 3 patients had unknown GEP classification. In this study, posterior tumors are defined as choroidal tumors that do not involve the iris or ciliary body. Anterior tumors are defined as tumors that involve the iris and/or ciliary body. A total of 11 patients had posterior tumors, and 8 patients had anterior tumors. When patients were grouped by either GEP classification (Table A1) or tumor location (Table A2), the AJCC (American Joint Committee on Cancer) stage showed statistically significant differences between groups. No patients withdrew or were lost to follow-up over the study period. Aqueous humor samples from five glaucoma (GLC) patients taken at the time of routine cataract surgery were included as the non-tumor control group.

2.2. Small-Extracellular-Vesicle Size and Concentration Profiling

Unprocessed AH was used for extracellular-vesicle and -particle (EVP) size and concentration profiling via Nanoparticle Tracking Analysis (NTA), with results shown in Figure A1. Due to sample availability, 15 pre-brachytherapy AH samples and 18 post-brachytherapy AH samples were used for this analysis. All nanoparticles' modal size was <150 mm, suggesting that sEVs (which range between 30 and 150 nm) are the major EV constituent in AH. The results demonstrate no significant difference in average particle counts per size, average modal size, or average concentration between pre- and post-brachytherapy samples (Figure A1A–C). Average modal size and particle concentration also did not demonstrate any significant differences in tumor location (Figure A1D,E) or GEP classes (Figure A1F,G) in pre-brachytherapy samples.

2.3. Tetraspanin Expression Profiling and Quantification

Total sEV counts and tetraspanin-based subpopulation profiles between the non-tumor control group and the UM group are shown in Figure A2. Total sEV counts between 5 glaucoma (GLC) and the 19 pre-brachytherapy UM samples showed no significant difference. CD63+ sEVs was the dominant subpopulation in both GLC samples and UM samples, with GLC having a significantly higher percentage. UM pre-brachytherapy samples exhibited a more diverse sEV subpopulation profile than GLC, with significantly increased percentages of CD9/63+, CD9/81+, CD63/81+, and CD9/81/63+ sEVs.

2.3.1. Comparison of Small Extracellular Vesicle Profiles between Paired Pre- vs. Post-Brachytherapy AH Samples

sEV counts are shown between the 19 paired pre- and post-brachytherapy AH samples in Figure 1. An increasing trend was observed for the total EV counts after therapy.

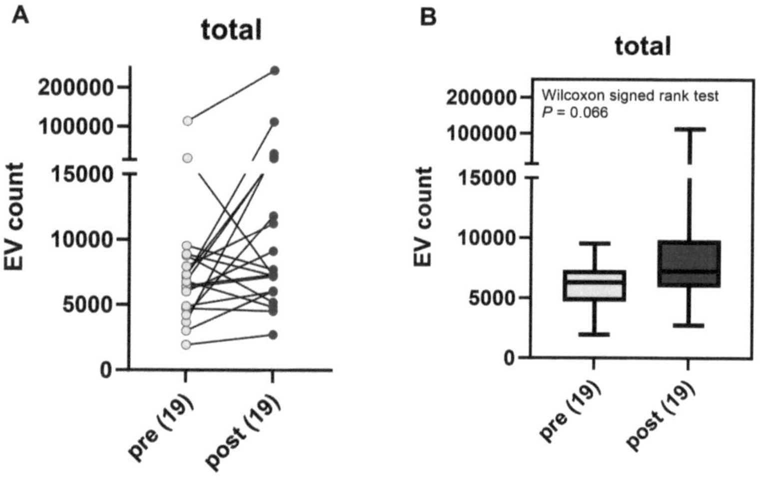

Figure 1. Comparison of total sEV counts pre- and post-brachytherapy. (**A**) Paired sEV counts for each sample. (**B**) Box-and-whisker plot of pooled sEV counts.

Figure 2 illustrates the changes in the percent composition of sEV subpopulations identified by expression profiles by using three tetraspanin markers (CD9+, CD63+, and CD81+). In comparison to pre-brachytherapy samples (Figure 2A), the mean percentage of CD63+ sEVs showed a significant decrease in post-brachytherapy samples ($p = 0.003$) (Figure 2A,B). On the other hand, the mean percentages of CD63/81+ sEVs and CD9/63/81+ sEVs showed a significant increase in post-brachytherapy samples ($p < 0.001$) (Figure 2A,B). Both paired pre- and post-brachytherapy results from each individual sample as well as pooled results for each of these three subpopulations demonstrate these trends (Figure 2C–E).

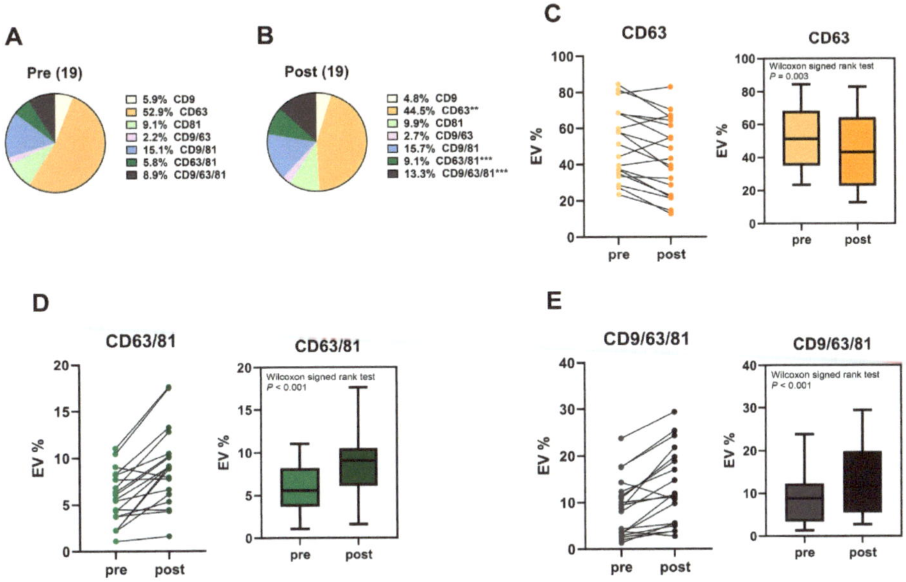

Figure 2. sEV subpopulation dynamics. (**A**,**B**) Comparison of percent composition of sEV subpopulations pre- and post-brachytherapy. Percentages of CD63+ (**C**), CD63/81+ (**D**), and CD9/63/81+ (**E**). sEV subpopulations pre- and post-brachytherapy, shown as paired percentage for each sample and pooled percentage for all samples. ** $p < 0.01$, *** $p < 0.001$.

2.3.2. Comparison of Small Extracellular Vesicle Profiles among Aqueous Humor Samples Stratified by Gene Expression Profile Class and Tumor Location

sEV tetraspanin expression profiles were compared in pre- and post-brachytherapy samples after stratifying by GEP class and tumor location. In pre-brachytherapy samples, there was no significant difference in total sEV counts or sEV subpopulation percentages between GEP classes and between anterior and posterior tumors (Figure A3). The subpopulation percentages of CD63+, CD63/81+, and CD9/63/81+ sEVs were compared for each GEP class between pre- and post-brachytherapy samples, as shown in Figure 3A–C. In GEP class 1 tumors, the mean percentages showed a statistically significant decrease from pre- to post-brachytherapy in CD63+ sEVs ($p = 0.002$) and a statistically significant increase in CD63/81+ sEVs ($p = 0.001$) and CD9/63/81+ sEVs ($p < 0.001$). GEP class 2 samples showed these same trends; however, the results were not significant. The subpopulation percentages of CD63+, CD63/81+, and CD9/63/81+ sEVs were compared for each tumor location between pre- and post-brachytherapy samples, as shown in Figure 3D–F. In posterior tumors, the mean percentages showed a statistically significant decrease from pre- to post-brachytherapy in CD63+ sEVs ($p = 0.042$) and a statistically significant increase in CD63/81+ sEVs ($p = 0.005$) and CD9/63/81+ sEVs ($p = 0.010$). In anterior tumors, the mean percentages showed a statistically significant increase in CD63/81+ sEVs ($p = 0.016$) and CD9/63/81+ sEVs ($p = 0.016$). There was a decrease in the mean percentage of CD63+ sEVs; however, it was not significant.

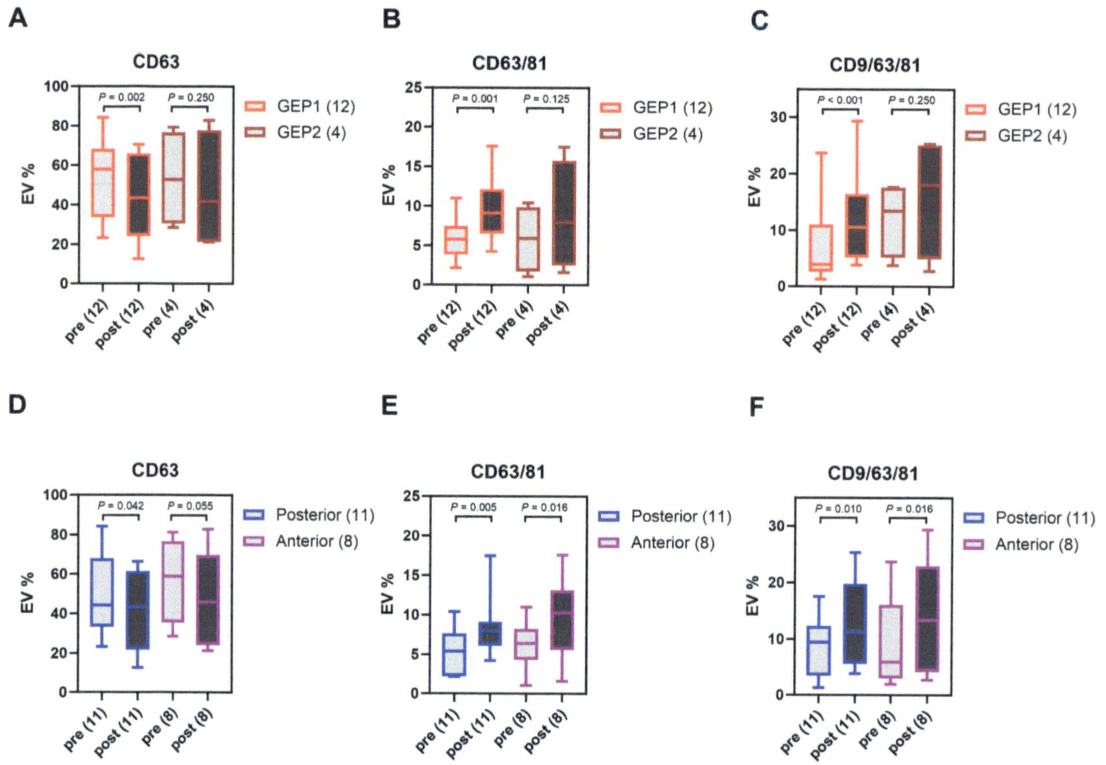

Figure 3. Comparison of percent composition of CD63+, CD63/81+, and CD9/63/81+ sEV subpopulations pre- and post-brachytherapy after GEP class stratification (**A–C**) and tumor location stratification (**D–F**).

3. Discussion

Herein, we present the first investigation into sEVs in UM patients with the use of single vesicle analysis. We demonstrate that sEVs can be detected in the AH of UM patients and their tetraspanin expression can be profiled by using single vesicle analysis. By comparing paired pre- and post- brachytherapy AH samples from 19 UM patients, the effect of radiation on sEV subpopulations was trended in all samples and further analyzed after sample stratification into GEP class and tumor location.

We have previously shown in UM patients that post-brachytherapy AH had significantly higher DNA and miRNA concentrations than pre-brachytherapy AH [22]. Higher tumor-derived nucleic acids have also been shown in plasma. In Francis et al., patients undergoing 3-day plaque brachytherapy had significantly more tumor-derived cell-free DNA in the plasma 48–72 h after plaque brachytherapy compared with less than 48 h after therapy. It was speculated in both studies that radiation from brachytherapy caused tumor cell necrosis and lysis, releasing DNA into the AH and blood [23].

Our results in this current study suggest that brachytherapy can similarly result in the release of sEVs into the AH. Two separate platforms, NTA NanoSight NS300 and ExoView R100, yielded consistent results of clear increasing trends in total sEV counts and concentrations in post-brachytherapy samples. While the results are consistent in showing an increase in sEVs post-brachytherapy, a limited sample size and high sample variability may be the cause of the statistical results not being significant. Two subpopulations, CD63/81+ and CD9/63/81+ sEVs, demonstrated significantly increased percentages in post-brachytherapy samples, suggesting radiation-induced release of these vesicles

(Figure 2). The increase in percentage of these two subpopulations is presumed to be the cause of the decrease in the percentage of the normally dominant CD63+ subpopulation, an AH-specific subpopulation reported in several pediatric ocular disease states [15].

When stratified based on GEP class, significantly increased percentages of CD63/81+ and CD9/63/81+ sEV subpopulations in post-brachytherapy samples were seen in GEP class 1 (low metastatic risk) tumors. While these same trends existed in GEP class 2 (high metastatic risk) tumors, the increase was not significant. Studies investigating differences in treatment response to brachytherapy between both GEP classes have not reached a consistent result. Some studies revealed that GEP class 1 tumors regress more rapidly [24,25], while others showed that GEP class 2 tumors had more rapid regression [26] or no statistically significant difference [27,28]. For this reason, we cannot conclude that inherent differences in responses to radiation between both classes of tumors result in more sEV release in GEP class 1 tumors. It is more likely that we could not achieve significance in GEP class 2 tumors due to a low sample size (12 GEP 1 pairs vs. 4 GEP 2 pairs). Tumor location does not influence the preferential release of these sEV subpopulations; the percentages of CD63/81+ and CD9/63/81+ sEVs demonstrated a significant increase post-brachytherapy in both anterior and posterior tumors (12 anterior vs. 7 posterior pairs).

The enrichment of CD63/81+ and CD9/63/81+ sEVs in post-brachytherapy samples suggests that these subpopulations may be tumor-derived. However, definitive tumor origin is not evidenced in this study, as these vesicles may have been the result of a radiation-induced change in the surrounding normal cells, such as brachytherapy-induced tissue necrosis resulting in sEV release.

In retinoblastoma eyes, the CD63/81+ sEV subpopulation was also hypothesized to be tumor-derived [18]. Given its association with both UM and retinoblastoma, further investigation into the cargo of these sEVs is warranted. Using UM cell lines, Tsering et al. showed that protein cargo derived from EVs may be involved in tumorigenesis and metastatic dissemination [7]. These studies demonstrate that investigating the cargo of tumor-derived EVs may increase our understanding of critical molecular processes of the tumor. These findings may have direct clinical benefits in the form of biomarkers of disease and drug targeting.

In conclusion, we present the phenotypic profiles of sEV subpopulations collected from AH pre- and post-brachytherapy in uveal melanoma patients. To our knowledge, this is the first time sEVs are reported from UM AH samples. Through the profiling of tetraspanin cell surface markers, we confirm that sEVs were present in all AH samples in UM patients. After brachytherapy, two subpopulations, CD63/81+ and CD9/63/81+ sEVs, constituted a significantly higher percentage of sEV distribution than before therapy. It is possible that CD63/81+ and CD9/63/81+ sEVs may be tumor-derived, although further studies are needed to identify the origin of these vesicles. We currently plan to expand our analysis of sEVs by using a multiplex bead-based flow cytometry assay with a panel that incorporates cancer-specific biomarkers. This approach would have the potential to improve our understanding of the origin of sEV subpopulations that are enriched after treatment.

4. Materials and Methods

This investigation was a case-series study at a tertiary care hospital (University of Southern California Roski Eye Institute). Samples were taken between August 2020 and May 2021.

4.1. Sample Collection

AH was collected from each patient before and after radioactive plaque placement, a form of localized radiation to treat UM. AH pre-brachytherapy samples were taken at the first surgery, before any significant radiation, and then again at the time of plaque removal, after a dose of 85 Gy to the tumor apex had been given (post-brachytherapy AH). As previously described in detail [29], clear corneal paracentesis was performed to extract up to 0.1 mL of AH by using a 32 gauge needle on a 1 cc syringe, as part of a routine procedure for anterior segment surgery at diagnosis or during treatment. Samples were transported on dry ice and stored at −80 °C until analysis.

4.2. Nanoparticle Tracking Analysis

By using 10 µL of unprocessed AH, NTA NanoSight NS300 Platform was used to evaluate the sizes and concentrations of extracellular vesicles and particles (EVPs). This platform is equipped with a 405 nm laser and sCMOS camera, which are used to record the Brownian movement of particles in suspension. The movement was then analyzed via the Stokes–Einstein equation to obtain the hydrodynamic radius and vesicle count for each modal size. NTA software 3.4 was used to perform data analysis, with the average of at least five camera recordings being presented. Results were shown as particle count per size distribution.

4.3. Single Particle-Interferometric Reflectance Imaging Sensor Analysis

Single sEV analysis was performed by using SP-IRIS-based ExoView R100 platform and ExoView Human Tetraspanin Kit (Unchained Labs, Pleasanton, CA, USA) as previously published [15,18]. Between 0.25 and 10 µL of unprocessed AH was diluted by using buffer A to a final volume of 40 µL. Then, 35 µL of each sample was incubated by using the ExoView Tetraspanin Chip at room temperature in an area free of vibrations or movement and sealed with tape to prevent drying out. The chips were then washed three times by using solution A from the kit and then incubated with immunocapture antibodies (anti-CD9 CF488, anti-CD81 CF555, and anti-CD63 CF647). Further information on antibodies can be found in the (Supplementary Material Figures S1 and S2). Antibodies were diluted as per the manufacturer's protocol (Unchained Labs) to a final concentration of 0.5 µg/µL. After 1 h incubation at room temperature and subsequent washing and drying, the chips were then imaged with the ExoView R100 reader by using ExoView Scanner version 3.2 acquisition software; data were analyzed with ExoView Analyzer version 3.2. The volume of each sample loaded was calculated to ensure that particle counts fell within the instrument's linear detection range (200 to 6000 particles per fluorescent channel). For the final analysis, particle counts were normalized to a standardized volume of 10 µL by using a dilution factor.

4.4. Statistical Analysis

Fisher's exact test was used to compare categorical variables (sex, eye, eye color, ciliary body involvement, and PRAME Status; Tables A1 and A2). Continuous variables were summarized as means ± standard error of mean (SEM) (all tables and figures). All continuous variables were non-normally distributed based on Shapiro–Wilk testing [30]. Non-normally distributed variables, such as EV counts and percentages, were compared by using non-parametric Wilcoxon Signed-Rank tests (paired samples; Figures 1–3) and Mann–Whitney U tests (EV count comparison (Figures A1–A3) and EV subpopulation percentages across GLC versus UM (Figure A2)) [31]. Analysis of variance (ANOVA) testing (Kruskal–Wallis test) and Dunn's multiple comparison tests were used for GEP classes and tumor locations (Figure A3). The ANOVA report can be found in the (Supplementary Material Table S1). Tetraspanin co-expression percentages were calculated based on the total number of fluorescent particles in the sample per SP-IRIS analysis. All statistical tests were 2-tailed, and $p < 0.05$ was considered statistically significant (Figures 1–3 and A1–A3; age, AJCC stage, and tumor stage in Tables A1 and A2). Analyses were conducted and plots obtained by using Prism 10 (GraphPad, La Jolla, CA, USA).

5. Patents

Drs. Jesse L. Berry and Liya Xu have filed a patent application entitled Aqueous humor cell free DNA for diagnostic and Prognostic evaluation of Ophthalmic Disease.

Supplementary Materials: The following supporting information can be downloaded at: https://www.mdpi.com/article/10.3390/ijms25116035/s1.

Author Contributions: L.X. and S.S. conceived the presented idea. S.S. and C.-C.P. carried out the data collection and data analysis. S.S. prepared the original draft. B.Y.X. contributed to resources. C.-C.P., P.N. and J.L.B. contributed to original draft editing. All authors have read and agreed to the published version of the manuscript.

Funding: This research was funded by Research to Prevent Blindness/Castle Biosciences Medical Student Eye Research Fellowship in Ocular Cancer (Recipient: Shreya Sirivolu; Mentor: Jesse Berry); USC Dean's Pilot; Children's Hospital Los Angeles Saban Research Institute (Research Career Development Award); The National Cancer Institute P30CA014089; The Knights Templar Eye Foundation; National Cancer Institute of the National Institute of Health Award Number K08CA232344; The Wright Foundation; Children's Oncology Group/St. Baldrick's Foundation, Danhakl Family Foundation; A. Linn Murphree, MD, Chair in Ocular Oncology; The Berle & Lucy Adams Chair in Cancer Research; The Larry and Celia Moh Foundation; and an unrestricted departmental grant from Research to Prevent Blindness.

Institutional Review Board Statement: This study was conducted in accordance with the Declaration of Helsinki and was approved by the Institutional Review Board of the Keck School of Medicine of the University of Southern California (HS-19-00293) on 14 June 2019 and at Children's Hospital Los Angeles (CHLA-20-00167) on 26 June 2020. This research conformed to the requirements of the United States Health Insurance Portability and Accountability Act.

Informed Consent Statement: Informed consent was obtained from all subjects involved in the study.

Data Availability Statement: De-identified original datasets have been uploaded as Tables A1 and A2.

Acknowledgments: We would like to acknowledge Seong Shin for the coordination of this study and Mark Reid from the Vision Center in CHLA for statistical correction, as well as USC Norris Library bioinformatics office.

Conflicts of Interest: Drs. Jesse L. Berry and Liya Xu have filed a patent application entitled Aqueous humor cell free DNA for diagnostic and Prognostic evaluation of Ophthalmic Disease. The funders had no role in the design of the study; in the collection, analyses, or interpretation of data; in the writing of the manuscript; or in the decision to publish the results.

Appendix A

Table A1. Univariate comparison of clinical characteristics between GEP1 and GEP2 UM patients.

Characteristic		GEP1, n = 12	GEP2, n = 4	*p*-Value
Sex (Fisher), n (%)				>0.999
	Females	7 (58.3)	3 (75.0)	
	Males	5 (41.7)	1 (25.0)	
Eye (Fisher), n (%)				0.569
	OD	7 (58.3)	1 (25.0)	
	OS	5 (41.7)	3 (75.0)	
Age at diagnosis, mean (±SD) (MWU)		53.3 (15.5)	58 (7.8)	0.425
Eye color (Fisher), n (%)				0.569
	Light (blue, gray, green, and hazel)	7 (58.3)	1 (25.0)	
	Dark (brown)	5 (41.7)	3 (75.0)	
Ciliary body involvement (Fisher), n (%)				>0.999
	Yes	6 (50.0)	2 (50.0)	
	No	6 (50.0)	2 (50.0)	

Table A1. *Cont.*

Characteristic		GEP1, n = 12	GEP2, n = 4	*p*-Value
AJCC stage (linear-by-linear association), n (%)				0.009
	I	5 (41.7)	0 (0)	
	IIA	5 (41.7)	1 (25.0)	
	IIB	2 (16.6)	1 (25.0)	
	IIIA, IIIB, IIIC	0 (0)	2 (50.0)	
PRAME status, known in 17 cases (Fisher), n (%)				0.450
	Negative	11 (91.7)	3 (75.0)	
	Positive	1 (8.3)	1 (25.0)	
Tumor Stage (linear-by-linear association), n (%)				0.009
	T1a	5 (41.7)	0 (0)	
	T1b	4 (33.3)	0 (0)	
	T2a	1 (8.3)	1 (25.0)	
	T2b	1 (8.3)	1 (25.0)	
	T3	1 (8.3)	1 (25.0)	
	T4	0	1 (25.0)	

AJCC, American Joint Committee in Cancer; Fisher, Fisher's exact test; GEP, gene expression profile; MWU, Mann–Whitney U test; OS, oculus sinister (left eye); OD, oculus dextrus (right eye); PRAME, preferentially expressed antigen in melanoma; SD, standard deviation.

Table A2. Comparison of clinical characteristics of UM patients with anterior and posterior tumors.

Characteristic		Posterior, n = 11	Anterior, n = 8	*p*-Value
Sex (Fisher), n (%)				0.633
	Females	6 (54.5)	6 (75.0)	
	Males	5 (45.4)	2 (25.0)	
Eye (Fisher), n (%)				0.370
	OD	4 (36.3)	5 (62.5)	
	OS	7 (63.7)	3 (37.5)	
Age at diagnosis, mean (±SD) (MWU)		57.2 (12.2)	54.0 (15.5)	0.888
Eye color (Fisher), n (%)				0.352
	Light (blue, gray, green, and hazel)	5 (45.4)	6 (75.0)	
	Dark (brown)	6 (54.5)	2 (25.0)	
GEP class (Fisher), n (%)				>0.999
	GEP1	6 (75.0)	6 (75.0)	
	GEP2	2 (25.0)	2 (25.0)	
AJCC stage (linear-by-linear association), n (%)				0.040
	I	8 (72.7)	0 (0)	
	IIA	1 (9.1)	5 (62.5)	
	IIB	1 (9.1)	2 (25.0)	
	IIIA, IIIB, IIIC	1 (9.1)	1 (12.5)	
PRAME status, known in 17 cases (Fisher), n (%)				>0.999
	Negative	7 (87.5)	7 (87.5)	
	Positive	1 (12.5)	1 (12.5)	
Tumor stage (linear-by-linear association), n (%)				0.011
	T1a	8 (72.7)	0 (0)	
	T1b	0 (0)	4 (50.0)	
	T2a	1 (9.1)	1 (12.5)	
	T2b	0 (0)	2 (25.0)	
	T3	1 (9.1)	1 (12.5)	
	T4	1 (9.1)	0 (0)	

AJCC, American Joint Committee in Cancer; Fisher, Fisher's exact test; GEP, gene expression profile; MWU, Mann–Whitney U test; OS, oculus sinister (left eye); OD, oculus dextrus (right eye); PRAME, preferentially expressed antigen in melanoma; SD, standard deviation.

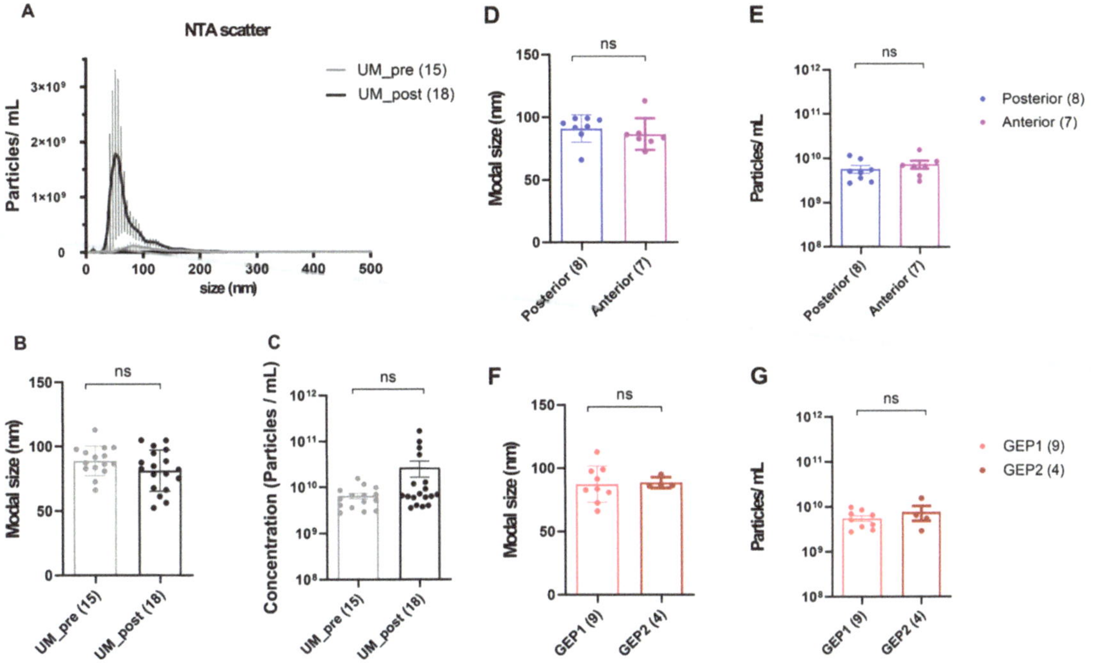

Figure A1. Characterization of EVPs from unprocessed AH by Nanoparticle Tracking Analysis. (**A–C**) Average particle counts versus size distribution, average modal size, and average particle concentration in pre- and post-brachytherapy samples. (**D**,**E**) Average modal size and particle concentration in anterior and posterior tumors. (**F**,**G**) Average modal size and particle concentration in GEP class 1 and GEP class 2 tumors. Error bars represent standard deviations obtained from each study group. Due to AH availability, full experimental cohort was not included in this analysis.

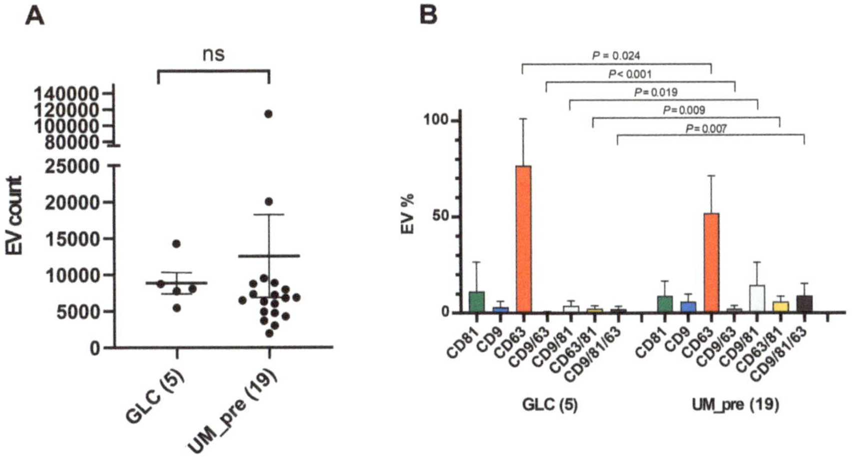

Figure A2. Quantitative comparison of sEV tetraspanin subpopulation expression profiles in glaucoma (GLC) and UM pre-brachytherapy (UM_pre) AH samples. (**A**) Mean total sEV counts compared between GLC and UM_pre AH samples. (**B**) Mean sEV subpopulation percentages compared between GLC and UM_pre AH samples.

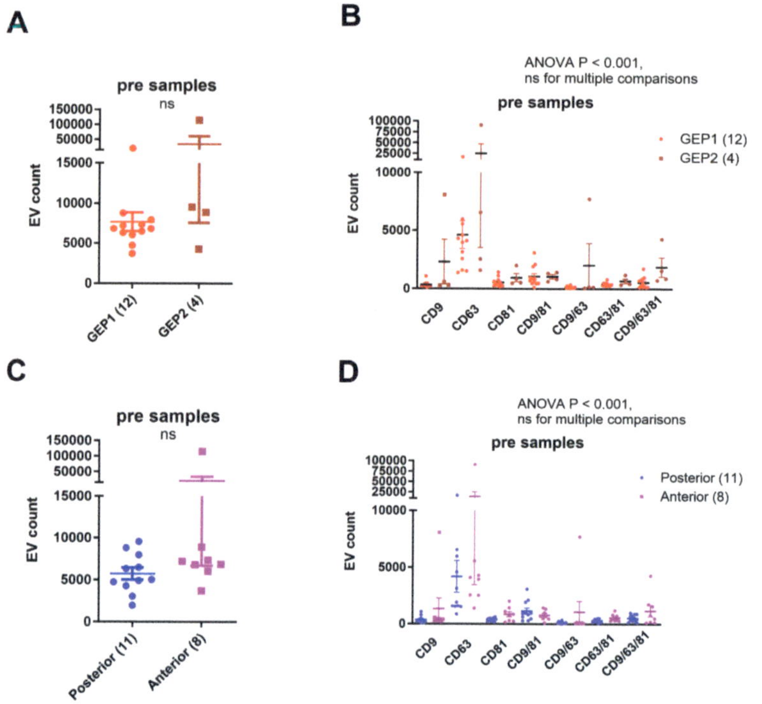

Figure A3. Comparison of pre-brachytherapy AH samples stratified into GEP class and tumor location. (**A**) Total EV counts in GEP class 1 and GEP class 2 tumors. (**B**) sEV tetraspanin expression profiles in GEP class 1 and GEP class 2 tumors. (**C**) Total EV counts in anterior and posterior tumors. (**D**) sEV tetraspanin expression profiles in anterior and posterior tumors.

References

1. Yu, B.; Zhang, X.; Li, X. Exosomes Derived from Mesenchymal Stem Cells. *Int. J. Mol. Sci.* **2014**, *15*, 4142–4157. [CrossRef] [PubMed]
2. Dismuke, W.M.; Challa, P.; Navarro, I.; Stamer, W.D.; Liu, Y. Human Aqueous Humor Exosomes. *Exp. Eye Res.* **2015**, *132*, 73–77. [CrossRef] [PubMed]
3. Whiteside, T.L. The Potential of Tumor-Derived Exosomes for Noninvasive Cancer Monitoring: An Update. *Expert. Rev. Mol. Diagn.* **2018**, *18*, 1029–1040. [CrossRef] [PubMed]
4. Whiteside, T.L. The Effect of Tumor-Derived Exosomes on Immune Regulation and Cancer Immunotherapy. *Future Oncol.* **2017**, *13*, 2583–2592. [CrossRef] [PubMed]
5. Li, S.; Yi, M.; Dong, B.; Tan, X.; Luo, S.; Wu, K. The Role of Exosomes in Liquid Biopsy for Cancer Diagnosis and Prognosis Prediction. *Int. J. Cancer* **2021**, *148*, 2640–2651. [CrossRef] [PubMed]
6. Fu, M.; Gu, J.; Jiang, P.; Qian, H.; Xu, W.; Zhang, X. Exosomes in Gastric Cancer: Roles, Mechanisms, and Applications. *Mol. Cancer* **2019**, *18*, 41. [CrossRef] [PubMed]
7. Tsering, T.; Laskaris, A.; Abdouh, M.; Bustamante, P.; Parent, S.; Jin, E.; Ferrier, S.T.; Arena, G.; Burnier, J.V. Uveal Melanoma-Derived Extracellular Vesicles Display Transforming Potential and Carry Protein Cargo Involved in Metastatic Niche Preparation. *Cancers* **2020**, *12*, 2923. [CrossRef] [PubMed]
8. Asleh, K.; Dery, V.; Taylor, C.; Davey, M.; Djeungoue-Petga, M.-A.; Ouellette, R.J. Extracellular Vesicle-Based Liquid Biopsy Biomarkers and Their Application in Precision Immuno-Oncology. *Biomark. Res.* **2023**, *11*, 99. [CrossRef] [PubMed]
9. Abdouh, M.; Floris, M.; Gao, Z.-H.; Arena, V.; Arena, M.; Arena, G.O. Colorectal Cancer-Derived Extracellular Vesicles Induce Transformation of Fibroblasts into Colon Carcinoma Cells. *J. Exp. Clin. Cancer Res. CR* **2019**, *38*, 257. [CrossRef]
10. Charrin, S.; le Naour, F.; Silvie, O.; Milhiet, P.-E.; Boucheix, C.; Rubinstein, E. Lateral Organization of Membrane Proteins: Tetraspanins Spin Their Web. *Biochem. J.* **2009**, *420*, 133–154. [CrossRef]
11. Hemler, M.E. Tetraspanin Functions and Associated Microdomains. *Nat. Rev. Mol. Cell Biol.* **2005**, *6*, 801–811. [CrossRef] [PubMed]

12. Yáñez-Mó, M.; Barreiro, O.; Gordon-Alonso, M.; Sala-Valdés, M.; Sánchez-Madrid, F. Tetraspanin-Enriched Microdomains: A Functional Unit in Cell Plasma Membranes. *Trends Cell Biol.* **2009**, *19*, 434–446. [CrossRef] [PubMed]
13. Andreu, Z.; Yáñez-Mó, M. Tetraspanins in Extracellular Vesicle Formation and Function. *Front. Immunol.* **2014**, *5*, 442. [CrossRef] [PubMed]
14. Jankovičová, J.; Sečová, P.; Michalková, K.; Antalíková, J. Tetraspanins, More than Markers of Extracellular Vesicles in Reproduction. *Int. J. Mol. Sci.* **2020**, *21*, 7568. [CrossRef] [PubMed]
15. Peng, C.-C.; Im, D.; Sirivolu, S.; Reiser, B.; Nagiel, A.; Neviani, P.; Xu, L.; Berry, J.L. Single Vesicle Analysis of Aqueous Humor in Pediatric Ocular Diseases Reveals Eye Specific CD63-Dominant Subpopulations. *J. Extracell. Biol.* **2022**, *1*, e36. [CrossRef] [PubMed]
16. Hoshino, A.; Kim, H.S.; Bojmar, L.; Gyan, K.E.; Cioffi, M.; Hernandez, J.; Zambirinis, C.P.; Rodrigues, G.; Molina, H.; Heissel, S.; et al. Extracellular Vesicle and Particle Biomarkers Define Multiple Human Cancers. *Cell* **2020**, *182*, 1044–1061.e18. [CrossRef] [PubMed]
17. Van Niel, G.; Charrin, S.; Simoes, S.; Romao, M.; Rochin, L.; Saftig, P.; Marks, M.S.; Rubinstein, E.; Raposo, G. The Tetraspanin CD63 Regulates ESCRT-Independent and Dependent Endosomal Sorting during Melanogenesis. *Dev. Cell* **2011**, *21*, 708–721. [CrossRef] [PubMed]
18. Pike, S.; Peng, C.-C.; Neviani, P.; Berry, J.L.; Xu, L. CD63/81 Small Extracellular Vesicles in the Aqueous Humor Are Retinoblastoma Associated. *Investig. Ophthalmol. Vis. Sci.* **2023**, *64*, 5. [CrossRef]
19. Kaliki, S.; Shields, C.L. Uveal Melanoma: Relatively Rare but Deadly Cancer. *Eye* **2017**, *31*, 241–257. [CrossRef]
20. Joh, S.; Kim, M.E.; Reilly, M.; Zhou, S.Y.; Kim, J.; Jennelle, R.L.; Berry, J.L. Outpatient Ocular Brachytherapy: The USC Experience. *Adv. Radiat. Oncol.* **2021**, *6*, 100737. [CrossRef]
21. Onken, M.D.; Worley, L.A.; Char, D.H.; Augsburger, J.J.; Correa, Z.M.; Nudleman, E.; Aaberg, T.M.; Altaweel, M.M.; Bardenstein, D.S.; Finger, P.T.; et al. Collaborative Ocular Oncology Group Report No. 1: Prospective Validation of a Multi-Gene Prognostic Assay in Uveal Melanoma. *Ophthalmology* **2012**, *119*, 1596–1603. [CrossRef] [PubMed]
22. Im, D.; Peng, C.-C.; Xu, L.; Kim, M.E.; Ostrow, D.; Yellapantula, V.; Bootwalla, M.; Biegel, J.A.; Gai, X.; Kuhn, P.; et al. Potential of Aqueous Humor as a Liquid Biopsy for Uveal Melanoma. *Investig. Ophthalmol. Vis. Sci.* **2022**, *63*, 1458. [CrossRef] [PubMed]
23. Francis, J.H.; Barker, C.A.; Brannon, A.R.; Canestraro, J.; Robbins, M.; Swartzwelder, C.E.; Levine, S.; Law, C.; Berger, M.F.; Shoushtari, A.; et al. Detectability of Plasma-Derived Circulating Tumor DNA Panel in Patients Undergoing Primary Treatment for Uveal Melanoma. *Investig. Ophthalmol. Vis. Sci.* **2022**, *63*, 17. [CrossRef] [PubMed]
24. Mruthyunjaya, P.; Seider, M.I.; Stinnett, S.; Schefler, A.; Seider, M.; Raufi, N.N.; Berry, D.; Harbour, J.W.; Berry, J.; Kim, J.; et al. Association between Tumor Regression Rate and Gene Expression Profile after Iodine 125 Plaque Radiotherapy for Uveal Melanoma. *Ophthalmology* **2017**, *124*, 1532–1539. [CrossRef] [PubMed]
25. Rao, R.C.; Khan, M.; Badiyan, S.N.; Harbour, J.W. Gene Expression Profiling and Regression Rate of Irradiated Uveal Melanomas. *Ophthalmic Surg. Lasers Imaging Retina* **2015**, *46*, 333–337. [CrossRef] [PubMed]
26. Sharma, A.; Mignano, J.E.; Duker, J.S. Relationship between Gene Expression Profile Class and Tumor Thickness Regression after Plaque Brachytherapy for Choroidal Melanoma. *Int. J. Retina Vitr.* **2022**, *8*, 47. [CrossRef] [PubMed]
27. Chappell, M.C.; Char, D.H.; Cole, T.B.; Harbour, J.W.; Mishra, K.; Weinberg, V.K.; Phillips, T.L. Uveal Melanoma: Molecular Pattern, Clinical Features, and Radiation Response. *Am. J. Ophthalmol.* **2012**, *154*, 227–232. [CrossRef] [PubMed]
28. Corrêa, Z.M.; Augsburger, J.J. Relationship between Rate of Posterior Uveal Melanoma Flattening Following Plaque Radiotherapy and Gene Expression Profile Class of Tumor Cells. *Investig. Ophthalmol. Vis. Sci.* **2014**, *55*, 556–559. [CrossRef] [PubMed]
29. Kim, M.E.; Xu, L.; Prabakar, R.K.; Shen, L.; Peng, C.-C.; Kuhn, P.; Gai, X.; Hicks, J.; Berry, J.L. Aqueous Humor as a Liquid Biopsy for Retinoblastoma: Clear Corneal Paracentesis and Genomic Analysis. *J. Vis. Exp. JoVE* **2021**, *175*, e62939. [CrossRef]
30. Ghasemi, A.; Zahediasl, S. Normality Tests for Statistical Analysis: A Guide for Non-Statisticians. *Int. J. Endocrinol. Metab.* **2012**, *10*, 486–489. [CrossRef]
31. Kim, H.-Y. Statistical Notes for Clinical Researchers: Nonparametric Statistical Methods: 1. Nonparametric Methods for Comparing Two Groups. *Restor. Dent. Endod.* **2014**, *39*, 235–239. [CrossRef] [PubMed]

International Journal of
Molecular Sciences

MDPI

Article

Serum Galectin-3 as a Non-Invasive Marker for Primary Sclerosing Cholangitis

Ganimete Bajraktari, Tanja Elger, Muriel Huss, Johanna Loibl, Andreas Albert, Arne Kandulski, Martina Müller, Hauke Christian Tews and Christa Buechler *

Department of Internal Medicine I, Gastroenterology, Hepatology, Endocrinology, Rheumatology, and Infectious Diseases, University Hospital Regensburg, 93053 Regensburg, Germany; ganimete1.bajraktari@stud.uni-regensburg.de (G.B.); tanja.elger@klinik.uni-regensburg.de (T.E.); muriel.huss@klinik.uni-regensburg.de (M.H.); johanna.loibl@klinik.uni-regensburg.de (J.L.); andreas.albert@klinik.uni-regensburg.de (A.A.); arne.kandulski@klinik.uni-regensburg.de (A.K.); martina.mueller-schilling@klinik.uni-regensburg.de (M.M.); hauke.tews@klinik.uni-regensburg.de (H.C.T.)

* Correspondence: christa.buechler@klinik.uni-regensburg.de

Abstract: Primary sclerosing cholangitis (PSC) is a serious liver disease associated with inflammatory bowel disease (IBD). Galectin-3, an inflammatory and fibrotic molecule, has elevated circulating levels in patients with chronic liver disease and inflammatory bowel disease (IBD). This study aims to clarify whether galectin-3 can differentiate between patients with IBD, PSC, and PSC-IBD. Our study measured serum galectin-3 levels in 38 healthy controls, 55 patients with IBD, and 22 patients with PSC (11 patients had underlying IBD and 11 patients did not), alongside the urinary galectin-3 of these patients and 18 controls. Serum and urinary galectin-3 levels in IBD patients were comparable to those in controls. Among IBD patients, those with high fecal calprotectin, indicating severe disease, exhibited lower serum and elevated urinary galectin-3 levels compared to those with low calprotectin levels. Serum galectin-3 levels were inversely correlated with C-reactive protein levels. PSC patients displayed higher serum and urinary galectin-3 levels than IBD patients, with the highest serum levels observed in PSC patients with coexisting IBD. There was no correlation between serum and urinary galectin-3 levels and laboratory indicators of liver injury in both IBD and PSC patients. In conclusion, this study demonstrates that serum and urinary galectin-3 levels can distinguish IBD from PSC patients, and also reveals higher serum galectin-3 levels in PSC-IBD patients compared to those with isolated PSC.

Keywords: galectin-3; aminotransferase; calprotectin; urine; primary sclerosing cholangitis; inflammatory bowel disease

Citation: Bajraktari, G.; Elger, T.; Huss, M.; Loibl, J.; Albert, A.; Kandulski, A.; Müller, M.; Tews, H.C.; Buechler, C. Serum Galectin-3 as a Non-Invasive Marker for Primary Sclerosing Cholangitis. *Int. J. Mol. Sci.* **2024**, *25*, 4765. https://doi.org/10.3390/ijms25094765

Academic Editor: Andrea Bernini

Received: 22 March 2024
Revised: 23 April 2024
Accepted: 26 April 2024
Published: 27 April 2024

1. Introduction

Primary sclerosing cholangitis (PSC) is a rare chronic biliary disease with a prevalence of approximately 6 per 100,000 person-years in the UK population [1]. This disease is progressive and is characterized by the destruction of the bile ducts, leading to cholestasis, liver fibrosis, and, ultimately, liver cirrhosis. Currently, there are no effective drugs identified to halt the progression of the disease. Diagnosing PSC can be challenging, as noted by Pria et al. [2]. The gold standard for initial non-invasive diagnostics is magnetic resonance cholangiopancreatography (MRCP) [3,4]. This cost-intensive imaging method also has uncertainties in the assessment of the bile ducts, which is why non-invasive markers are needed. Furthermore, it is advised that patients with IBD undergo annual liver function screening, irrespective of symptoms [5].

PSC is frequently linked to inflammatory bowel disease (IBD), with about 70% of PSC patients also suffering from IBD [5].The two main entities of IBD are Crohn's disease (CD) and ulcerative colitis (UC) [6–8]. Despite extensive research identifying genetic factors, microbial dysbiosis, immune dysfunction, and environmental influences as contributors to IBD pathogenesis, its exact cause remains unclear [9].

Non-invasive diagnostic and prognostic biomarkers for PSC are currently an area of unmet need [10]. Biomarkers such as anti-neutrophil cytoplasmic antibodies (ANCAs) are suboptimal in their performance [10]. One study from Australia showed that more than 80% of IBD patients had a positive ANCA result and a significant proportion had proteinase 3 antibodies. However, no specific ANCA pattern predicted a specific IBD subtype or PSC [11]. Anti-glycoprotein 2 combined with ANCAs to serine proteinase 3 (PR3-ANCA) are prognostic in PSC for poor survival [12] but have still to be tested in larger cohorts.

Galectin-3 is expressed by various human cells, especially immune cells, adipocytes, epithelial cells, and endothelial cells, and serum galectin-3 levels are elevated in obesity [13–15]. Galectin-3 is an inflammatory and fibrotic molecule. Inflammation is known to be an important pathogenic event in liver injury, and reactive oxygen species (ROS) cause the induction of pro-inflammatory genes. Excessive production of ROS and overexpression of these pro-inflammatory proteins play an important role in the progression of liver disease severity [16,17]. Galectin-3 activates myofibroblasts, which produce excess extracellular matrix, resulting in scar formation. Mice deficient in galectin-3 were protected from carbon tetrachloride-induced liver fibrosis and non-alcoholic steatohepatitis (NASH) [18,19]. It has been suggested that galectin-3 plays a causal role in tissue fibrosis. In the liver, galectin-3 expression is induced in cirrhosis. Higher hepatic and serum galectin-3 levels have been reported in patients with chronic liver disease of various etiologies compared to liver-healthy controls [18,20–25]. In liver cirrhosis, hepatocyte galectin-3 levels are strongly induced, whereas in patients with cholestasis, mostly galectin-3 expression of Kupffer cells is upregulated. In the normal liver, galectin-3 is expressed in bile duct cells, and the galectin-3 protein levels of bile duct cells also seem to be increased in cholestasis and cirrhosis [26,27].

This has led to the development of galectin-3-blocking agents as anti-fibrotic drugs. Currently, the effects of galectin-3 inhibitors are being investigated in patients with liver fibrosis. However, the drugs tested so far have not been able to improve the fibrosis stage in patients with NASH [28].

Dextran sodium sulfate (DSS) is frequently used to induce IBD in mice by adding it to their drinking water. DSS disrupts the epithelial barrier, allowing intestinal bacteria to invade the mucosa, leading to immune cell infiltration and inflammation. A leaky gut leads to liver and bile duct inflammation, which can progress to cirrhosis. Impaired liver function and intestinal inflammation contribute to intestinal dysbiosis, which can further exacerbate the disease [29,30] (Figure S1). In the DSS model, a comparison between wild-type and galectin-3-deficient mice revealed that galectin-3 plays a role in resolving inflammation. Furthermore, acute DSS-induced colitis was ameliorated by galectin-3 treatment [31]. In both acute and chronic models of colitis, a peritoneal injection of recombinant galectin-3 significantly reduced colonic IL-6 levels [32].

It is worth noting that serum galectin-3 levels in patients with UC showed a negative correlation with endoscopic and histological parameters of colitis. Additionally, elevated fecal galectin-3 levels were found to be an indicator of UC remission [31].

Recent studies observed increased serum galectin-3 levels in IBD, which were unable to distinguish between active disease and remission in both UC and CD patients [33,34]. This suggests that galectin-3 may not be a reliable biomarker for disease activity in IBD. It has also been reported that serum galectin-3 levels did not differ between IBD patients and controls [35]. To summarize, current evidence is inconclusive about systemic levels of galectin-3 and its association with disease activity in IBD.

Urinary proteins are increasingly recognized as biomarkers for a range of diseases [36]. Elevation of galectin-3 in urine may serve as a marker for progressive kidney injury and renal fibrosis. Additionally, it has been identified as a biomarker for heart failure and various cancers [36–38].

Renal manifestations are present in approximately 6% of patients with IBD. The most frequent type of renal involvement is nephrolithiasis [39,40]. IBD patients with renal involvement had a lower glomerular filtration rate and higher serum creatinine levels compared to IBD patients with normal renal function [40]. Nephrolithiasis is twice as common in patients with chronic liver disease in comparison to healthy individuals [41].

Galectin-3 in serum can indicate liver diseases. However, it has not been evaluated whether it can distinguish between PSC and IBD. Therefore, in this study, the serum galectin-3 levels of healthy controls, patients with IBD, and patients with PSC were measured to address this issue. In addition, the potential of urinary galectin-3 as a biomarker for PSC was analyzed.

2. Results

2.1. Serum and Urinary Galectin-3 of IBD Patients

The cohort included 55 IBD patients, 22 PSC patients (11 patients had underlying IBD and 11 patients did not), and 38 controls. The controls and patients had a comparable age and gender distribution (Table 1). The PSC patients had higher serum creatinine and a lower glomerular filtration rate (GFR) than the IBD patients. Aspartate aminotransferase (AST), gamma glutamyl transferase (gamma GT), alkaline phosphatase (AP) and bilirubin were elevated in PSC compared to IBD patients (Table 1). The urinary protein/creatinine levels of controls and patients with IBD or PSC were similar. The cutoff value for proteinuria is 0.2 [42], showing that our controls and patients had normal renal function.

Table 1. Characteristics of the study groups. The PSC cohort included 11 patients with PSC and IBD and 11 patients with PSC without underlying IBD (PSC_{woIBD}). The IBD cohort does not include patients with PSC. Data are reported as median, minimum, and maximum values. The Model for End Stage Liver Disease (MELD) score was documented for PSC patients. Urine protein/creatinine ratio of 18 controls has been determined. The statistical test used: Kruskal–Wallis Test (alanine aminotransferase (ALT), alkaline phosphatase (AP), aspartate aminotransferase (AST), gamma glutamyl transferase (gamma GT), glomerular filtration rate (GFR), not determined (n.d.)). * $p < 0.05$, ** $p < 0.01$, *** $p < 0.001$.

Characteristics	IBD	PSC	Controls
Number (females/males)	55 (23/32)	22 (6/16)	38 (20/18)
Age (years)	47 (19–70)	52 (18–70)	55 (23–78)
BMI (kg/m^2)	25.1 (15.5–40.4)	24.8 (18.0–31.8)	n.d.
C-reactive protein (mg/L)	3 (0–144)	2 (0–51)	n.d.
Creatinine (mg/dL)	0.83 (0.51–1.25) *	1.02 (0.68–3.94) *	n.d.
GFR (mL/min)	100 (61–136) *	86 (12–135) *	n.d.
Fecal calprotectin (µg/g)	49 (0–3889)	35 (0–999)	n.d.
AST (U/L)	25 (10–41) *	27 (15–177) *	n.d.
ALT (U/L)	20 (7–63)	27 (5–89)	n.d.
Gamma GT (U/L)	26 (11–74) *	54 (10–345) *	n.d.
AP (U/L)	64 (43–142) ***	117 (57–537) ***	n.d.
Bilirubin (mg/dL)	0.50 (0.15–1.90) **	0.70 (0.20–14.00) **	n.d.
MELD Score	n.d.	6 (6–20)	n.d.
Urine protein/creatinine	0.05 (0–14.89)	0.08 (0–4.60)	0.05 (0 –0.20)

In the control cohort, serum galectin-3 was negatively correlated with age (r = −0.363, p = 0.025). No such association was observed for urinary galectin-3 (r = 0.439, p = 0.101). Serum galectin-3 was higher in women than in men (p = 0.054), whereas urinary galectin-3 was similar in both sexes (p = 0.814).

In IBD patients, serum and urine galectin-3 did not correlate with age (r = −0.080, p = 0.562 and r = 0.008, p = 0.954, respectively) and BMI (r = −0.054, p = 0.705 and r = 0.057, p = 0.708, respectively). Males and females had similar serum (p = 0.986) and urinary (p = 0.821) galectin-3 levels.

Serum and urine galectin-3 in IBD were not correlated (r = −0.083, $p = 0.566$). The IBD cohort included 36 patients with CD and 19 patients with UC. Serum and urinary galectin-3 were similar between these two groups (Figure 1A,B).

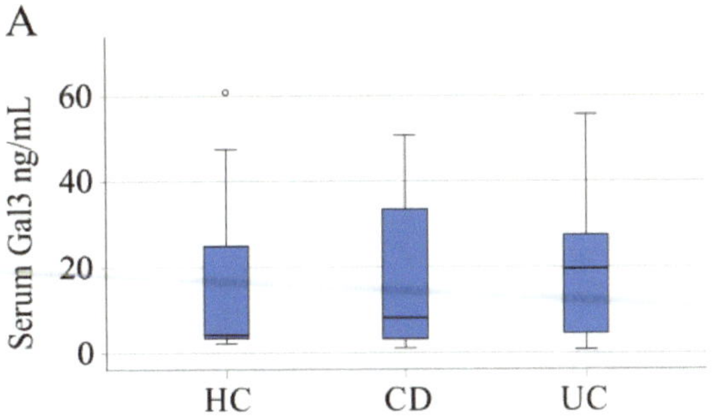

Figure 1. Serum and urinary galectin-3 (Gal3) of healthy controls (HC) and patients with Crohn's disease (CD) and ulcerative colitis (UC). (**A**) Serum Gal3; (**B**) urinary Gal3. Small circles in the figures are outliers.

Stratifying IBD patients by fecal calprotectin levels showed no significant difference in serum galectin-3 levels across groups ($p = 0.168$; Figure 2A). Specifically, 27 patients had fecal calprotectin levels < 50 µg/g, 11 had levels between 50 and 150 µg/g, 8 had levels between 150 and 500 µg/g, and 7 had levels > 500 µg/g, with data for 2 patients undocumented. Nonetheless, a trend toward lower serum galectin-3 levels was observed in patients with high fecal calprotectin compared to those with low levels ($p = 0.056$). Of note, urinary galectin-3 levels were significantly different ($p = 0.024$) and increased in patients with high calprotectin (Figure 2B). Serum creatinine ($p = 0.639$) and GFR ($p = 0.073$) were not changed with increasing levels of fecal calprotectin.

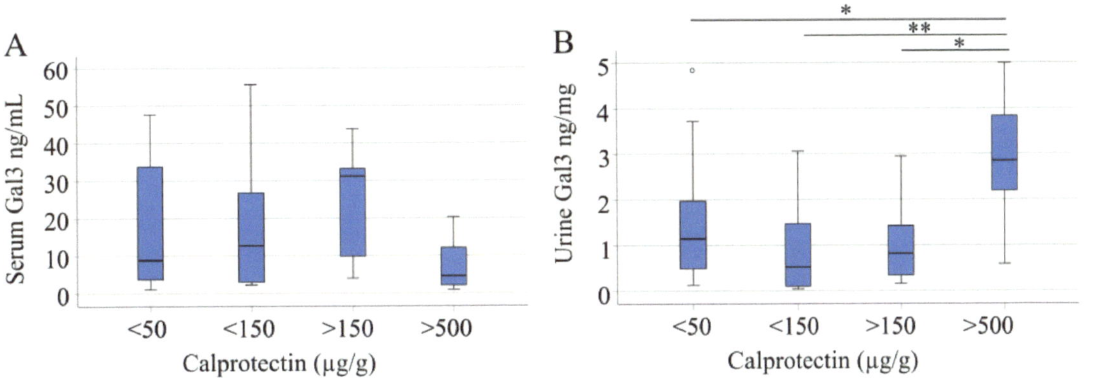

Figure 2. Serum and urinary galectin-3 (Gal3) in relation to fecal calprotectin: (**A**) serum Gal3; (**B**) urinary Gal3. * $p < 0.05$, ** $p < 0.01$. The small circle in the (**B**) is an outlier.

In IBD, galectin-3 in serum and urine did not correlate with creatinine, GFR, and fecal calprotectin. Serum galectin-3 was negatively correlated with CRP (Table 2). ALT, AST, gamma GT, and AP correlated with neither the serum nor urinary galectin-3 of IBD patients ($p > 0.05$ for all).

Table 2. Spearman correlation of serum and urinary galectin-3 with creatinine, glomerular filtration rate, C-reactive protein, and fecal calprotectin in IBD.

Creatinine	Glomerular Filtration Rate	C-Reactive Protein	Fecal Calprotectin
Serum Galectin-3			
r = 0.049 p = 0.727	r = 0.002 p = 0.989	r = −0.284 p = 0.043	r = −0.102 p = 0.463
Urinary Galectin-3			
r = −0.203 p = 0.161	r = 0.126 p = 0.387	r = 0.107 p = 0.474	r = 0.090 p = 0.539

Galectin-3 has been described as a diagnostic biomarker for tumors [43]. In our IBD patients, 10 patients (of the 52 patients where this was documented) developed tumors during therapy. Serum galectin-3 ($p = 0.296$) was similar between IBD patients with and without tumors. The urinary galectin-3 of patients with tumors tended to be increased ($p = 0.141$) (Figure S2).

Intestinal fistulas are mostly a complication of CD [44], but IBD patients with fistulas (14 patients) had similar serum ($p = 0.323$) and urine ($p = 0.441$) galectin-3 levels to patients without fistulas.

2.2. *Serum and Urinary Galectin-3 of PSC Patients*

PSC is a liver disease associated with IBD [5], and serum galectin-3 was higher in PSC compared to IBD patients and healthy controls (Figure 3A). The area under the receiver operating characteristic (AUROC) for discrimination of PSC and IBD was 0.673.

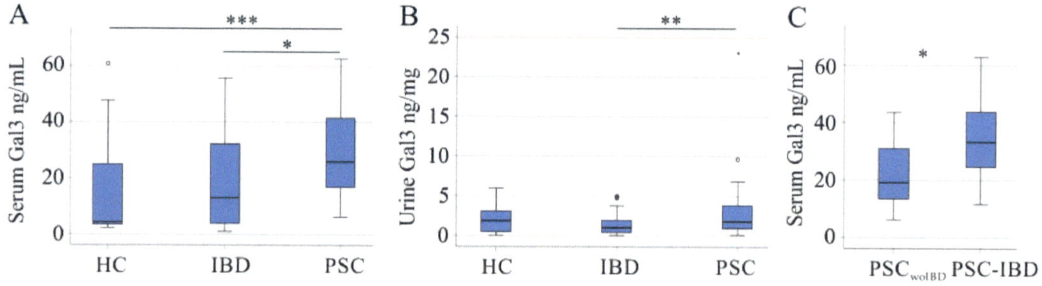

Figure 3. Serum and urinary galectin-3 (Gal3) of healthy controls (HC), patients with inflammatory bowel disease (IBD) and patients with primary sclerosing cholangitis (PSC). (**A**) Serum Gal3; (**B**) urinary Gal3; (**C**) serum Gal3 of PSC patients without IBD (PSC_{woIBD}) and PSC patients with IBD (PSC-IBD). * $p < 0.05$, ** $p < 0.01$, *** $p < 0.001$. Small circles and asterisk in the figure are outliers.

Urinary galectin-3 of PSC patients was induced in comparison to IBD patients but was similar to healthy controls (Figure 3B). The AUROC for discrimination of PSC and IBD patients was 0.696.

The 11 PSC-IBD patients had higher serum galectin-3 than the 11 PSC patients without underlying IBD (PSC_{woIBD}) ($p = 0.033$) (Figure 3C). Urinary galectin-3 was comparable ($p = 0.751$).

Galectin-3 serum levels of PSC_{woIBD} patients were similar to IBD patients, and were significantly higher in PSC-IBD (Figure 4A). The AUROC for discrimination of PSC-IBD from IBD was 0.760. PSC_{woIBD} and PSC-IBD patients had similar calprotectin, CRP, creatinine, and GFR ($p > 0.05$).

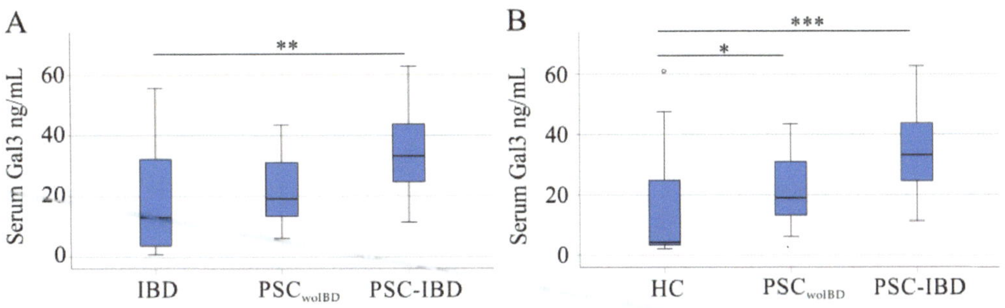

Figure 4. Serum galectin-3 (Gal3) of healthy controls (HC), patients with inflammatory bowel disease (IBD), patients with primary sclerosing cholangitis without underlying IBD (PSC_{woIBD}), and PSC-IBD patients: (**A**) Serum Gal3 of IBD, PSC_{woIBD}, and PSC-IBD patients; (**B**) serum Gal3 of HC, PSC_{woIBD}, and PSC-IBD patients. * $p < 0.05$, ** $p < 0.01$, *** $p < 0.001$. Small circle in the figure is an outlier.

Galectin-3 serum levels of PSC_{woIBD} and PSC-IBD were higher compared to healthy controls (Figure 4B).

In PSC patients, galectin-3 in both serum and urine did not correlate with creatinine, GFR, CRP, and fecal calprotectin (Table 3). ALT, AST, gamma GT, AP, and the MELD score correlated with neither the serum nor urinary galectin-3 of PSC patients ($p > 0.05$ for all).

Table 3. Spearman correlation coefficients for the correlation of serum and urinary galectin-3 with creatinine, glomerular filtration rate, C-reactive protein, and fecal calprotectin in patients with PSC.

Creatinine	Glomerular Filtration Rate	C-Reactive Protein	Fecal Calprotectin
Serum Galectin-3			
r = 0.091 p = 0.803	r = 0.035 p = 0.914	r = −0.335 p = 0.287	r = 0.105 p = 0.759
Urinary Galectin-3			
r = −0.145 p = 0.592	r = −0.049 p = 0.858	r = −0.145 p = 0.592	r = −0.138 p = 0.623

In PSC-IBD patients, a negative correlation of serum galectin-3 and CRP (r = −0.652, $p = 0.041$) was observed. All other laboratory measures did not significantly correlate with the serum or urinary galectin-3 of PSC-IBD and PSC_{woIBD} patients ($p > 0.05$ for all).

3. Discussion

In our study, we analyzed galectin-3 levels in serum and urine, finding that serum galectin-3 levels can distinguish between IBD and PSC-IBD patients. Notably, PSC-IBD patients exhibit higher serum galectin-3 levels compared to both PSC without IBD (PSC_{woIBD}) and IBD patients alone, suggesting a pathophysiological and clinical relevance.

Experimental studies in mice with colitis have mostly provided evidence for a protective role of galectin-3 [31,32,45]. Current evidence about systemic levels of galectin-3 in IBD and its association with disease activity is, however, inconclusive [31,33–35]. In our cohort, serum and urinary galectin-3 levels were similar in IBD patients and controls, and did not differ between UC and CD patients. IBD patients with high fecal calprotectin levels showed a decrease in serum galectin-3 and a concomitant increase in urinary galectin-3. Creatinine and GFR did not change in patients with massive inflammation, and renal disease did not appear to cause this change. The negative correlation of serum galectin-3 with CRP in IBD is consistent with a decrease in serum galectin-3 levels in active disease. Reduced serum galectin-3 in patients with higher disease activity has been reported by Volarevic et al. [31], but, e.g., Yu et al. did not identify different galectin-3 serum levels of IBD patients with active and inactive disease [34]. However, it is difficult to compare the disease severity of patients enrolled in different studies. Thus, some of our IBD patients may have more severe disease and correspondingly lower serum galectin-3 levels compared to patients enrolled in other studies. The median fecal calprotectin level of our patients was 50 µg/g and was higher than that of the patients in the study by Volarevic et al. [31], where most patients had calprotectin levels below 25 µg/g. It remains to be seen whether fecal calprotectin levels determined by different assays can be compared.

PSC is often associated with IBD [5,46]. The diagnosis of PSC remains challenging and there is currently no non-invasive biomarker available [46,47]. As serum galectin-3 is elevated in patients with chronic liver disease [20], it can be hypothesized that serum galectin-3 may discriminate between IBD and PSC. Serum and urinary galectin-3 levels were indeed higher in PSC compared to IBD patients. Serum galectin-3 was also elevated in PSC patients compared to healthy controls, in agreement with previous studies [20,21,25].

It is of particular interest that serum and urinary galectin-3 levels are not associated with elevated ALT and AST levels, yet are clearly associated with PSC.

PSC patients with underlying IBD had higher serum galectin-3 than patients with isolated PSC. Previous studies suggested a protective role of galectin-3 in IBD [32,48], and higher levels in PSC-IBD may thus contribute to less severe IBD, commonly described in patients with PSC-IBD [5]. Whether serum galectin-3 can be used as a biomarker to differentiate patients with isolated PSC from patients with PSC-IBD is a matter for future research. PSC_{woIBD} patients have a more rapid progression of liver fibrosis compared to PSC-IBD patients, who have a higher risk of malignancy [49].

Galectin-3 plays a role in several diseases and has been described as a biomarker for different types of cancer, liver disease, kidney disease, and heart failure [36,43]. In IBD, serum galectin-3 was not increased in patients who developed tumors during therapy. Urinary galectin-3 was slightly higher, but this effect was not significant. The morbidity pattern among middle-aged and older individuals in Germany was evaluated in 2023. This study identified clinically meaningful multimorbidity classes (their prevalence is given in brackets) such as the arthrosis/inflammation/mental illness class (21%), the hypertension/metabolic class (22%), and the cardiovascular/cancer class (2%) [50]. Although not documented for our cohorts, those patients likely had a similar or even higher range of comorbidities [51]. However, further studies are needed to evaluate a possible confounding effect of the above-mentioned comorbidities on serum and urinary galectin-3 in IBD and PSC.

In females, higher serum galectin-3 levels were observed compared to men [52], and this was also evident in our control cohort. A sex-specific difference of serum galectin-3 level was not detected in our patients. The study by Boer et al. including approximately 8000 subjects from the general population described a positive correlation of plasma galectin-3 with age [52], whereas this association was negative in our controls. In our patient cohort, serum galectin-3 levels were not found to be related to age. Although higher galectin-3 levels were present in overweight/obese subjects, galectin-3 did not correlate with BMI in IBD patients [15,52]. Age, sex, and BMI are, therefore, not regarded as confounding factors for the analysis of galectin-3 in the serum and urine of IBD and PSC patients.

A previous study suggested that higher plasma galectin-3 levels are due to impaired renal excretion of galectin-3 [53]. In the IBD cohort, the high serum galectin-3 of patients with active disease was indeed related with low galectin-3 levels in urine. Otherwise, we could not detect correlations between serum and urinary galectin-3 levels. In PSC patients, both serum and urinary galectin-3 levels were elevated. The urinary protein/creatinine ratio is a measure of proteinuria [42] and was normal in our patients. This shows that impaired renal galectin-3 excretion is not the main cause of higher serum galectin-3 levels.

This study has limitations. The samples were collected at only one time point, and the number of patients with PSC_{woIBD} and PSC-IBD, which are rare diseases, was small. Additionally, comorbidities such as diabetes or cardiovascular diseases were not documented. Urine and blood samples have been collected at any time during the day, which is an advantage for clinical tests. The significance of our study is to suggest studies evaluating continuous monitoring of galectin-3 for the diagnosis of PSC. A potential pathogenic role of galectin-3 in PSC should also be evaluated.

4. Materials and Methods

4.1. Patients and Control Cohorts

From 6 December 2021 to 31 January 2023, patients diagnosed with IBD or PSC (11 patients had underlying IBD and 11 patients did not) at the Department of Internal Medicine I (University Hospital of Regensburg) were recruited. Diagnosis of IBD and PSC was based on histologic, endoscopic, and clinical criteria [54–56]. In the IBD cohort, 12 patients were treated with corticosteroids, 19 with mesalazine, 17 with anti-interleukin 12/23 antibodies, 18 with anti-TNF antibodies, and 7 with azathioprine. The 22 PSC patients in our study were treated with ursodeoxycholic acid. Patients with coagulopathy were excluded from the study. The serum and urine of patients, as well as controls, were collected and stored at −80 °C until use. For this retrospective study, the controls were students, hospital staff, and partners of the patients who lived in the same area as the patients with IBD.

4.2. Measurement of Galectin-3, Urinary Creatinine, and Urinary Protein

The ELISA to measure urinary galectin-3 was from AOBIOUS INC. (Cat. No.: E0497h; Gloucester, MA, USA). Urine was used undiluted. For analysis of serum galectin-3, serum was diluted 1:3-fold in Reagent Diluent as recommended by the provider of the ELISA (Cat. No.: DY1154; R&D Systems, Wiesbaden-Nordenstadt, Germany). Urinary creatinine was measured by the creatinine parameter assay kit (Cat. No.: KGE005; R&D Systems) in 1:20-fold diluted urine. Urinary galectin-3 levels relative to urinary creatinine levels were used for calculations. Pierce™ BCA Protein Assay Kits (Cat. No.: 23225, Thermo Fisher Scientific, Waltham, MA, USA) were used to determine urinary protein concentrations using 1:10-diluted spot urine samples of our patients and controls.

4.3. Collection of Urine

Urine was collected during the day, aliquoted into appropriate portions, and stored at −80 °C.

4.4. Statistical Analysis

Data are shown as boxplots and outliers are marked as circles and asterisks. A receiver operating characteristic curve, Mann–Whitney U-test, Kruskal–Wallis Test, and Spearman correlation were the statistical tests used (SPSS Statistics 26.0 program, IBM, Leibniz Rechenzentrum, München, Germany). A value of $p < 0.05$ was regarded as significant.

5. Conclusions

This study demonstrates that serum galectin-3 levels differentiate between IBD, PSC, and PSC-IBD patients. Notably, individuals with both PSC and IBD exhibit elevated serum galectin-3 levels compared to those with PSC alone, suggesting potential pathophysiological implications. Thus, serum galectin-3 emerges as a promising novel biomarker for PSC. PSC is difficult to diagnose, and no specific biomarker that could support the diagnosis of PSC has been identified [10,57]. Liver function tests may be useful to start an investigation, but there are no specific laboratory markers that help diagnose PSC [58]. Although patients with PSC may be initially assessed with ultrasound or computed tomography, cholangiography based on magnetic resonance cholangiography or endoscopic retrograde cholangiopancreatography evaluation is usually required to make a definitive diagnosis. Liver biopsy is only needed in suspected cases of small duct PSC when the cholangiogram is normal, or to confirm an overlap with autoimmune hepatitis [57]. Biomarkers are non-invasive and comparatively cheap analysts without the drawbacks of liver biopsy. However, a group of markers related to the different aspects of a disease are requested for clinical purposes. Galectin-3 may become one marker of a biomarker panel for the diagnosis of PSC. Moreover, increased galectin-3 in PSC may also initiate studies to evaluate galectin-3-blocking agents in cholestatic liver diseases.

Supplementary Materials: The following supporting information can be downloaded at https://www.mdpi.com/article/10.3390/ijms25094765/s1.

Author Contributions: Conceptualization, C.B.; formal analysis, G.B. and C.B.; investigation, G.B., resources, T.E., M.H., J.L, A.A., and H.C.T.; writing—original draft preparation, C.B.; writing—review and editing, G.B., T.E., M.H., J.L., A.A., A.K., M.M., H.C.T., and C.B. All authors have read and agreed to the published version of the manuscript.

Funding: This research received no external funding.

Institutional Review Board Statement: This study was approved by the Ethics Committee of the University Hospital Regensburg (protocol no. 19-1309-101, approval date: 20 February 2019; protocol no. 21-2390-101, approval date: 19 May 2021) and all participants gave written informed consent. The study was conducted under the updated guidelines for good clinical practice and the updated Declaration of Helsinki.

Informed Consent Statement: Informed consent was obtained from all subjects involved in this study.

Data Availability Statement: Original research data can be obtained on request.

Acknowledgments: The support of Elena Underberg, Birgit Meier, Elisabeta Sepsy, Maria Eichinger, Tanja Fererberger, Stefanie Sommersberger, and Stefan Gunawan is greatly acknowledged.

Conflicts of Interest: Arne Kandulski (scientific presentations and scientific advisory activities): Roche Pharma AG, Eisai GmbH, Abbvie Germany AG, Janssen-Cilag GmbH, MSD Sharp & Dohme GmbH, Boston Scientific Corp., Fujifilm Germany, Micro-Tech Germany, Bayer Pharma AG Germany. Hauke Christian Tews (scientific presentations and scientific advisory activities): Abbvie Germany AG, Janssen-Ciag GmbH, Celltrion, Bristol Myers Squibb, Pfizer Pharma GmbH. Martina Müller (travel grants, scientific presentations): United European Gastroenterology, Abbvie Germany, Falk foundation, Germany. The remaining authors declare that the research was conducted in the absence of any commercial or financial relationships that could be construed as a potential conflict of interest.

References

1. Liang, H.; Manne, S.; Shick, J.; Lissoos, T.; Dolin, P. Incidence, prevalence, and natural history of primary sclerosing cholangitis in the United Kingdom. *Medicine* **2017**, *96*, e7116. [CrossRef] [PubMed]
2. Pria, H.D.; Torres, U.S.; Faria, S.C.; Velloni, F.G.; Caiado, A.H.M.; Tiferes, D.A.; D'Ippolito, G. Practical Guide for Radiological Diagnosis of Primary and Secondary Sclerosing Cholangitis. *Semin. Ultrasound CT MR* **2022**, *43*, 490–509. [CrossRef] [PubMed]
3. Schramm, C.; Eaton, J.; Ringe, K.I.; Venkatesh, S.; Yamamura, J.; MRI Working Group of the IPSCSG. Recommendations on the use of magnetic resonance imaging in PSC-A position statement from the International PSC Study Group. *Hepatology* **2017**, *66*, 1675–1688. [CrossRef]
4. Dave, M.; Elmunzer, B.J.; Dwamena, B.A.; Higgins, P.D. Primary sclerosing cholangitis: Meta-analysis of diagnostic performance of MR cholangiopancreatography. *Radiology* **2010**, *256*, 387–396. [CrossRef] [PubMed]
5. Mertz, A.; Nguyen, N.A.; Katsanos, K.H.; Kwok, R.M. Primary sclerosing cholangitis and inflammatory bowel disease comorbidity: An update of the evidence. *Ann. Gastroenterol.* **2019**, *32*, 124–133. [CrossRef]
6. Brown, S.J.; Mayer, L. The immune response in inflammatory bowel disease. *Am. J. Gastroenterol.* **2007**, *102*, 2058–2069. [CrossRef] [PubMed]
7. Gajendran, M.; Loganathan, P.; Catinella, A.P.; Hashash, J.G. A comprehensive review and update on Crohn's disease. *Dis. Mon.* **2018**, *64*, 20–57. [CrossRef]
8. Gajendran, M.; Loganathan, P.; Jimenez, G.; Catinella, A.P.; Ng, N.; Umapathy, C.; Ziade, N.; Hashash, J.G. A comprehensive review and update on ulcerative colitis. *Dis. Mon.* **2019**, *65*, 100851. [CrossRef] [PubMed]
9. Lee, S.H.; Kwon, J.E.; Cho, M.L. Immunological pathogenesis of inflammatory bowel disease. *Intest. Res.* **2018**, *16*, 26–42. [CrossRef]
10. Nguyen, H.H.; Fritzler, M.J.; Swain, M.G. A Review on Biomarkers for the Evaluation of Autoimmune Cholestatic Liver Diseases and Their Overlap Syndromes. *Front. Mol. Med.* **2022**, *2*, 914505. [CrossRef]
11. Lee, W.I.; Subramaniam, K.; Hawkins, C.A.; Randall, K.L. The significance of ANCA positivity in patients with inflammatory bowel disease. *Pathology* **2019**, *51*, 634–639. [CrossRef]

12. Wunsch, E.; Norman, G.L.; Milkiewicz, M.; Krawczyk, M.; Bentow, C.; Shums, Z.; Mahler, M.; Lopens, S.; Reinhold, D.; Franke, A.; et al. Anti-glycoprotein 2 (anti-GP2) IgA and anti-neutrophil cytoplasmic antibodies to serine proteinase 3 (PR3-ANCA): Antibodies to predict severe disease, poor survival and cholangiocarcinoma in primary sclerosing cholangitis. *Aliment. Pharmacol. Ther.* **2021**, *53*, 302–313. [CrossRef]
13. Dong, R.; Zhang, M.; Hu, Q.; Zheng, S.; Soh, A.; Zheng, Y.; Yuan, H. Galectin-3 as a novel biomarker for disease diagnosis and a target for therapy (Review). *Int. J. Mol. Med.* **2018**, *41*, 599–614. [CrossRef] [PubMed]
14. Krautbauer, S.; Eisinger, K.; Hader, Y.; Buechler, C. Free fatty acids and IL-6 induce adipocyte galectin-3 which is increased in white and brown adipose tissues of obese mice. *Cytokine* **2014**, *69*, 263–271. [CrossRef]
15. Weigert, J.; Neumeier, M.; Wanninger, J.; Bauer, S.; Farkas, S.; Scherer, M.N.; Schnitzbauer, A.; Schaffler, A.; Aslanidis, C.; Scholmerich, J.; et al. Serum galectin-3 is elevated in obesity and negatively correlates with glycosylated hemoglobin in type 2 diabetes. *J. Clin. Endocrinol. Metab.* **2010**, *95*, 1404–1411. [CrossRef] [PubMed]
16. Alsahli, M.A.; Almatroodi, S.A.; Almatroudi, A.; Khan, A.A.; Anwar, S.; Almutary, A.G.; Alrumaihi, F.; Rahmani, A.H. 6-Gingerol, a Major Ingredient of Ginger Attenuates Diethylnitrosamine-Induced Liver Injury in Rats through the Modulation of Oxidative Stress and Anti-Inflammatory Activity. *Mediat. Inflamm.* **2021**, *2021*, 6661937. [CrossRef]
17. Wang, W.; Gao, W.; Zhu, Q.; Alasbahi, A.; Seki, E.; Yang, L. TAK1: A Molecular Link Between Liver Inflammation, Fibrosis, Steatosis, and Carcinogenesis. *Front. Cell Dev. Biol.* **2021**, *9*, 734749. [CrossRef]
18. Henderson, N.C.; Mackinnon, A.C.; Farnworth, S.L.; Poirier, F.; Russo, F.P.; Iredale, J.P.; Haslett, C.; Simpson, K.J.; Sethi, T. Galectin-3 regulates myofibroblast activation and hepatic fibrosis. *Proc. Natl. Acad. Sci. USA* **2006**, *103*, 5060–5065. [CrossRef] [PubMed]
19. Jeftic, I.; Jovicic, N.; Pantic, J.; Arsenijevic, N.; Lukic, M.L.; Pejnovic, N. Galectin-3 Ablation Enhances Liver Steatosis, but Attenuates Inflammation and IL-33-Dependent Fibrosis in Obesogenic Mouse Model of Nonalcoholic Steatohepatitis. *Mol. Med.* **2015**, *21*, 453–465. [CrossRef]
20. Gudowska, M.; Gruszewska, E.; Cylwik, B.; Panasiuk, A.; Rogalska, M.; Flisiak, R.; Szmitkowski, M.; Chrostek, L. Galectin-3 Concentration in Liver Diseases. *Ann. Clin. Lab. Sci.* **2015**, *45*, 669–673.
21. Wanninger, J.; Weigert, J.; Wiest, R.; Bauer, S.; Karrasch, T.; Farkas, S.; Scherer, M.N.; Walter, R.; Weiss, T.S.; Hellerbrand, C.; et al. Systemic and hepatic vein galectin-3 are increased in patients with alcoholic liver cirrhosis and negatively correlate with liver function. *Cytokine* **2011**, *55*, 435–440. [CrossRef] [PubMed]
22. Hsu, D.K.; Dowling, C.A.; Jeng, K.C.; Chen, J.T.; Yang, R.Y.; Liu, F.T. Galectin-3 expression is induced in cirrhotic liver and hepatocellular carcinoma. *Int. J. Cancer* **1999**, *81*, 519–526. [CrossRef]
23. Cervantes-Alvarez, E.; Limon-de la Rosa, N.; Vilatoba, M.; Perez-Monter, C.; Hurtado-Gomez, S.; Martinez-Cabrera, C.; Argemi, J.; Alatorre-Arenas, E.; Yarza-Regalado, S.; Tejeda-Dominguez, F.; et al. Galectin-3 is overexpressed in advanced cirrhosis and predicts post-liver transplant infectious complications. *Liver Int.* **2022**, *42*, 2260–2273. [CrossRef] [PubMed]
24. Bacigalupo, M.L.; Manzi, M.; Rabinovich, G.A.; Troncoso, M.F. Hierarchical and selective roles of galectins in hepatocarcinogenesis, liver fibrosis and inflammation of hepatocellular carcinoma. *World J. Gastroenterol.* **2013**, *19*, 8831–8849. [CrossRef]
25. An, Y.; Xu, S.; Liu, Y.; Xu, X.; Philips, C.A.; Chen, J.; Mendez-Sanchez, N.; Guo, X.; Qi, X. Role of Galectins in the Liver Diseases: A Systematic Review and Meta-Analysis. *Front. Med.* **2021**, *8*, 744518. [CrossRef]
26. Butscheid, M.; Hauptvogel, P.; Fritz, P.; Klotz, U.; Alscher, D.M. Hepatic expression of galectin-3 and receptor for advanced glycation end products in patients with liver disease. *J. Clin. Pathol.* **2007**, *60*, 415–418. [CrossRef]
27. Shimonishi, T.; Miyazaki, K.; Kono, N.; Sabit, H.; Tuneyama, K.; Harada, K.; Hirabayashi, J.; Kasai, K.; Nakanuma, Y. Expression of endogenous galectin-1 and galectin-3 in intrahepatic cholangiocarcinoma. *Hum. Pathol.* **2001**, *32*, 302–310. [CrossRef]
28. Slack, R.J.; Mills, R.; Mackinnon, A.C. The therapeutic potential of galectin-3 inhibition in fibrotic disease. *Int. J. Biochem. Cell Biol.* **2021**, *130*, 105881. [CrossRef]
29. Kim, S.I.; Choi, M.E. TGF-beta-activated kinase-1: New insights into the mechanism of TGF-beta signaling and kidney disease. *Kidney Res. Clin. Pract.* **2012**, *31*, 94–105. [CrossRef]
30. Yang, C.; Merlin, D. Unveiling Colitis: A Journey through the Dextran Sodium Sulfate-induced Model. *Inflamm. Bowel Dis.* **2024**, izad312. [CrossRef]
31. Volarevic, V.; Zdravkovic, N.; Harrell, C.R.; Arsenijevic, N.; Fellabaum, C.; Djonov, V.; Lukic, M.L.; Simovic Markovic, B. Galectin-3 Regulates Indoleamine-2,3-dioxygenase-Dependent Cross-Talk between Colon-Infiltrating Dendritic Cells and T Regulatory Cells and May Represent a Valuable Biomarker for Monitoring the Progression of Ulcerative Colitis. *Cells* **2019**, *8*, 709. [CrossRef] [PubMed]
32. Lippert, E.; Stieber-Gunckel, M.; Dunger, N.; Falk, W.; Obermeier, F.; Kunst, C. Galectin-3 Modulates Experimental Colitis. *Digestion* **2015**, *92*, 45–53. [CrossRef] [PubMed]
33. Frol'ova, L.; Smetana, K., Jr.; Borovska, D.; Kitanovicova, A.; Klimesova, K.; Janatkova, I.; Malickova, K.; Lukas, M.; Drastich, P.; Benes, Z.; et al. Detection of galectin-3 in patients with inflammatory bowel diseases: New serum marker of active forms of IBD? *Inflamm. Res.* **2009**, *58*, 503–512. [CrossRef] [PubMed]
34. Yu, T.B.; Dodd, S.; Yu, L.G.; Subramanian, S. Serum galectins as potential biomarkers of inflammatory bowel diseases. *PLoS ONE* **2020**, *15*, e0227306. [CrossRef] [PubMed]
35. Cibor, D.; Szczeklik, K.; Brzozowski, B.; Mach, T.; Owczarek, D. Serum galectin 3, galectin 9 and galectin 3-binding proteins in patients with active and inactive inflammatory bowel disease. *J. Physiol. Pharmacol.* **2019**, *70*, 95–104. [CrossRef] [PubMed]

36. Tews, H.C.; Elger, T.; Grewal, T.; Weidlich, S.; Vitali, F.; Buechler, C. Fecal and Urinary Adipokines as Disease Biomarkers. *Biomedicines* **2023**, *11*, 1186. [CrossRef] [PubMed]
37. Ahmad, T.; Rao, V.; Chunara, Z.; Mahoney, D.; Jackson, K.; Hodson, D.; Tarleton, C.; Thomas, D.; Chen, M.; Jacoby, D.; et al. Urine Galectin-3 Levels Identify High Risk Renal Dysfunction in Patients with Heart Failure. *J. Card. Fail.* **2017**, *23*, S32. [CrossRef]
38. Balasubramanian, K.; Vasudevamurthy, R.; Venkateshaiah, S.U.; Thomas, A.; Vishweshwara, A.; Dharmesh, S.M. Galectin-3 in urine of cancer patients: Stage and tissue specificity. *J. Cancer Res. Clin. Oncol.* **2009**, *135*, 355–363. [CrossRef] [PubMed]
39. Ambruzs, J.M.; Larsen, C.P. Renal Manifestations of Inflammatory Bowel Disease. *Rheum. Dis. Clin. N. Am.* **2018**, *44*, 699–714. [CrossRef]
40. Dincer, M.T.; Dincer, Z.T.; Bakkaloglu, O.K.; Yalin, S.F.; Trabulus, S.; Celik, A.F.; Seyahi, N.; Altiparmak, M.R. Renal Manifestations in Inflammatory Bowel Disease: A Cohort Study During the Biologic Era. *Med. Sci. Monit.* **2022**, *28*, e936497. [CrossRef] [PubMed]
41. Porter, I.E., 2nd; Palmer, W.C.; Parker, A.S.; Hodge, D.O.; Diehl, N.N.; Haley, W.E. Prevalence of Nephrolithiasis in Patients with Chronic Liver Disease: A Case-Control Study. *J. Clin. Exp. Hepatol.* **2018**, *8*, 375–379. [CrossRef] [PubMed]
42. Kaminska, J.; Dymicka-Piekarska, V.; Tomaszewska, J.; Matowicka-Karna, J.; Koper-Lenkiewicz, O.M. Diagnostic utility of protein to creatinine ratio (P/C ratio) in spot urine sample within routine clinical practice. *Crit. Rev. Clin. Lab. Sci.* **2020**, *57*, 345–364. [CrossRef]
43. Hara, A.; Niwa, M.; Noguchi, K.; Kanayama, T.; Niwa, A.; Matsuo, M.; Hatano, Y.; Tomita, H. Galectin-3 as a Next-Generation Biomarker for Detecting Early Stage of Various Diseases. *Biomolecules* **2020**, *10*, 389. [CrossRef] [PubMed]
44. Jiang, H.; Shen, J.; Ran, Z. Epithelial-mesenchymal transition in Crohn's disease. *Mucosal Immunol.* **2018**, *11*, 294–303. [CrossRef] [PubMed]
45. Tsai, H.F.; Wu, C.S.; Chen, Y.L.; Liao, H.J.; Chyuan, I.T.; Hsu, P.N. Galectin-3 suppresses mucosal inflammation and reduces disease severity in experimental colitis. *J. Mol. Med.* **2016**, *94*, 545–556. [CrossRef] [PubMed]
46. Rabiee, A.; Silveira, M.G. Primary sclerosing cholangitis. *Transl. Gastroenterol. Hepatol.* **2021**, *6*, 29. [CrossRef]
47. Banales, J.M.; Inarrairaegui, M.; Arbelaiz, A.; Milkiewicz, P.; Muntane, J.; Munoz-Bellvis, L.; La Casta, A.; Gonzalez, L.M.; Arretxe, E.; Alonso, C.; et al. Serum Metabolites as Diagnostic Biomarkers for Cholangiocarcinoma, Hepatocellular Carcinoma, and Primary Sclerosing Cholangitis. *Hepatology* **2019**, *70*, 547–562. [CrossRef] [PubMed]
48. Jovanovic, M.; Simovic Markovic, B.; Gajovic, N.; Jurisevic, M.; Djukic, A.; Jovanovic, I.; Arsenijevic, N.; Lukic, A.; Zdravkovic, N. Metabolic syndrome attenuates ulcerative colitis: Correlation with interleukin-10 and galectin-3 expression. *World J. Gastroenterol.* **2019**, *25*, 6465–6482. [CrossRef]
49. Rennebaum, F.; Demmig, C.; Schmidt, H.H.; Vollenberg, R.; Tepasse, P.R.; Trebicka, J.; Gu, W.; Ullerich, H.; Kabar, I.; Cordes, F. Elevated Liver Fibrosis Progression in Isolated PSC Patients and Increased Malignancy Risk in a PSC-IBD Cohort: A Retrospective Study. *Int. J. Mol. Sci.* **2023**, *24*, 15431. [CrossRef]
50. Amirzada, M.; Buczak-Stec, E.; Konig, H.H.; Hajek, A. Multimorbidity patterns in the German general population aged 40 years and over. *Arch. Gerontol. Geriatr.* **2023**, *114*, 105067. [CrossRef]
51. Chen, B.; Collen, L.V.; Mowat, C.; Isaacs, K.L.; Singh, S.; Kane, S.V.; Farraye, F.A.; Snapper, S.; Jneid, H.; Lavie, C.J.; et al. Inflammatory Bowel Disease and Cardiovascular Diseases. *Am. J. Med.* **2022**, *135*, 1453–1460. [CrossRef]
52. de Boer, R.A.; van Veldhuisen, D.J.; Gansevoort, R.T.; Muller Kobold, A.C.; van Gilst, W.H.; Hillege, H.L.; Bakker, S.J.; van der Harst, P. The fibrosis marker galectin-3 and outcome in the general population. *J. Intern. Med.* **2012**, *272*, 55–64. [CrossRef]
53. Meijers, W.C.; Schroten, N.F.; Ruifrok, W.P.; Assa, S.; Dokter, M.M.; Damman, K.; Gansevoort, R.T.; Van Gilst, W.H.; Sillje, H.H.; De Boer, R.A. Urinary and plasma galectin-3 in heart failure—Insights in renal handling. *Eur. Heart J.* **2013**, *34*, P4243. [CrossRef]
54. Kucharzik, T.; Dignass, A.; Siegmund, B. Aktualisierung der S3-Leitlinie Colitis ulcerosa 2019. *Z. Gastroenterol.* 2019; 57, 1279–1280. [CrossRef]
55. Sturm, A.; Maaser, C.; Calabrese, E.; Annese, V.; Fiorino, G.; Kucharzik, T.; Vavricka, S.R.; Verstockt, B.; van Rheenen, P.; Tolan, D.; et al. ECCO-ESGAR Guideline for Diagnostic Assessment in IBD Part 2: IBD scores and general principles and technical aspects. *J. Crohns Colitis* **2019**, *13*, 273–284. [CrossRef] [PubMed]
56. EASL Clinical Practice Guidelines on sclerosing cholangitis. *J. Hepatol.* **2022**, *77*, 761–806. [CrossRef] [PubMed]
57. Tornai, D.; Ven, P.L.; Lakatos, P.L.; Papp, M. Serological biomarkers for management of primary sclerosing cholangitis. *World J. Gastroenterol.* **2022**, *28*, 2291–2301. [CrossRef]
58. Morgan, M.A.; Khot, R.; Sundaram, K.M.; Ludwig, D.R.; Nair, R.T.; Mittal, P.K.; Ganeshan, D.M.; Venkatesh, S.K. Primary sclerosing cholangitis: Review for radiologists. *Abdom. Radiol.* **2023**, *48*, 136–150. [CrossRef]

MDPI AG
Grosspeteranlage 5
4052 Basel
Switzerland
Tel.: +41 61 683 77 34

International Journal of Molecular Sciences Editorial Office
E-mail: ijms@mdpi.com
www.mdpi.com/journal/ijms

www.ingramcontent.com/pod-product-compliance
Lightning Source LLC
Chambersburg PA
CBHW041344210926
51850CB00013B/2978
9783725824120